CONTENTS

# THE QUESTION OF ANIMAL CULTURE

*Edited by*

Kevin N. Laland and

Bennett G. Galef

HARVARD UNIVERSITY PRESS

*Cambridge, Massachusetts*

*London, England*

*2009*

*Library of Congress Cataloging-in-Publication Data*

The question of animal culture / edited by Kevin N. Laland and Bennett G. Galef.
   p.   cm.
  Includes bibliographical references and index.
  ISBN 978-0-674-03126-5
1. Animal behavior.  2. Social behavior in animals.  3. Learning in animals.  4.
Psychology, Comparative.  I. Laland, Kevin N.  II. Galef, Bennett G.
QL751.Q84 2009
591.56—dc22    2008030549

# ACKNOWLEDGMENTS

This book would not have been possible without the hard work of many people who have helped in various ways. We would especially like to thank Celia Heyes, for reading the entire manuscript and giving detailed and extremely constructive critical feedback on the text, Willemijn Spoor and Lewis Dean for putting together the reference list and sorting out anomalies in it, Laurel Fogarty for compiling the index, Pat Bateson and Marc Hauser for providing endorsements, and Ann Downer-Hazell at HUP and Meredith Phillips at Westchester Book Group for their support, professionalism, and efficiency. We are indebted to them all.

# 1

## INTRODUCTION

KEVIN N. LALAND AND BENNETT G. GALEF

Chimpanzees from the Gombe Stream Reserve in Tanzania use sticks to fish for termites, while those that live in the Taï Forest in the Ivory Coast do not; conversely, the use of a stone hammer to crack open nuts is observed at Taï but not at Gombe. Some orangutans in Borneo make leaf-bundle "dolls," others use tools as sexual stimulants, and still others blow raspberries at bedtime. Groups of white-faced capuchin monkeys in Costa Rica exhibit peculiar social conventions that are not seen in other capuchin populations, such as sniffing each other's hands and placing fingers in each other's mouths. Humpback whales from different regions sing different songs, and some female dolphins and their daughters use sponges as tools while foraging.

At first sight, such reports of behavioral differences among species members that live in different locations are evocative of human cultural variation. Just as people from different regions of the world eat different foods, have varying customs, and speak different languages, some animals also appear to have local traditions. Much circumstantial and some experimental evidence suggests that as in human societies, these traditions are learned from others and are handed down from one generation to the next. But are the similarities between animal "cultures" and those of humans meaningful or superficial?

No one who reads about animal behavior can be unaware of the recent spate of articles in prominent scientific journals, newspapers, and news magazines that argue that differences in the behavioral repertoires of animals living in different locales provide evidence that they, like humans, are cultural beings. Those with a slightly deeper interest in the possibility of culture in animals are probably also aware that many experts in behavioral development are unconvinced by the data that field biologists claim support the view that animal and human culture are fundamentally similar.

The question whether the traditions of animals and the culture of humans are truly similar is contentious, and psychologists, primatologists, behavioral ecologists, and anthropologists often hold somewhat different positions. Until now, these conflicting perspectives were to be found only in a widely scattered and often esoteric literature. Consequently, anyone who sought a comprehensive overview of the various opinions had to undertake a demanding search of the primary literature to find relevant materials. *The Question of Animal Culture* is designed to make that task far easier. The book's contributors, each an established authority on either social learning or a related field, address the question whether, in their opinions, animals have culture. The authors were asked to provide a précis of the data that they find most relevant to the issue, and to emphasize their interpretation of those data with respect to the question of the ways in which animal and human cultures are similar or different. By carefully choosing contributors who represent the full range of perspectives on this issue, we hope to have provided an up-to-date, comprehensive overview of the controversy.

## A Brief History of the Animal Culture Debate

The idea that animals might acquire important components of their behavioral repertoires by copying others has a long history that dates back to Aristotle, who provided the first evidence of social learning of song in birds. Charles Darwin was aware of animal traditions, noting in *The Descent of Man* (1871, p. 161) that "apes are much given to imitation . . . and the simple fact previously referred to, that after a time no animal can be caught in the same place by the same sort of trap, shews [*sic*] that animals learn by experience, and imitate each other's caution." Early evolutionists, including Alfred Wallace, George Romanes, Conwy Lloyd Morgan, and James Baldwin, placed great emphasis on learned traditions as a source of adaptive behavior. In spite of his belief that the human brain required a special explanation, Wallace (1870) did not regard the handing on of skills and habits from one generation to the next as restricted to humans, and he saw a great deal of similarity between the processes that underlay the construction of nests by birds, deemed to be learned, in part, through imitation, and the building of shelters by humans. Romanes (1884) regarded imitation as the critical means by which animals, particularly mammals, refine their instincts. Morgan stressed birdsong dialects and traditional food preferences in animals as suggesting continuity of mental abilities between humans and other animals (C. L. Morgan 1896a). Further, C. L. Morgan (1896b), Baldwin (1896), Spalding (1873), and Osborne (1896) independently suggested that organisms could survive ecological challenges by virtue of their acquired knowledge and skills, frequently learned from others, and that this would then channel natural selection to favor unlearned versions of the same adaptive behavior.

Over the last century field researchers have reported many cases of the spread of novel foraging behaviors in natural animal populations. Lefebvre and Palameta (1988) document many "possible socially transmitted foraging behaviors" in a variety of vertebrates, going back to 1887, when Carpenter reported the putatively socially transmitted habit of cracking oysters with stones in crab-eating macaques. More familiar examples include the drinking of cream from milk bottles by some European birds (Fisher and Hinde 1949) and the spread of food-washing techniques in Japanese macaques (Kawai 1965). Such behavioral innovations appear to have spread too quickly to be explained plausibly by population genetic, ecological, or demographic factors and have been assumed to spread through social learning. However, in general, researchers have rarely been able to substantiate the claim that such diffusions are actually the product of social (as opposed to asocial) learning, and this has left the assumption that the behaviors are spread socially open to criticism.

The modern debate over animal culture began in earnest in Japan a little more than half a century ago. Inspired by Imanishi's claim that culture is widespread in animals, Japanese researchers began to document traditions in free-living, but often provisioned, primate populations (Kawai, 1965; de Waal 2001). The most famous among these is the washing of sweet potatoes by Japanese macaques.

In September 1953 Satsue Mito first saw Imo, an 18-month-old, female Japanese macaque, wash a dirt-covered sweet potato in a small freshwater stream on Koshima Island in the Sea of Japan. A dozen years later, when the first publication appeared in the West that described the pattern of diffusion through Imo's troop of the habit of washing dirt from sweet potatoes before eating them, its author referred to this behavior and other unique patterns of behavior seen on Koshima as "precultural" (Kawai 1965), much as Kawamura, the author of an earlier article on socially learned behaviors of macaques in Japan, had referred to the behavioral variants he described as "sub-cultures" (Kawamura 1959). The implication was, as Kawai made explicit in a more recent

3

publication (Hirata et al. 2001, p. 489), that "we must not overestimate the situation and say that 'monkeys have culture' and then confuse it with human culture." At the same time, the use of the word "culture," even prefixed as it was, implied some unusual degree of correspondence between monkey and human behavior, be it homologous or analogous.

For several years after Kawai's publication, most researchers who studied behaviors that observation suggested had been socially transmitted through a population referred to the behavioral phenomena that they were interested in as "precultural" (e.g., Menzel 1973a), "protocultural" (Menzel et al. 1972), or "traditional" (e.g., Beck 1974; Strum 1975). Possibly because all the socially transmitted behaviors studied during this period, like many human cultural traditions, functioned primarily to increase the efficiency with which bearers of a tradition could extract resources from the environment, primatologists seemed reluctant to think of such animal traditions as equivalent to human traditions, which often have important social and symbolic as well as practical functions.

In 1978 McGrew and Tutin reported the first evidence of a tradition involving an apparently arbitrary pattern of behavior, the grooming handclasp, prevalent in a troop of chimpanzees at Kasoge in western Tanzania but never observed at Gombe, a mere 50 kilometers distant. McGrew and Tutin (1978) argued forcefully that handclasp grooming satisfied many of the criteria used to identify cultural patterns in humans and that use of the term "culture" to refer not only to handclasp grooming but also to other population-specific behaviors of chimpanzees was justified. McGrew and Tutin's article was the first in the modern era to directly address the question of the relationship between the traditions of animals and the culture of humans. McGrew and Tutin appeared to initiate a trend. Increasingly the prefixes were dropped as talk of "preculture" and "protoculture" changed to discussion of "culture," particularly when speaking of chimpanzees (Goodall 1986; Nishida 1987; McGrew 1992; Boesch 1993a; Wrangham, McGrew et al. 1994).

McGrew went on to document substantial differences in the behavioral repertoires of populations of chimpanzees scattered across Africa. He set out to study this behavioral variation systematically and to make detailed comparisons between sites. He found a number of different behavior patterns, ranging from foraging to sexual, aggressive, and even medicinal behavior, that varied systematically among chimpanzee populations, and he argued that these were passed across generations as learned traditions.

The variation in chimpanzee behavioral repertoires and McGrew's interpretation of this variation as cultural received considerable attention through his influential book *Chimpanzee Material Culture* (1992) and Wrangham, McGrew, and colleagues' (1994) edited volume *Chimpanzee Cultures*. McGrew's argument was seconded in widely read popular books, notably Frans de Waal's *The Ape and the Sushi Master*, that presented further evidence of humanlike cognition, emotions, ethics, and culture in other primates, especially chimpanzees (de Waal 2001).

Not everyone was convinced by these arguments. Critics, notably psychologists Bennett Galef (1992, 2003b) and Michael Tomasello (1994, 1999a), took issue with claims of animal culture, primarily on two levels. First, they criticized the data; any claim of culture demanded clear demonstration that putative traditions are a consequence of social learning. Critics pointed out that the observed behavioral differences between populations of chimpanzees could be the result of variation in ecological resources between sites (see Galef this book). Second, they suggested that parallels between animal and human culture rested on superficial analogies rather than on homologies in cognitive processing (see Galef this book; Tomasello this book). In particular, Galef and Tomasello insisted that human culture was supported by imitation and teaching, different psychological mechanisms than those that supported animal traditions. Tomasello (1994) further suggested that imitation and teaching were critical for traditions to exhibit the "ratchet effect" (Tomasello 1994, this book) that produced an increase in the complexity or efficiency of technology over time that was never observed in animal traditions. With publication of the articles by Galef and Tomasello, the debate over animal cultures began in earnest.

Meanwhile, biologists had begun to use the term "culture" in a broad manner. John Tyler Bonner (1980, p.9), in his widely read book *The Evolution of Culture in Animals,* defined culture as "the transfer of information by behavioral means" and was willing to describe invertebrates as exhibiting rudimentary culture. He traced the increasing complexity of acquired information transmission from simple imprinting mechanisms through crude forms of social learning in birds and mammals to "imitation" in chimpanzees and then to full-blown human culture. Similarly, in their book *Genes, Mind, and Culture* Charles Lumsden and Edward Wilson (1981) attributed culture to some 10,000 species, including even some bacteria. Lumsden and Wilson deemed any extragenetic form of acquired information transmission "cultural." Mundinger (1980) took a slightly more restricted line, describing as culture vocal learning in passerine birds, a label that stuck (Catchpole and Slater 1995). For Mundinger, culture simply implied social learning.

In the late 1990s experimental evidence of imitation by chimpanzees began to appear (Whiten et al. 1996; Whiten 1998), which some regarded as undermining the animal culture skeptics' position (Whiten et al. 1999; Whiten this book; van Schaik this book). The case for chimpanzee culture was given a major boost by a remarkable international collaborative effort among nine leading primatologists, each of whom had spent many years studying chimpanzee behavior (Whiten et al. 1999). These researchers collated behavioral information from seven long-term field studies of chimpanzees at different sites across Africa. This mammoth undertaking revealed patterns of variation far more extensive than had previously been documented for any animal species other than humans. Sixty-five categories of behavior were described, 42 of which exhibited significant variability across sites.

Although some of this variation was attributed to differences in the availability of resources (absence of algae fishing can be explained by the rarity of algae at some sites), some behavior patterns, including tool use, grooming, and courtship behaviors, were common in some communities but absent in others, and this distribution had no apparent ecological explanation. Moreover, the repertoire of such traditional behavior patterns in each chimpanzee community was highly distinctive, a phenomenon characteristic of human cultures but previously undiscovered in any nonhuman species.

Whiten and colleagues' (1999) systematic analysis of multiple sites, documentation of the absence, as well as the presence, of behaviors, and recording of frequencies of behavioral variants were important improvements in the scale and rigor of analyses of animal traditions. On the basis of their data, Whiten and colleagues (1999) felt comfortable titling their article "Cultures in chimpanzees." Whiten (2005, this book) later stressed that it was no coincidence that our nearest relatives exhibit the traditions most like those of human culture of all animals, and he argued that chimpanzee and human cultures result from homologous processes.

Whiten and colleagues' (1999) analysis precipitated a series of articles that applied similar methods to other species (van Schaik, Ancrenaz et al. 2003; Perry, Panger et al. 2003; Krützen et al. 2005). Collectively these papers implied that differences in the behavioral repertoires of many large-brained mammals living in different locales provided evidence that they were cultural beings. That articles proposing animal culture were often published in highly prestigious journals *(Nature, Science, Proceedings of the National Academy of Sciences USA)* illustrates the attention that the topic of animal culture could now garner. A conference on the topic of animal traditions, followed by an edited volume (Fragaszy and Perry 2003a), drew further attention to the field.

In an article titled "Orangutan cultures and the evolution of material culture," van Schaik, Ancrenaz, and colleagues (2003) identified 24 putative cultural variants (including feeding techniques and social signals) in six populations of orangutans, with each population again characterized by a distinctive repertoire of traditional behaviors. Primatologists who studied free-living populations of orangutans provided additional support for their interpretation of this variation as reflecting socially transmitted traditions by demonstrating correlations between geographic proximity and cultural similarity of populations and between opportunities for social learning and size of cultural repertoire. There is no doubt that van Schaik, Ancrenaz, and colleagues' (2003) use of the term "cultural" implied homology with human culture: "The presence in orangutans of humanlike skill (material) culture pushes back its origin in the hominoid lineage to about 14 million years ago, when the orangutan and African ape clades last shared a common ancestor" (p. 105).

At about the same time, researchers who were studying capuchin monkeys published results of a major, long-term collaborative study of white-faced capuchin monkeys *(Cebus capucinus)* that revealed behavioral variation in the social conventions of 13 social groups throughout Costa Rica (Perry, Panger et al. 2003). Several striking and often bizarre social conventions were candidates for traditional status, including hand sniffing, sucking of body parts, and placing fingers in the mouths of other monkeys. What is particularly compelling about these data is that it is all but impossible to attribute variation in such conventions to ecological differences among sites. However, Perry, Panger and colleagues (2003) carefully avoid describing these traditions as culture (see Perry this book).

In parallel to the debate over interpretation of primate foraging traditions, material culture, and social conventions, similar controversies were starting to develop over vocal traditions in birds, dolphins, and whales. The existence of socially transmitted vocal dialects in birds had been known since Marler (1952), and geographic variation in the songs of many passerines has been documented, notably white-crowned sparrows and chaffinches (Marler and Tamura 1964; Catchpole and Slater 1995). From the 1970s evidence began to appear for vocal traditions in mammals, particularly cetaceans (Caldwell and Caldwell 1972; Janik and Slater 1997). Much of the research on vocal traditions in cetaceans has focused on bottlenose dolphins (*Tursiops* spp.) and humpback whales *(Megaptera novaeangliae)* (Janik and Slater 1997). For example, all males in a humpback whale population share a song that changes gradually during the singing season, a change much too rapid to be explained by changes in genotype (Payne and Payne 1985). Most striking, off the east coast of Australia, a song was observed to change in 2 years to one previously heard only off the west coast of Australia, possibly as a result of movement of a few individuals from west to east (Noad et al. 2000).

Claims of cetacean social learning have also been made in domains other than vocalization, particularly foraging and migratory traditions, and have moved the topic of culture to the center of cetacean research. A review titled "Culture in whales and dolphins" (Rendell and Whitehead 2001) lists a broad range of traits that can be interpreted as cultural, including killer whales *(Orcinus orca)* beaching themselves during foraging and bottlenose dolphins using sponges to grub for prey.

As this brief historical account reveals, over the last two decades there has been a profound change in the frequency with which scientists who write about population-specific behaviors in animals refer to the phenomena they discuss as "culture." The growing number of long-term behavioral studies of primate and cetacean populations, detailed comparisons of the behavioral repertoires of different populations, and documentation of diversity in animals' use of tools, foraging patterns, vocalizations, and modes of social interaction have brought to the fore behavioral variation in animals that many researchers view as similar to human culture. However, the species that are most commonly put forward as culture bearing (primates and cetaceans) are often among the most difficult animals to study. Several have endangered or threatened status, and for a variety of ethical and practical reasons, at least in the field (but see Matsuzawa et al. 2001), investigation of purportedly traditional behaviors of most is largely restricted to observational studies. As a consequence, the evidence that advocates of animal cultures are able to muster is largely circumstantial in nature. Although, in theory, one major component of the controversy over animal culture could be resolved by experimental manipulations, for instance, the translocation of individuals between populations or of populations between sites, as has been successfully used to demonstrate traditional behavior in fishes (Warner 1988, 1990), in reality, it is not possible to apply this methodology to chimpanzees or humpback whales. One ramification of these methodological constraints is that the case for animal culture rests largely on judgments of plausibility (Laland and Hoppitt 2003), about which opinions vary considerably, as this book demonstrates.

Part of the disagreement over animal culture reflects definitional issues. Biologists (see, for instance, Whitehead or Laland, Kendal, and Kendal's chapters in this book) seemingly tend to employ less exacting definitions than do anthropologists (as exemplified by Perry or Hill's contributions), and psychologists often take an intermediate position between the two. Some researchers deem a species cultural if it exhibits socially transmitted traditions, while others raise the bar to demand, for instance, teaching, group-specific norms, or ethnic markers. The range of definitions adopted reflects, in part, variability in the questions that researchers from different disciplines address.

As Perry (this book) points out, in the main, sociocultural anthropologists have not yet engaged in the animal culture debate: "Cultural anthropologists are so dismissive of the notion of animal 'culture' that it is difficult to find one who thinks it worth his time to articulate his objections in print." Given that human culture is widely regarded as the "type specimen" for animal culture, the lack of input from the very researchers who dedicate their lives to its investigation is surely a major omission. In this respect, the contributions of Hill and Perry in this book are particularly valuable.

However, there is more to the debate over animal culture than squabbles over definitions. A large part of the controversy concerns the kinds of evidence sufficient to establish that differences in the behavior of geographically separate populations of a species result from social learning rather than from genetic differences between populations or differences in the way diverse ecologies shape behavioral development of individuals. Here researchers differ in the degree to which they are willing to rely on circumstantial evidence and plausibility arguments, and laboratory experimentalists and field researchers often take different sides.

In addition, researchers disagree over whether human culture and animal cultures are fundamentally different or fundamentally similar (or perhaps more accurately, in what ways human and animal cultures are similar to or different from one another). Naturally, animal culture advocates stress the similarities of animal traditions to human culture, focusing on common characteristics, such as behavioral variation underpinned by social learning, group-specific repertoires, or the diffusion of innovations, while skeptics stress the many differences, such as social learning mechanisms, evidence of cumulative culture, and norms.

As noted earlier, the range of perspectives on the question of animal culture has produced a widely dispersed and, at times, esoteric literature. Nonetheless, the field has important implications for both our understanding of the continuity of animal and human minds and the way in which we characterize *Homo sapiens*. The goal of this book is to capture the current breadth of opinion and to get to the heart of the issues. We are fortunate in having recruited essentially all of today's major players in the debate about the nature of culture and thus can provide the reader with a comprehensive, concise, and accessible overview of the current state of the field.

## The Structure of This Book

Because of the nature of this book, the participants rather than the issues shaped it. However, the editors asked each author to address, at least in passing, one or more of a small number of questions: What is the most useful way to conceptualize culture? If you feel that a definition of culture is helpful, what is yours? Which animals, if any, exhibit culture? What data and which methods provide the best evidence for culture in animals? Which commonly employed methods or commonly cited data fail to provide such evidence? In what ways are animal and human cultures similar and different? What is the source of the controversy over animal culture, and what would be required to resolve that controversy?

The authors' contributions have been organized sequentially from the strongest advocates of animal culture to the strongest skeptics. In structuring the book in this manner, we acknowledge that the reduction of a multifaceted debate to a single dimension results in a somewhat arbitrary placement of individuals in the debate. Nonetheless, we persisted with the ordering because no matter how crude it is, it does serve to place the various contributions in context and makes it very easy to see the breadth of opinion.

In chapter 2 Frans de Waal, author of numerous books on chimpanzee social behavior, and his student, Kristin Bonnie, present their view that "there is good evidence for culture in many mammals, fish, and birds."

The chapter opens with a spirited opposition to the idea that the products of only a limited set of mechanisms of transmission qualify as cultural. The authors instead advocate a functional, biological perspective in which mechanisms are secondary to social relationships.

The chapter presents observations and experimental findings on brown capuchin monkeys and chimpanzees that, together with data on other aspects of primate behavior, support the Bonding- and Identification-Based Observational Learning (BIOL) model first proposed in de Waal's (2001) book *The Ape and the Sushi Master*. Instead of being dependent on external rewards, "BIOL is a form of learning born out of the desire to belong and fit in." Young individuals ("the apprentices") identify with a certain model ("the master"), whom they copy, often without receiving extrinsic rewards for doing so. Observations such as the inheritance of rank positions, culturally learned communication, handclasp grooming, and other arbitrary conventions in various primates are regarded by de Waal and Bonnie as providing evidence for affiliation and relationship-dependent forms of learning, consistent with BIOL. For these authors, social learning is more than just individual learning in a social context; it is subject to powerful social modifiers and motivators.

Chapter 3 is by William McGrew, a chimpanzee primatologist and, as indicated earlier, the first researcher to make the case that a nonhuman animal, the common chimpanzee, possesses culture. McGrew remains among the strongest proponents of the view that chimpanzees are cultural animals. Drawing on his three books (*Chimpanzee Material Culture* [1992], *Great Ape Societies* [1996], and *The Cultured Chimpanzee* [2004]) and numerous other publications on the topic, McGrew describes variation in chimpanzee behavior across different populations in Africa and argues that this variation cannot be explained by individual learning or genetic or environmental influences. Rather, this rich diversity in social and material culture reflects socially learned traditions, in many respects more similar to cultural variation in humans than is the behavior of other animals.

The title of his chapter, "Ten Dispatches from the Chimpanzee Culture Wars, plus Revisiting the Battlefronts"[1] betrays the hostile reception that the notion of animal culture evoked among anthropologists. McGrew outlines the criteria by which he believes a species can legitimately be categorized as cultural, which extend beyond behavioral diversity, social learning, and tradition. He challenges the assertions (Tomasello 1994) that cumulative culture is uniquely human and a definite feature of human culture and that language and culture are both isomorphic and inseparable to conclude that "mounting evidence gives a rationale for cultural primatology."

Chapter 4 by Carel van Schaik focuses on orangutan culture. Van Schaik is a long-term student of traditions in orangutans and has catalogued orangutan behaviors that vary systematically across sites (van Schaik, Ancrenaz et al. 2003). In a robust defense of the "method of elimination," van Schaik describes why he feels that there are no realistic alternatives to accounts of orangutan behavioral variation in terms of culture. He also describes simple statistical analyses that support interpretation of this variation as socially transmitted traditions, for instance, correlations between geographic proximity and cultural similarity and between opportunities for social learning and size of cultural repertoire. For van Schaik, the attribution of culture requires multiple traditions, interpopulation variation, and group-typical behavioral repertoires.

Andrew Whiten studies social learning and imitation, particularly in chimpanzees and human children. He is also the first author of the primary article on culture in chimpanzees (Whiten et al. 1999), regarded by many as a methodological breakthrough in its pioneering use of detailed comparisons of the behavior of chimpanzees between sites across Africa. In chapter 5 Whiten presents the findings of his "method of exclusion" (ethnographic) approach to chimpanzee behavioral variation, which has become the standard method within the field, subsequently echoed in studies of orangutans, dolphins, and monkeys. On the basis of extensive experimental and comparative evidence, Whiten concludes that the "cultural" credentials of chimpanzees exceed those of other species capable of traditional behavior, with chimpanzees (and perhaps one or two other species) possessing multiple diverse traditions. Whiten describes experimental data (including studies of imitation and of transmission chains), in addition to comparative analyses of chimpanzee behavior across populations in Africa, that he argues collectively make a compelling, if circumstantial, case for chimpanzee culture.

Chapter 6 is by Hal Whitehead, a biologist who has investigated whale behavior for many years, with a focus on their social systems, migration, and "culture." Whitehead is also the author of *Sperm Whales: Social Evolution in the Ocean* (2003a), in which he argues that whales have culture, and co-author, with Luke Rendell, of a highly cited target article in *Behavioral and Brain Sciences* that put the topic of cetacean culture on the scientific map (Rendell and Whitehead 2001). In chapter 6

Whitehead argues, on the basis of behavioral variation in wild populations, that there is good evidence for culture, that is, socially learned traditions, in several species of whales and dolphins.

Humpback and sperm whales, in particular, Whitehead suggests, exhibit complex vocal traditions, characterized by specific dialects in local populations (clans). Whitehead argues that these patterns cannot be explained by ecological differences together with individual learning, because clans use the same areas, or by genetic factors, because genetically unrelated animals perform clan-specific behaviors. However, Whitehead recognizes the need for new tools to address these issues, and much of his chapter is devoted to the development of novel statistical methods that use similarity matrices to isolate cultural variation in animals. For Whitehead, the key question relating to animal culture is not whether a particular behavior is socially learned, but rather how much of the variation in a behavioral pattern is determined by social leaning.

Chapter 7, by Brooke Sargeant and Janet Mann, focuses on within-population variation in dolphin foraging behavior. Janet Mann is a director of the Shark Bay Dolphin Research Project and heads a longitudinal study, the Dolphin Mother-Infant Behavioral Ecology Project, initiated in 1988, that investigates calf development, female reproduction, genetics, ecology, and behavior. Together with her student Brooke Sargeant, Mann has identified a number of distinctive foraging behaviors in a dolphin population and has catalogued the behavioral repertoires of numerous individuals, including many mother-daughter pairs. Sargeant and Mann argue that a small number of dolphin foraging behaviors may meet stringent definitions of cultural traditions. They hold that if the main criterion for animal culture is reliance on social learning, many species are likely to be deemed "cultural." Like Whiten, they present a variety of sources of evidence, ranging from experimental studies suggesting imitation to parent-offspring correlations in behavior in the field, that collectively make an impressive plausibility argument for dolphin culture.

Chapter 8 is by Kevin Laland, Jeremy Kendal, and Rachel Kendal. Laland has carried out extensive laboratory-based research into animal social learning in fish, birds, rodents, and primates, as well as mathematical analyses of cultural evolution and gene-culture coevolution. Laland and colleagues favor a broad and minimalist definition of culture, reflecting the continuity between humans and other animals, and are sympathetic to the idea that a range of vertebrates may possess culture. At the same time, they are critical of much evidence put forward in favor of primate and cetacean culture. Laland has claimed that currently the experimental evidence of culture in fish is stronger than that in chimpanzees, since the best evidence for culture is found in species most amenable to experimental manipulation. However, Laland and colleagues are also critical of the arguments of those who are resistant to the notion of animal culture, and they suggest that the case against animal culture is often mediated by an anthropocentric bias. They maintain that the disparity of views over animal culture reflects the paucity of methodological tools available to researchers in this field, and they draw attention to some new mathematical and statistical methods that potentially could resolve the debate by allowing social learning to be identified inferentially.

In chapter 9[2] Michael Tomasello, a lifelong student of primate cognition and leading skeptic concerning claims of animal culture, stresses that the psychological mechanisms that underlie human culture and animal traditions are quite different. He criticizes claims of imitation in chimpanzees (although in his postscript Tomasello acknowledges recent work showing imitation in chimpanzees). For Tomasello, animal social learning is primarily reliant on simple mechanisms, such as local enhancement and emulation, that cannot support cumulative culture, while human culture is based largely on imitation and teaching.

According to Tomasello, the differences in the psychological processes that underpin social learning in humans and other animals are sufficient to explain why animal traditions do not have the "ratcheting" (cumulative) property that characterizes human culture. Tomasello identifies three key characteristics of human cultural traditions: *universality* (some traditions are practiced by virtually everyone), *uniformity* (people exhibit a high degree of similarity), and *history* (cultural traditions are passed faithfully between generations and accumulate knowledge). He concludes that although chimpanzee behavioral traditions exhibit strong evidence for universality, there is only weak evidence for uniformity or history. Tomasello hypothesizes that compared with chimpanzee "cultures," the greater universality, uniformity, and history of human culture are a manifestation of higher fidelity of information transmission among humans, reflecting differences in the psychological mechanisms employed.

Psychologist Bennett Galef has spent a long career engaged in laboratory experiments on social learning in rats and quail. Galef (1992) was one of the first to argue in print that the evidence of culture in primates is seriously flawed and not sufficient to support the claims of those who would call traditions in animals cultural.

with the problem of transmission fidelity (e.g., Tomasello 1994), but he places more emphasis on a line of skepticism that has come to prominence more recently, that human (but not animal) behavior is regulated by norms. Sterelny argues that the roots of norm-guided action will be much more elusive than the roots of fidelity. "Fidelity is empirically tractable . . . Normativity is much less scrutable because it has no overt ethnographic behavioral signature." For Sterelny, animal societies are not cooperative enough to exemplify the earlier stages of the coevolution of norms, cooperation, and the division of labor. He regards human culture as an emergent property of other evolved human traits rather than a specific adaptation.

Sterelny argues that the idea of social learning, juxtaposed with asocial learning, is problematic since it neglects the important role of niche construction, by which he means information transmitters shaping the learning experience of the observer, which breaks down the social-asocial dichotomy. Teaching is just one of several ways that parents can shape the learning environment of their young. From the niche-construction perspective, the evolutionary transition from agents that learn mostly by individual exploration of their environment to agents that live in social worlds with stabilized traditions need not involve the transformation of the individual cognitive equipment of the agents. Social learning is not an individual trait but an interaction.

Sterelny criticizes social learning researchers for failing to distinguish between the content of socially transmitted information (by which he means what the information is about) and the channel through which the learning agent has access to that information (which may be social or asocial). This has implications for interpreting the animal culture literature. For instance, other contributors to this book suggest that interspecific variation in social behavior (from songs to handclasp grooming) provides the most compelling examples of culture since it is unlikely to reflect variation in ecological resources. To the contrary, Sterelny regards such examples as comparatively weak since they provide evidence only that the animals can learn social facts rather than that such facts are learned socially. When we are considering ecological skills, Sterelny accepts the method of exclusion as a decent, though overly conservative, indicator of social learning. However, when the information source is the agent's social environment, Sterelny complains that the method of exclusion cannot show that agents learn in a distinctive socially mediated or socially enhanced manner.

It will be apparent from this brief overview that there is indeed a broad range of opinions over how to interpret "animal culture," and a rich set of issues is at stake. We encourage you to study the remaining chapters in detail and to make up your own mind.

## Notes

1. Reprinted, with permission, from F. B. M. de Waal and P. L. Tyack, *Animal Social Complexity* (Cambridge, MA: Harvard University Press, 2003).

2. Reprinted, with permission, from R.W. Wrangham, W.C. McGrew, F.B.M. de Waal, and P.G. Heltne, *Chimpanzee Cultures* (Cambridge, MA: Harvard University Press, 1994). Unfortunately, because of previous commitments, McGrew and Tomasello were unable to write new chapters for this volume but both kindly consented to the reprinting of their earlier chapters and provided updates of them.

ACKNOWLEDGMENTS

Kevin N. Laland would like to thank the Biotechnology and Biological Sciences Research Council and Bennett G. Galef the National Sciences and Engineering Research Council of Canada for financial support.

# 2

## IN TUNE WITH OTHERS:

## THE SOCIAL SIDE OF PRIMATE CULTURE

FRANS B. M. DE WAAL AND KRISTIN E. BONNIE

> Imanishi (1952) . . . asserts that instinct is an inherited be-
> havior and thus is something opposite to culture, which rep-
> resents acquired behavior. If it is dogmatic to regard all
> animal behavior as instinctive, it is equally dogmatic to re-
> gard all human behavior as cultural, says Imanishi.
>
> Itani and Nishimura 1973, p. 27

There is nothing more circular than saying that we, humans, are the product of culture if culture is at the same time the product of us. Natural selection has produced our species, including our cultural abilities, and hence these abilities fall squarely under biology. This inevitably raises the question whether natural selection may have produced similar abilities in more than one species.

That this controversial issue was first broached in the East rather than the West is not surprising, given how tightly the culture concept is interwoven with claims of human uniqueness. Plato's "great chain of being," which places humans above all other animals, is absent from Eastern belief systems, according to which all living things are spiritually connected (Asquith 1996; de Waal 2003a). As far back as 1952 Kinji Imanishi, a Japanese anthropologist, wrote an essay that challenged the human-animal divide. He inserted a fictional debate between a wasp, a monkey, an evolutionist, and a layman in which the possibility was raised that other animals might have culture. The proposed definition was straightforward: if individuals learn from one another, their behavior may, over time, become different from that in other groups, thus creating a characteristic culture.

*Figure 2.1* More than half a century after the sweet-potato-washing habit spread among Japanese macaques on Koshima Island, they are still doing it even though the current population has never known the innovator. Nowadays transmission is mostly from mother to offspring, although this was not so in the early years of the habit. The infant clinging to its mother may learn to associate sweet potatoes with the ocean simply by picking up dropped pieces. Photograph by Frans de Waal.

This approach brought culture down to its lowest common denominator: the social rather than genetic transmission of behavior. It was confirmed within a few years by observations of Japanese macaques *(Macaca fuscata)* washing sweet potatoes on Koshima Island (Figure 2.1). The thorough reports of Kawai (1965), Watanabe (1994), Hirata et al. (2001), and others show that sweet-potato washing spread among the monkeys in a manner consistent with the troop's social relationships. The first individuals to show the behavior after Imo, the juvenile female who initiated it, were her mother and age peers. Even though the Koshima study is a historic report that lacks controls, there are good reasons to consider it the first documented case of an innovation that became a tradition (de Waal 2001).

If animal groups vary with respect to a single behavior pattern, such as sweet-potato washing, there is perhaps no reason to employ the loaded "culture" label. "Group-specific behavior" or "tradition" will do. That things might not be so simple with regard to our closest relatives was first intimated by McGrew's (1992) review of tool use among wild chimpanzees. Since then a steady stream of new observations indicates an entire slew of cross-group variants in certain species, as amply evident in the present book. This justifies terminology that goes beyond mere "tradition." Our own definition of culture reflects this broader perspective:

> Culture is a way of life shared by the members of one group but not necessarily with the members of other groups of the same species. It covers knowledge, habits, and skills, including underlying tendencies and preferences, derived from exposure to and learning from others . . . The way individuals learn from each other is secondary, but that they learn from each other is a requirement. Thus, the "culture" label does not apply to knowledge, habits, or skills that individuals can and will readily acquire on their own.
>
> (de Waal 2001, p. 31).

The "culture" label thus befits any species in which one community can readily be distinguished from another on the basis of socially transmitted behavior (cf. Menzel 1973a; Bonner 1980). So far, there is good evidence for culture in many mammals, fish, and birds. Nevertheless, scientists trained in disciplines in which mechanisms are paramount, such as experimental psychology, sometimes resist the idea of animal culture by insisting on specific forms of social transmission—such as teaching and imitation—that most animals may not exhibit (e.g., Premack and Premack 1994; Tomasello 1994). They wonder if the learning mechanisms of humans and other animals are truly "homologous," that is, derived from common ancestry. Inasmuch as the fundamentals of learning, such as association and conditioning, are widespread, they are likely homologous. Beyond this, the homology concept is difficult or impossible to apply. This concept was developed for anatomical traits, which are easily defined and compared, and has been successfully extended to the muscle movements of facial expressions (Preuschoft and van Hooff 1995). But the question whether human and ape cognition are homologous will remain unanswerable until we have far more precise definitions and tests of the underlying capacities.

An alternative approach to culture, illustrated in the definition given earlier, considers learning mechanisms as secondary. This more evolutionary approach focuses on the effects and function of culture rather than the specific cognitive mechanisms that support it. In the same way that the definition of respiration does not specify whether the process takes place through lungs or gills, or the definition of locomotion does

21

not specify whether it is accomplished with legs or wings, the concept of cultural propagation does not need to specify how organisms acquire behavior from each other.

As a result of these differing emphases, the debate about animal culture sometimes resembles one between two deaf men, with one insisting on acquired behavioral variants and the other on specific cognitive capacities. The present chapter explores mechanisms of cultural transmission, but not from the angle of cognitive complexity. We rather seek to explore the social nature of cultural learning, reviewing observations and experiments on brown capuchin monkeys *(Cebus apella)* and chimpanzees *(Pan troglodytes)*. We believe that there is far more to social learning than simply individual learning in a social setting, as others have suggested. Social learning has unique dynamics of its own.

## No Imitation without Identification
### Three Theories of Social Learning

Japanese macaques have rubbed pebbles together for a quarter century on Arashiyama, a mountain that overlooks Kyoto. It is a peculiar behavior whose main feature is that it produces noise. The behavior is absent from nearby monkey troops. It is unknown exactly how the monkeys learn this so-called stone handling from each other: young monkeys must copy it from their mothers without ever being rewarded (Huffman 1996). The primate literature contains myriad examples of traditions of which the reinforcement is unclear (reviewed by de Waal 2001), which raises the possibility that primate social learning stems at least partly from conformism and a motivation to act like others.

To give this process a name and emphasize the role of social models, such as mothers and peers, we will use the acronym BIOL, which stands for Bonding- and Identification-based Observational Learning (de Waal 2001). Instead of being dependent on external reinforcement, BIOL is a form of learning born out of the desire to belong and fit in. Young individuals identify with certain models, which they copy in an often playful, imperfect, and exploratory fashion. These models act as "masters" to the naïve "apprentice" (Matsuzawa et al. 2001). Rewards are secondary, although one could argue that the copying of others is intrinsically rewarding.

This model conflicts with traditional learning theory. When Galef (1992, p. 171) claimed that "although imitation might introduce some novel behavior into the repertoire of members of a population, through time this behavioral novelty would be maintained, modified, or extinguished depending on its effectiveness in acquiring rewards," he expressed the prevailing view that although social partners can influence behavior, learning is ultimately decided by tangible rewards. However, we know from mirror-neuron studies that monkeys do not need any rewards to match the observed actions of others to their own behavior; that is, mirror neurons respond similarly during an action performed by the monkey itself, such as grasping, and while watching another monkey perform a similar action. These "monkey see, monkey do" neurons do the same for entire chains of actions and their predicted outcomes. In other words, the *intentions* of others seem to be encoded from observed motor sequences. Thus social animals are hardwired to be in tune with each other at the level of both actions and goals (Fogassi et al. 2005; Fadiga and Craighero 2007).

Although these findings are usually discussed in the context of empathy, they bear on behavioral copying as well. In their simplest manifestations both mimicry and imitation depend on the degree to which the subject "maps" the model's body movements onto its own (de Waal 2007). The predisposition to do so has high survival value in relation to group life. Primates are nomadic and hence need to sleep when others sleep, play when others play, and forage or hunt when others forage or hunt. Experiments show that satiated primates, like many other animals, begin eating again when they see others eat (Addessi and Visalberghi 2001; Ferrari et al. 2005; Dindo and de Waal 2007), scratch themselves when they see others scratch themselves (Nakayama 2004), and yawn in response to a video of a yawning conspecific (Anderson et al. 2004). This phenomenon is known as social facilitation. Novel behavior, too, is copied, at least by apes. Examples include the imitation of an unusual limping walk by juvenile groupmates of an injured adult male (de Waal 1982) and successful "do-as-I-do" experiments involving human models (Custance et al. 1995; Myowa-Yamakoshi and Matsuzawa 1999). A nice illustration of how unimportant tangible rewards are for this copying tendency comes from the nut-cracking attempts of young chimpanzees. During the first 5 years of their lives, they lack the strength and coordination to crack nuts with stones but continue to re-create the actions of their mothers without a single success (Matsuzawa et al. 2001).

De Waal (1998) proposed "identification" with others as the motivation behind behavioral copying. Like Preston and de Waal's (2002) characterization of perception-action mechanisms for empathy, identification entails bodily mapping the self onto the other (or the other onto the self), resulting in shared representations with the other. This obviously requires a motivation to do so, in which motivation is thought to increase with social closeness and bodily similarity—such as with members of one's own species and gender. It is hardly surprising, therefore, that after the initial skepticism about imitation in nonhuman primates, based on their failure to copy complex human actions (Tomasello, Kruger et al. 1993), two kinds of studies have yielded more promising results, namely, those concerning (a) human-raised apes watching a human model (Tomasello, Savage-Rumbaugh et al. 1993; Bjorklund et al. 2000) or (b) apes raised by their own kind watching a conspecific model (Whiten et al. 2005; Horner et al. 2006; Whiten this book). In both cases identification with the model species is facilitated by rearing history. It is also not surprising that when young chimpanzees learn to use a wand to fish for ants, daughters copy their mothers more faithfully than do sons (Lonsdorf et al. 2004). The tendency to copy another seems to vary with identification, therefore—a motivational issue not to be confused with cognitive capacity.

Table 2.1 compares the predictions of BIOL (i.e., social learning is guided by social relations, and extrinsic rewards are not essential) with predictions from traditional learning theory, according to which behavior will be extinguished in the absence of extrinsic rewards, as well as with predictions from vicarious learning theory (e.g., Bandura 1977), according to which seeing another individual achieve benefits will make observers attend to the model and aim for the same benefits.

### Testing BIOL on Capuchin Monkeys

Brown capuchin monkeys are gregarious New World primates whose social structure is characterized by a loose dominance hierarchy and marked tolerance among unrelated individuals within the group. Although experimental evidence for social learning within the genus has been mixed, that individual behavior is influenced by others is certain. There are ample suggestions that behavior is socially transmitted in the field (Perry, Baker et al. 2003; Perry this book), and in an earlier experiment we found that these monkeys can learn which of two tokens yields the better reward from merely watching a partner exchange tokens for food (Brosnan and de Waal 2004). Capuchins thus provide an interesting model to address questions regarding social influences on learning.

**Table 2.1  Three theories about the role of reward in social learning**

| Theory | Who requires reward? Model | Who requires reward? Observer | Prediction Effect(s) of tangible reward | References |
|---|---|---|---|---|
| Vicarious reinforcement | Yes | No | • Attracts attention to model and facilitates acquisition by observer<br>• Demonstrates adaptive (rewarding) value of modeled behavior | Bandura (1977)<br>Palameta and Lefebvre (1985)<br>Akins and Zentall (1998) |
| Traditional learning | No | Yes | • Required to maintain a socially acquired behavior<br>• Absence of reward leads to extinction | Galef (1992)<br>Galef et al. (1986)<br>Heyes et al. (1993) |
| Bonding- and Identification-based Observational Learning (BIOL) | No | No | • Rewards not required<br>• Emphasis is on social ties and attention to specific others | de Waal (1998, 2001)<br>Matsuzawa et al. (2001) |

*Figure 2.2*   A capuchin model opens one out of three boxes of different color and marking, while the test subject, behind mesh, stands upright to get a better look at the procedure. After this, the subject will be presented with a rearrangement of the same three boxes. Drawing from video still by Frans de Waal.

In order to study the role of food rewards in social learning, and more specifically to explore the competing predictions outlined in Table 2.1, we paired 13 adult capuchin monkeys with a familiar conspecific model, with each subject observing two or three models throughout the experiment, thus creating 34 unique model-subject pairs. Each pair was temporarily separated from its social group and brought into a mobile test chamber consisting of two equally sized areas separated by mesh. The monkeys could see and hear their partner but had no access to the other's space. We designed a simple test—three opaque boxes with hinged lids that were opened by the monkeys without training. Models were taught to open only one of the three boxes, and we were interested in whether the subjects, who were naïve to the contents of the box, would copy the models' choices (Figure 2.2).

On each trial the model was first presented with three boxes and allowed to open one. After the subjects observed this, the same three boxes were rearranged (forcing the subjects to attend to the color and pattern of the box rather than its relative position) and presented to the subject, who was allowed to make only one choice. Each test session consisted of 12 trials with the same boxes. We analyzed the proportion of trials in which subjects chose the same box as the model under the three following conditions:

*Neither Rewarded:* All boxes were empty for both model and subject.

*Model Rewarded:* Only the model was rewarded for opening the trained box. The subject's boxes were empty.

*Both Rewarded:* The model was rewarded for opening the trained box, and so was the subject.

All subjects proceeded through each condition in this order, completing three sessions with each model before moving on to the next condition. We analyzed the data for the first two models observed by each subject and found that subjects copied the choice made by the model at a level significantly above 33.3 percent chance under all conditions: they did so on average in 44.6 percent, 46.1 percent, and 70.8 percent of trials in, respectively, the Neither Rewarded, Model Rewarded, and Both Rewarded conditions (Bonnie and de Waal 2007; Figure 2.3). Interestingly, we found no significant difference in performance between the Neither Rewarded and Model Rewarded conditions. Apparently, observing the model collect a reward does not enhance a capuchin monkey's inclination to copy the model, as predicted by vicarious reinforcement theory.

To see how often subjects would open a baited box by themselves, we added a Control condition in which a partner was present but not involved in the task. In these Control conditions the subject's random chance of finding a reward was the same as in the Both Rewarded condition, 33.3 percent, but subjects were significantly less successful at finding a reward in the Control than in the Both Rewarded condition, which suggests that model demonstrations did in fact enhance the subjects' performance.

These results support the BIOL model in that capuchin monkeys do not need to be rewarded, and do not in fact need to see another individual being rewarded, to copy the behavior of others. Even with no rewards in sight and no previous experience with baited boxes (as in the Neither Rewarded condition), there was still behavioral matching. Reinforcement also played a role, however. Monkeys chose the same box as the model significantly more in the Both Rewarded condition than in any other condition. Their performance increased noticeably, reaching around 90 percent correct on choices made in the final session.

A second prediction from the BIOL model is that a close social tie between two individuals will stimulate learning through observation (see also Coussi-Korbel and Fragaszy 1995). For each model-subject

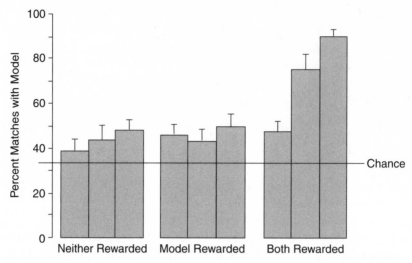

*Figure 2.3*   Mean (+ SEM) proportion of trials in which 13 capuchin monkeys copied the choices demonstrated by a conspecific model in the Neither Rewarded, Model Rewarded, and Both Rewarded conditions. All subjects were tested in that order in 12 trials per session and 3 sessions per condition. As can be seen, subjects were on average above the chance level of 33.3 percent even in the absence of any food rewards (as in the Neither Rewarded condition). Their performance improved dramatically when they themselves had a chance of gaining a reward. From Bonnie and de Waal (2007).

pair, we calculated a measure of affiliation, the Relationship Quality Index (RQI), used previously to qualify mother-infant relationships in our capuchin colony (Weaver and de Waal 2003). RQI is defined as the relative hourly rate of exchanged grooming and contact sitting (affiliation) divided by the relative hourly rate of agonistic exchanges and avoidance (aggression) occurring between two individuals. Relative affiliation and aggression rates for each pair were calculated by dividing the rate for each dyad by the average rate among all dyads in the group. We used behavioral data extracted from a 2-year database of regularly repeated 30-minute observations of the group, during which affiliative and aggressive behaviors were sampled while the monkeys were confined to the indoor area of their living quarters, free to associate with whomever they wanted. A dyad with a positive RQI has more observed instances of affiliation than of conflict and avoidance, whereas dyads with a negative RQI have a higher rate of conflict and avoidance than of affiliation.

For each subject, we categorized each model observed as having an RQI above or below the group median. We labeled these categories, respectively, AFF (predominantly affiliative) and AGG (predominantly aggressive and conflictual). Subjects for whom all models were of the same type were excluded from analysis. For the remaining subjects ($N=10$), we calculated the proportion of trials with either AFF or AGG models during which the subjects copied the model's choice. Doing so separately for each reward condition, we found that across all three sessions in the Model Rewarded condition, subjects copied the choices of AFF models $58.8\pm18.5$ percent (mean $\pm$ standard deviation) of the time, which was significantly above the $45.0\pm15.1$ percent copying of AGG models (paired comparison with randomization test, $P=0.032$, two-tailed; Manly 1997). No such difference was found for the other two conditions ($P>0.10$).

Thus if rewards are available to the model only, subjects tend to match the model's choice, especially if they have a close tie with her. Possibly it is in these closer, more tolerant relationships that subjects follow the model's discoveries of food with the least inhibition. Competitive tendencies may interfere in the more distant relationships. If the model finds no food, on the other hand, tolerance versus competitiveness is not an issue, and hence the social relationship is less relevant. Similarly, when the subject has a chance to find its own food, the subject's competition with the model lessens because the baited boxes in front of the subject become the primary concern. These findings fit the BIOL model in that attention paid to others varies with previously established relationships; that is, competitiveness between two individuals interferes with social learning, whereas social tolerance promotes it (cf. van Schaik 2003).

## Social Culture
### Social Organization

Compared with variation in how primates deal with the environment (e.g., foraging for food, tool use), little attention has been paid to social culture, which we might define as the transmission of social positions, preferences, habits, and attitudes. This is a more elusive topic than material culture. In human culture, for instance, it is easy to tell if people eat with knife and fork or with chopsticks, but to notice if a culture is egalitarian or hierarchical, warm or distant, is much harder to capture in behavioral measures.

A well-documented primate example of social culture is the inheritance of rank positions in macaques and baboons. The future position in the hierarchy of a newborn female can be predicted with almost 100 percent certainty on the basis of her mother's rank. Despite its stability, the matrilineal system depends on learning. Early in life the young monkey finds out against which opponents she can expect help from her mother and sisters: she will end up dominating the parties she is supported against. Experiments that have manipulated the presence of family members have found that when support dwindles, dominant females are unable to maintain their positions (Chapais 1988). In other words, the kin-based hierarchy is maintained for generation after generation through social rather than genetic transmission.

The same applies to the affiliative network. De Waal (1996b) found that rhesus monkey *(M. mulatta)* daughters copy their mothers' association preferences. Even when they have grown fully independent and are approaching motherhood themselves, they spend much time with the daughters of their mothers' friends. We do not know exactly how friendships are being transmitted across the generations, but the simplest way would be that when two mothers sit down to groom and relax, their daughters take the opportunity to play nearby. Being playmates early on, these youngsters then develop an association for the rest of their lives. Given these processes, imagine that females in a particular group begin to strengthen ties outside their own families. Over time this trend will become more and more deeply embedded because their daughters will start doing the same; hence a different social culture will be born.

Cultural effects on social behavior have been documented in relation to conflict and conflict resolution. One experiment managed to turn monkeys into pacifists. Juveniles of two different macaque species were placed together, day and night, for 5 months. Rhesus monkeys, known as quarrelsome and violent, were housed with the more tolerant and easygoing stump-tailed macaques *(M. arctoides)*. Stump-tailed monkeys easily reconcile with their opponents after fights by holding each others' hips, whereas reconciliations are rare among rhesus monkeys. Because the mixed-species groups were dominated by the stump-tailed monkeys, physical aggression was rare. The atmosphere was relaxed, and after a while juveniles of the two species played together, groomed together, and slept in large, mixed huddles. Most important, the rhesus monkeys developed peacemaking skills on a par with those of their more tolerant groupmates. The two species were separated at the end of the experiment, but even then the rhesus monkeys maintained a threefold higher reconciliation rate after fights with conspecifics than is typical of their species (de Waal and Johanowicz 1993).

Not unlike rhesus monkeys, baboons have a reputation of being fiercely competitive. Sapolsky and Share (2004) produced the first field evidence that these monkeys can deviate from this characterization. Wild olive baboons *(Papio anubis)* developed an exceptionally pacific social tradition that outlasted the individuals that established it. For years Sapolsky (1994) had documented how these baboons on the plains of the Masai Mara in Kenya wage wars of nerves, whereby the stress of conflict compromises their rivals' immune systems and increases the level of blood cortisol. An accident of history, however, selectively wiped out all the male bullies of his main study troop. As a result, the number of aggressive incidents dropped dramatically. This by itself was not very surprising. It became more interesting when it was discovered that the behavioral change was maintained for a decade. Baboon males migrate after puberty, and the study group had experienced a complete turnover of males during the intervening decade. Nevertheless, compared with neighboring troops, the affected troop upheld its reduced aggression, increased friendly behavior, and exceptionally low stress levels. The conclusion from this natural experiment is that like human societies, each animal society has its own ecological and behavioral history, which determines its prevalent social style.

### Culturally Learned Communication

Expressions of emotions appear in every member of a species in similar or identical form even if opportunities for learning have been scant. As a parallel to deaf and blind children who, despite deprived learning opportunities, exhibit all human facial expressions in emotionally appropriate contexts (Eibl-Eibesfeldt 1989), a deaf female chimpanzee at the Arnhem Zoo seemed to utter all the varied calls of her species in the right context (de Waal 1982).

It is often assumed, therefore, that the production of communication signals is little affected by learning in primates (but see Taglialatela et al. 2003). The correct reading and interpretation of signals, on the other hand, seems open to environmental influences. For example, responsive-ness to communication signals varies with exposure to species-typical stimuli and opportunities for associative learning (Mason 1985), and the appropriate response to alarm calls by juvenile primates increases with age and experience (Cheney and Seyfarth 1990).

To the general rule that the production of communication displays is less influenced by learning than their appraisal, one important exception exists, however. This is the culturally transmitted communication dis-plays of the great apes, that is, displays that individuals learn from each other. The result of transmission through learning is that a group may de-velop a set of communication displays shared by all of its members yet distinct from the displays found in other groups. Thus most bonobos *(Pan paniscus)* in the San Diego Zoo show a ritual unknown in other bonobo groups, captive or wild. During grooming they customarily clap their hands or feet together or tap their chests with their hands. One bonobo will sit down in front of another, clap her hands a couple of times, and then start grooming the other's face, alternating this with more hand clap-ping. The behavior seems to function in the same way as the spluttering and tooth clacking of grooming chimpanzees, which express enthusiasm for the task. This makes the San Diego Zoo the only place in the world where one can actually hear apes groom. When new individuals are intro-duced, they pick up the habit in about 2 years (de Waal 1988).

Other examples of group-specific communication derive from a com-parison of vocalizations across zoo groups of chimpanzees (Marshall et al. 1999), as well as from field studies on chimpanzees across Africa (Whiten et al. 1999). The latter report includes communication displays such as leaf clipping in courtship, the "rain dance," and handclasp groom-ing (discussed later). Recently, yet another custom was reported for wild chimpanzees, the so-called social scratch. In this gesture one individual rakes the hand back and forth across the body of another, usually scratching the other with the nails. It seems the typical "you scratch my back, I'll scratch yours" gesture, but however familiar this sounds, in wild chimpanzees the social scratch is limited to a single community (Nakamura et al. 2000). This behavior is thought to arouse pleasure in the recipient and to initiate a grooming session. In relation to the role of reinforcement, discussed earlier, it is intriguing that the main reward of this behavior goes to its recipient, as opposed to its performer.

Cultural communication patterns tend to be nonfacial and nonvocal, perhaps because of the apes' limited control over face and voice, especially at emotionally charged moments. Also in humans, facial expressions seem less culturally variable than manual gestures. It is perhaps due to this bias that so few good examples of culturally transmitted communication exist for monkeys, because one of the striking differences between monkey and ape visual communication is the virtual absence of free-hand gestures in monkeys (de Waal 2003b). The observations of white-faced capuchin monkeys *(C. capucinus)* by Perry (this book) may constitute an exception in that they concern group-specific interaction patterns, not unlike the handclasp or hand-clap grooming described for apes.

### Handclasp Grooming

In 1992 we first saw two chimpanzees at the Yerkes Primate Center's Field Station clasp their hands together while grooming (Figure 2.4). The two were sitting in a metal climbing structure grooming each other when one female, Georgia, unexpectedly took the hand of an older female and lifted both of their hands high into the air. They thus sat in a perfectly symmetrical A-frame posture, each with their free hand grooming the pit of the other's lifted arm. The great advantage was that the custom seemed to be in its early stages. Rather than encountering a wild community of chimpanzees that has been doing it for hundreds or perhaps thousands of years, here we had a group in which the behavior was initially extremely rare (only a dozen instances were seen in the entire first year of daily observation) and was always initiated by the same individual, Georgia, the presumed inventor. The posture strongly resembled the so-called handclasp grooming (HCG) reported for wild chimpanzees in the Mahale Mountains, Tanzania (McGrew and Tutin 1978).

A unique property of the handclasp grooming posture is that it is not required for grooming the armpit of another individual. Chimpanzees that do not perform HCG are no less hygienic in the underarm area than those who do. Thus it appears to yield no obvious benefits or rewards to the groomers. It has been proposed that handclasp grooming and even specific aspects of the posture itself act to symbolize a close relationship between the grooming pair. Indeed, the intimate nature of the posture involves a degree of cooperation and trust among the partners (McGrew and Tutin 1978; de Waal and Seres 1997). This has led to the hypothesis that the nature of the relationship between two individuals will predict the development of HCG between them. While we would

*Figure 2.4* One group of chimpanzees at the Yerkes Primate Center shows the handclasp grooming posture, also observed in a few wild communities. Here the posture is shown between the adult sister and the mother of the behavior's originator. Photograph by Frans de Waal.

expect to see HCG among kin or close affiliates, we would not expect HCG between two individuals who are rivals.

To explore this prediction, we analyzed retrospectively 11 years of data (from 1992 to 2003) on the occurrence of HCG in one group of chimpanzees at the Yerkes Field Station (a second group in identical surroundings at the same facility never showed the posture). During this period nearly 300 instances of HCG were observed by members of our team. The pattern spread gradually until all adults regularly performed HCGs (Figure 2.5).

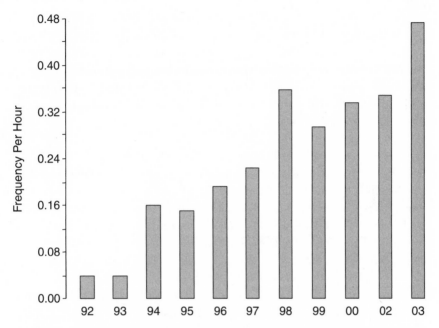

*Figure 2.5*  Rate per hour of observation of all handclasp grooming bouts from 1992 through 2003. The rate increased steadily and significantly throughout the study period. From Bonnie and de Waal (2006).

We were able to define for each year the degree of affiliation between all possible grooming pairs within the group. To do so, we looked at the proportion of scan samples collected during 90-minute observations of the group in its outdoor enclosure in which dyads were engaged in contact sitting, sitting within arms' reach, grooming, and mutual grooming. From these data a dyadic matrix was created of average proximity patterns (contact sitting and sitting within arm's reach). A second matrix included handclasp grooming, a binary variable regarding the occurrence or nonoccurrence of HCG within a dyad.

For each year the proximity matrix was correlated with the HCG matrix, and the results were compared with 5,000 random permutations of the same matrices (cf. Dow and de Waal 1989). Proximity was found to correlate positively with HCG, which confirmed our prediction that dyads with a higher rate of affiliative exchange were more likely to develop HCG than others. More than half of the top 25 percent of dyads in

35

terms of affiliation were observed to engage in HCG. In contrast, of dyads that fell within the lowest quartile of affiliation, only 15 percent were observed to engage in HCG at some point during the study period. In addition, in nearly all dyads formed, at least one individual had been previously observed to handclasp groom. We concluded that affiliation and individual experience determine the spread of handclasp grooming (Bonnie and de Waal 2006).

### Learning of Arbitrary Conventions

Social cultures, like material ones, require controlled experiments to explore how behaviors are transmitted within a group. Thus far, such experiments have included mimicking by apes of arbitrary gestures, body movements, and actions on objects that were demonstrated by humans (Tomasello, Savage-Rumbaugh et al. 1993; Custance et al. 1995; Myowa-Yamakoshi and Matsuzawa 1999; Call 2001) and the previously mentioned co-rearing of two different macaque species (de Waal and Johanowicz 1993). However, since all of these studies have relied on cross-species interaction, they have limited ecological relevance. No study has successfully generated a new social convention in an established group. Indeed, in a study in which one chimpanzee was trained to employ arm raising and other arbitrary gestures to gain food from a human, the behavior failed to be adopted by other members of its group (Tomasello et al. 1997). Even though this was a relatively short, single attempt, the authors took their negative finding to mean that nonhuman species lack the capacity to observationally learn the significance of arbitrary actions, a capacity considered fundamental to human culture.

Working with the same chimpanzees at the Yerkes Primate Center Field Station, we seeded in each of two groups a different endpoint to a complex action chain involving familiar objects. Although the chimpanzees were ultimately rewarded for completing the task, the study was not a tool task (as was Whiten et al. 2005), nor was the connection with food obvious at first. Indeed, the chimpanzees showed little interest in the apparatus and did not perform the desired behavior throughout baseline sessions conducted before training of a model. Only after they had observed another chimpanzee complete the series of actions did many individuals adopt unambiguously the method specific to the group in which they lived.

*Figure 2.6* Chimpanzees in a large outdoor compound are rewarded by an experimenter positioned on an observation tower for depositing tokens into either of two receptacles on the right. From Bonnie et al. (2007). Drawing by Devyn Carter.

We provided chimpanzees with two dozen tokens constructed from PVC pipe—objects that are familiar enrichment items for our chimpanzees and have been used in previous experiments (Brosnan and de Waal 2005). Tokens were scattered throughout the compound, a large outdoor area to which the chimpanzees have free access. In addition, two unique "receptacles," a bucket and a chute, in which tokens could be placed were available. The experimenter stood on a platform 6 meters above the scene so as not to bias the chimpanzees toward the location of either receptacle (Figure 2.6).

To obtain a reward, chimpanzees needed to complete a five-step behavioral sequence that included searching for a token in the compound, picking it up and transporting it to the receptacles, inserting (and letting go of) the token into either the bucket or chute, and finally looking up toward the experimenter for a reward. A high-ranking adult female from each group was trained to deposit tokens in one of the two receptacles, although both receptacles were always available and both yielded equal rewards to all.

As a result of observing a conspecific model interact with the tokens and receptacles in a meaningful way, two different but equally rewarding conventions spread to become traditions in both groups. Because both receptacles were available and yielded equal rewards, it is unlikely that all individuals in a single group would have discovered just one of the receptacles on their own. Thus we have demonstrated that chimpanzees are capable of perceiving the benefits associated with an initially meaningless action chain performed by a conspecific model, and of duplicating the entire chain so as to gain the same reward (Bonnie et al. 2007).

## Conclusion

To understand the mechanisms of social transmission that underlie animal (and human) culture, we need to move beyond the learning paradigms developed on individually tested laboratory animals. Not that individual experience plays no role. Rewards for individual performance definitely exert a major influence on the speed of behavioral acquisition and the maintenance of behavior. But this is usually not how social learning starts. It starts with paying attention to others, often literally being "in their face" (Figure 2.7), watching every move they make. It is here that identification with the other and being socially close matter. One might downplay this as merely the attentional level, but there is more to it. There are many examples of unrewarded behavior that is nevertheless copied from others (e.g., handclasp grooming, the opening of boxes in our monkey experiment). This means that acting like others has its own intrinsic motivation, regardless of external reinforcement. Social learning is more, therefore, than individual learning in a social context: it is subject to powerful social modifiers and motivators.

A good example, apart from the multitude of examples provided in this chapter, is the "conformism" suggested by Whiten et al. (2005), who

*Figure 2.7* Watching adults is a favorite activity of young chimpanzees and is thought to be the way they gain knowledge about sources of food and feeding techniques, such as this female's way of picking grubs out of rotten wood. Photograph by Frans de Waal.

found indications that even if chimpanzees discover an alternative solution to a problem, they nevertheless converge on the "culturally" prevailing solution. We have found additional suggestions for this effect, which, if confirmed, would nicely fit the BIOL model but not the other models of social learning.

This has implications for how cultural learning research on primates is being conducted. Instead of using human models, for example, we conduct all of our experiments with conspecific models. And instead of looking at social learning as mainly a cognitive issue, we have ample evidence that it is very much part of social relationships, requiring bonding, identification, and tolerance. Even though traditional learning paradigms undoubtedly need to be part of any cultural learning framework, we suggest that social relationships and close observation of certain "role models" will need to be part of it as well, at least for the many social mammals that live in highly individualized societies.

ACKNOWLEDGMENTS

We are grateful for constructive comments from Kevin Laland and Victoria Horner. We thank the technicians, animal care staff, and undergraduate students who helped during our projects. The capuchin work was supported by a National Science Foundation grant (IBN-0077706) to the first author, an NSF Graduate Research Fellowship to the second author, and a grant from the National Institutes of Health (RR-00165) to the Yerkes National Primate Research Center. The center is fully accredited by the American Association for Accreditation of Laboratory Animal Care.

# TEN DISPATCHES FROM THE CHIMPANZEE CULTURE WARS, PLUS POSTSCRIPT (REVISITING THE BATTLEFRONTS)

W. C. MCGREW

## Introduction

Culture applies equally to being yogurt or watching *kabuki*. A culture-bearer may be a petri dish or an *imam*. Herein lies the problem: The label refers to a wide range of phenomena. Somewhere in this range lie boundaries of uncertainty. The fermentation of bacteria is easy, but the fermentation of ideas raises issues. What is culture? When did it emerge? How do we know it? Who has it?

Add a heavy dollop of strong feelings to this mix, and controversy is assured. The title of this essay is taken from the current preoccupations of anthropology (and the social sciences in general), which is riven by bitter struggle. The "culture wars" in "cultural studies" are about essentials and jurisdiction, and, ultimately, about identity. The same issues affect what is called here "the chimpanzee culture wars." (For a recent and somewhat detached view of the human version, see Kuper 1999.)

Wars have battles, and from battlegrounds come dispatches, which are meant to be timely and terse. Here, these 10 dispatches are very much the latter, as depth gives way to breadth. Imagine them as being light enough to be flown by carrier pigeon from far-flung battlefields, in anthropology, biology, linguistics, psychology, and so on, amid shot and shell.

Put another way, the aim of this chapter is to look closely at cultural primatology. (Though young, the phrase already has two meanings. Human attitudes, beliefs, and treatment of apes, monkeys, and prosimians, better termed ethno-primatology, is *not* covered here. See Wheatley 1999.) Cultural primatology is analogous to cultural anthropology, as a subset of investigators interested in the culture (as opposed to the anatomy, ecology, genetics, or physiology) of nonhuman primates. However, whereas cultural primatologists assume that nonhuman primates are cultural creatures, most cultural anthropologists instead presume that only humans have culture. Both cannot be right.

More than 40 populations of wild chimpanzees across Africa make and use tools, from Tanzania in the east to Senegal in the far west. No two populations appear to have the same technological profile. Some use probes of vegetation to harvest social insects; others use stone hammers and anvils to crack open nuts. In neither case is the geography of tool use explained by absence of raw materials or targeted tasks. Even when these are constant, variations in style occur that cannot be explained by ecological factors. Such variation in humans is called material culture and ends up in museums. What to do when it occurs in apes?

## Chimpanzee Culture? Absurd!

In this dispatch, from the battlefield of ethnology, culture is taken to be both universally and uniquely human, that is, all *Homo sapiens* have it and only *Homo sapiens* has it. Thus, culture is both a necessary and sufficient condition for humanity. (Modern humanity, that is. There is a problem of what to do with archaic forms like Neanderthals. Whether they were cultural or not evokes some debate.)

Some examples will show how difficult is this stand to maintain: Celestial navigation is neither universal nor unique to humans, as humans living in closed-canopy tropical forests do not show it, but migrating songbirds do. Constructed shelters may be universal to all human societies, but these are not unique, as shown by beaver lodges and orangutan beds. Writing is uniquely human, but not all humans know of it, as there are many non-literate cultures (or were, prior to contact by outsiders). So, what is left to humanity as both universal and unique? How about the space shuttle, or for that matter, the wheel? Or, the computer, or for that matter, the abacus? Unfortunately for this line of argument, most foraging peoples, whether Inuit or San, have none of these. Either they must be excluded from cultured humanity, or we must look deeper.

A skeptical pragmatist might point to seemingly obvious truths, e.g., that all humans clearly depend on culture, while just as clearly other species do not. (This is sometimes expressed cleverly as culture being humanity's ecological niche.) To scupper this proposition, we would need to find only one human society lacking culture, but we cannot, as the ethnographic record shows. However, the same logic applies to chimpanzee populations; more than 40 have been studied, and all seem to have culture. Therefore, which species is more dependent? Neither one. Moreover, as de Waal (2001) has argued, there is much evidence that chimpanzees do need culture, as unacculturated apes may be at best incompetent, and at worst, dead. Textbooks in social sciences, especially in anthropology, are full of such assertions, presented as "obvious" truths, but these need querying.

A skeptical mentalist might object that all that we know of nonhuman culture is based on behavior, which is the least notable aspect of culture. Much more interesting is the knowledge that underlies and informs behavior. Even more challenging is the meaning attributed to the knowledge that drives behavior. And finally, there are the emotions that color the meanings that pervade the knowledge that is manifest in behavior.

How can any sensible person imagine that all of this exists in animals? Surely, it is said, this is what distinguishes a mating system (pair-bonded gibbons) from the institution of marriage (wedded Mennonites). Surely this is what distinguishes an optional taboo (the English do not eat horses) from obligatory carnivory (tarsiers eat no plants). Surely this is what distinguishes a rite of passage (Maasai initiation of young men) from puberty (adolescent male chimpanzees challenging their elders).

There are several answers to this question. One is that all that we know of knowledge, meaning, and emotion is based on behavior. We have no direct access to human or any other minds, so all is inference. Whether or not we are any better at divining human than nonhuman minds is debatable. We share the perceptual world of our fellow humans (advantage), but we are also susceptible to their mendacity (disadvantage).

It may be that the chimpanzee mind is wholly devoid of knowledge, meaning, and emotion, or it may be that some or all of these phenomena are there, in distinctively apish form. As with our fellow humans, we can only try to draw valid inferences. It would seem that cultural anthropology would be of great help to cultural primatology in this task. At the very least, the former could tell the latter what evidence would suffice in principle, so that primatologists can seek it in practice. There is a history of such aid (flirtation?) in sociocultural anthropology, dating back to Kroeber (1928), but including also the works of Ruth Benedict, Marvin Harris, and G. P. Murdock.

The prospect of chimpanzee culture is absurd only if it is unimaginable.

*Figure 3.1*   Party of Gombe chimpanzees of mixed age and sex.

## Chimpanzee Culture? Of Course!

In this dispatch from the battlefront of ethology, the problem is not exclusion of other species from the cultured, but rather finding the limits of inclusion. In his groundbreaking book, *The Evolution of Culture in Animals,* Bonner (1980) was willing to grant candidacy to slime-molds. That is, depending on definition (see Dispatch 7, Culture Is by Definition), culture is present not only in primates, or mammals, or homiotherms, or vertebrates, or invertebrates, but also in organisms lacking a nervous system! (This is *not* just word play nor is such inclusiveness limited to culture. See Strassmann and colleagues 2000 for the case for altruism in a protozoan.)

Textbook examples abound. In the same year as Bonner, Mundinger (1980) made the focused case for animal culture based on vocal learning in passerine birds. Invoking the research of pioneers such as Marler on white-crowned sparrows, Mundinger argued that characteristics such as plasticity, diffusion, tradition, and innovation were met. Hundreds of studies have extended these results and teased out the mechanisms of song learning and transmission (e.g., West and King 1996). In mammals, California sea otters crack mollusks on stone anvils but their Alaskan cousins do not. Among the Californian population, most crack while floating on their backs at the surface, but some take their tools underwater!

The problem with these (and others, e.g., Galapagos finch, bower-bird, black rat) marvels of natural history among nonprimates is that they seem to be mostly "one-trick ponies." That is, sea otters are great at anvil use, but do little else that is not species-typical. European black-birds are wonderfully creative singers, but the rest of their behavioral repertory seems stereotyped by comparison. If one of the hallmarks of culture is its comprehensiveness, then single-trait candidates are bound to fall short.

None of these caveats applies to Japanese monkeys, however. Not only do these primates show an impressive array of cultural patterns, but also some of these habits have been tracked by primatologists for decades. Furthermore, their interpretation as cultural has been asserted as cultural from the outset, by Imanishi and his intellectual successors (de Waal, 2001). It is no accident that sweet-potato washing, wheat sluicing, hot-spring bathing, and so on are to be found in every introductory textbook. We marvel at photographs of snow-topped monkeys immersed to their necks in hot springs and wonder at their ingenuity.

Yet there is a problem with most (but not all; see Nakamichi et al. 1998) of these classic cases. Their origins lie in human facilitation; the monkeys were lured to the beach or into the pools by artificially providing them with domesticated plant foods. This enhancement takes nothing away from what happened next, how the habits spread by horizontal or vertical transmission or continued to elaborate or became fixed. But one must always wonder: How many of these habits would have occurred without human assistance?

Within behavioral biology, crediting chimpanzees with culture or not is polarized. Ethologists (Boesch, Goodall, Nishida, Wrangham), who study chimpanzees observationally in nature, tend to say yes. Comparative psychologists (Galef, Premack, Tomasello), who study apes or rodents experimentally in captivity, tend to say no. Less of a dichotomy exists among researchers on capuchin monkeys, with field-workers (Boinski, Fedigan, Panger, Perry) largely affirmative but lab workers (Anderson, Fragaszy, Visalberghi, de Waal) of more mixed opinion. In any case, methodology and degree of direct experience are often confounded in investigators.

Notably absent (with a few exceptions, e.g., Boehm) are researchers in cultural primatology who were educated in cultural anthropology. Most primatologists are committed to the neo-Darwinian evolutionary paradigm, as natural scientists. Most anthropologists are not so committed, and think of themselves as social scientists, if they consider themselves to be scientists at all. This is a recipe for misunderstanding.

Illumination may come from a completely different mammalian order, the Cetacea. Despite the obvious logistical difficulties of studying aquatic culture, recent research on dolphins and whales is productive and provocative (Rendall and Whitehead 2001). Cetaceans do things that no ape has been seen to do: chimpanzees form small coalitions; dolphins form larger coalitions of coalitions. Cetaceans do things that we primates cannot even imagine: we can only guess at the networked communicative capacities of an echo-locating pod of orcas.

## Culture Is Not Behavioral Diversity

Who can fail to be moved by the richness of human cultural diversity? Nowadays, every city has a wealth of ethnic cuisines, so that even a humble onion can turn up on a plate in a wonderful variety of forms. Satellite television brings this cross-cultural variation into our living rooms. So, it is argued that if we find behavioral diversity in apes, then they must be cultural.

True, field primatologists realized in the 1970s that we could no longer speak of The Chimpanzee. (If pressed for a milestone, one might point to Menzel's symposium on precultural behavior in primates, held at the 1972 International Congress of Primatology.) Instead, every study, whether at Bossou, Gombe, Mahale, or Taï, seemed to reveal a new twist, if not a new behavioral pattern. This eye-catching reporting continues, and the ethnography piles up, so it is hard any more to keep straight which population of apes does what. Surely, there is now a need for a Chimpanzee Relations Area File, else how can anyone keep up with the information?

However, if diversity means differences across equivalent sets, and such sets are hierarchical, there is a potential for confusion across levels. Individual variation is easy to see in primates, as Imanishi and his students showed for wild Japanese monkeys, and Köhler even earlier showed for captive chimpanzees. In chimpanzees, each alpha male has his own style. But individual differences are usually seen as something for personality psychologists to study, not as matters of culture.

At the other extreme, species differ in behavior, and sibling species like chimpanzee *(Pan troglodytes)* and bonobo *(Pan paniscus)* provide a pastiche of similarities and differences (Stanford 1999). But these are not cultural matters, any more for primates than they would be for congeneric lion and leopard, and are usually left to comparative ethology or psychology. (Such boundaries are more fuzzy for artifacts found where and when anatomically modern humans and Neanderthals coexisted. Who made what may or may not be cross-cultural diversity.)

More to the cultural point is diversity at the level of community, population, or subspecies. In chimpanzees, neighboring unit-groups at Mahale show different versions of the grooming hand-clasp (McGrew et al. 2001). The separate but nearby populations of Mahale and Gombe have contrasting cultural profiles (Nishida 1987). Far western African chimpanzees *(P. t. verus)* are nut crackers, while the other sub-species *(P. t. troglodytes* and *P. t. schweinfurthii)* show none (McGrew et al. 1997). All of this diversity could be cultural.

However, lots of species show behavioral diversity. In nonapes, hamadryas baboons on opposite sides of the Red Sea in Arabia and Ethiopia differ in group size and composition (Kummer 1995). In nonprimates, sea otter lithic technology varies along the Pacific coast. In nonmammals, scrub jay family structure differs from Florida to Arizona to California (Woolfenden and Fitzpatrick 1984). Whether any or all of this variation is cultural or merely reflects environmental constraints is a matter of investigation, but not assumption.

Kummer (1971) also pointed out a third way that behavior may vary, in addition to nature (environmental dictate) or nurture (social learning).

47

Organisms also may learn much individually from interaction with the nonsocial environment, by trial and error. If such learning (e.g., predator avoidance) occurs in parallel, even simultaneously across individuals, it may appear social when it is not. No one needs to show us that the sun is hot; we all learn this for ourselves at a young age.

A more vexing problem in judging behavioral diversity is quality of data. Anecdotes are of little use, as a single event can be a coincidence, subject mistake, observer error, or hoax. At best, an anecdote alerts us to a possibility. Equally limited in usefulness is idiosyncrasy. This eliminates all the problems of anecdote, but an act done by only one individual, however often, can hardly be cultural, as it is asocial. Even a habit, that is, behavior done repeatedly by several group members, is but a hypothesis. Only customs, that is, acts performed normatively by appropriate subsets within a group (e.g., cooks cook, soldiers fight, elders advise), are evidence of culture. Such distinctions are often ignored by eager reporters and may make ethnology difficult (Whiten et al. 1999).

Finally, behavioral diversity is neither a necessary nor sufficient condition for culture. Kwakiutl eat elephant seal but Bantu eat elephant. This is diversity but need not be culture, as no Kwakiutl ever met an elephant, nor Bantu an elephant seal. All the world now consumes carbonated cola drinks. This is global uniformity, but it is still culture. The custom has spread, not the geographical distribution of Cola trees.

Thus, in seeking culture, behavioral diversity is just a possible starting point.

## Culture Is beyond Social Learning

In its broadest sense, social learning occurs when information gained from others of the same species alters one's behavior, thoughts, attitudes, and feelings (although only the first of these is observable!). This contrasts with information gained from the rest of the animate or inanimate world. So, does social learning equal culture?

Social learning occurs in all classes of vertebrates and in several kinds of invertebrates, which lack true brains. In the short term, honey bees learn from their fellow workers where to find nectar. In the longer term, an octopus may learn permanently to avoid a predator, from one observation of another doing so.

Given how widespread is social learning, many investigators have focused on its mechanisms, that is, on *how* rather than *what*. The array of

*Figure 3.2*    Adult male plays with infant *(right)* while mother grooms her juvenile daughter *(left)*, Gombe.

possible means (and their associated jargon terms) by which information is transferred is daunting (Whiten and Ham 1992; Byrne and Russon 1998). These distinctions among mechanisms are differently emphasized, depending on the species being studied. Comparative psychologists studying nonhumans spend much time on thresholds (Is stimulus enhancement enough?) or alternatives (Is emulation *really* imitation?) or rubicons (No imitation, no culture!). This can be confusing to nonspecialists.

On the other hand, sociocultural anthropologists studying humans pay little attention to mechanism, being more interested in *what* is passed on. When information is available on processes, it turns out that most customs are transmitted by a mélange of passive observational learning. For example, Aka pygmies of Congo learn most of the 50 most important activities of daily life by watching others, not by being instructed (Hewlett and Cavalli-Svorza 1986).

Of the cognitive mechanisms of social learning, teaching is deified by some comparative psychologists. It is said to be essential to culture and unique to humans, and so becomes a hallmark of "true" culture. If teaching is defined as acts by a tutor with the goal of improving the performance of a pupil, then much social learning by humans may not qualify. Formal teaching is absent in most traditional societies, except for specific contexts such as initiation rites. This makes sense: Teaching is costly, in terms of time, energy, and emotion. It is the mode of last resort for social learning, when simpler means fail. Teaching is arguably a curse, not a blessing, made necessary by a large, plastic, and expensive brain (McGrew 2001).

Riddle: When is social learning not really social learning? Answer: When human-reared apes are given "honorary" human status for the purpose of developmental cognitive studies. Thus, our closest relatives are put into experimental settings where humans are their models (and caretakers and surrogate parents and kin). Then, their ability to learn socially from human models is compared to similarly aged human children. The apes cannot win in such a setup. If they perform well, it is dismissed as "enculturation," that is, upgraded ability that is not generalizable to nonenculturated apes. If they perform badly, their inferiority is confirmed. Such an experimental design is sometimes termed "cross-fostering," but of course it is not, since no human child is ever taken from its kind and turned over to apes for rearing. It makes an interesting thought experiment: Who would show more social learning and cultural superiority, a human infant reared by an ape, or an ape infant reared by a human? Arguably the artificiality of both conditions means that little can be learned about evolved processes of social learning or culture from them.

In summary, there are two alternatives: If culture equals social learning, then many creatures, e.g., octopus, guppy, and lizard, must be granted cultural status. If culture is more than social learning, then we must look elsewhere for essential criteria. On these grounds, it seems sensible to consider social learning as necessary but not sufficient for culture.

## Tradition Is Not Enough

Tradition is continuity over time. More precisely, tradition is vertical transmission of information across generations, from old to young. Rarely, when innovation is youthful, vertical transmission may go in the reverse direction. This occurred with the first spread of sweet-potato washing by Japanese monkeys, but it seems to be rare (Hirata et al. 2001).

Examples of traditions in animals abound. Every year, wildebeests migrate across the Serengeti, whooping cranes winter at Padre Island, monarch butterflies flit to Chihuahua, salmon surge up the Tweed. Some of these traditions have gone on since before the human species emerged. In some cases, especially with migratory birds, we can monitor how off-spring retrace the routes of their parents, or even ancestors. In a few cases, we have detailed information: At Gombe, the termite fishing of wild chimpanzees has been recorded over four generations of the "F" family, starting with Goodall's observation of the matriarch Flo in the 1960s.

For human beings, the central role of tradition is clear. All human so-cieties emphasize origin myths, however fanciful; attention to tradition, such as appeal to ancestors, is a human universal (Brown 1991). People who tell stories to sociocultural anthropologists often stress that they have always done things a certain way. Conversely, people who fail to keep traditions may be severely punished.

So, does culture equal tradition, and tradition equal culture? No, it is not that simple. First, some information is transmitted genetically across generations. This is deucedly difficult to establish in the wild, where vari-ables of nature and nurture are confounded, even for behavior. Do gen-erations of warthogs wear down a path to a water hole because it is their cultural inclination, or because it is the most energetically efficient or least predator-risky route? Even if we were lucky enough to be there to see them tread a new trail, would it be from whimsy or from changed (but unseen to us) environmental contingencies?

Second, not all transmission of information is vertical. Some is hori-zontal, within generations and across peers. The power of human "popu-lar culture" is impressive—ask any teenager. But horizontal transmission of culture is more than fad. Opie and Opie (1987) showed that some as-pects of children's culture, such as jumping-rope rhymes, were maintained for centuries by horizontal transmission, child to child. Thus, there is tra-ditional culture, but no intergenerational transfer.

Finally, even if we learn from our elders, and pass on those customs to our successors, those traditions need not be cultural. Whether the models are kin, companion, or even stranger need make no difference. I learned to fish from my uncle, but I now realize that he had idiosyncratic techniques. I make pineapple upside-down cake using the recipe that is a

*Figure 3.3* Two adolescent male chimpanzees play with adolescent olive baboon at Gombe. (This is apparently unique to Gombe.)

family tradition, but it is unlike the same dish made by the other matrilines in my culture. These are surely traditions, but whether or not they are cultural, in the rich sense, depends on definition. After all, most people are not anglers, and who knows how many folks have ever made such a cake, or passed it on to their descendants? How normative does a behavioral (or cognitive or emotional) pattern have to be to be cultural?

Some particular aspects of tradition as culture have been singled out as crucial, such as the ratchet effect (Tomasello 1999b). Here, information not only spreads but accumulates; thus, with each transmission, either vertically or horizontally, new "mutations" (memes?) enrich the message. We stand on the shoulders of our predecessors, making progress. This is said to be uniquely human, and so is presented as both a necessary and sufficient condition for culture. It is neither. Putting aside the problems of misreplication and maladaptation, racheting is neither unique nor universal. Since its invention in 1956, wheat sluicing or washing by Japanese monkeys has elaborated and diversified (Hirata et al. 2001). Imo's initial technique now looks crude, and successive generations have left it behind.

Equally, the evidence for ratcheting in the human ethnographic literature is slim. Most ethnographers of traditional societies report stasis, not dynamic change.

So, even if tradition is a necessary condition for culture, at least in the long term, it is not a sufficient condition, at least as the term is used here.

## Culture without Language

Language is everyone's favorite example of human culture. Each of us imprints upon what we hear in the cradle; some of us go on to learn more than one language. It is all a matter of exposure. No one ever suggests that the brains of infants born in Patagonia might be more genetically receptive to hearing Patagonian than Danish. Instead, we see the results of an inadvertent but global experiment in cross-fostering. A Korean newborn adopted by a Canadian grows up to speak English, or French, or both, but not Korean.

Thus, it was no surprise that the first published response to a claim of a social custom in chimpanzees (McGrew and Tutin 1978) was to deny it on the basic grounds that nonlinguistic creatures could not have culture (Washburn and Benedict 1979). This belief that language and culture are isomorphic is widespread.

For humans, the evidence is too strong to make a claim about the relationship between language and culture. All known human societies have both, so there is no informative variance. One could enable the other or vice versa, or both could be a by-product of a third phenomenon (e.g., big-brained intelligence), or each could be independently derived (e.g., language from vocal communication and culture from extractive technology). We would need to have cultures without language or languages without culture to test the relationship. We could seek correlations between linguistic variables, such as vocabulary size, and cultural variables, such as technological complexity, but this seems not to have been done.

Or, it may be that apes are helpful models. Despite the huge and contentious published literature in "pongo-linguistics," there is no consensus (Savage-Rumbaugh 1998). There seems to be a positive correlation between time spent with chimpanzees and conviction that they are capable of language (but note the careful wording!). Thus, the most dismissive critics have spent no time with apes. On the other hand, there is another

53

correlation among chimpologists, so that the strongest claims come from the researchers using the most artificial systems of linguistic communication, e.g., Yerkish (Savage-Rumbaugh 1998). Until an open-minded linguist is willing to go to the field and take wild chimpanzees as they are found, we are unlikely to know more. This seems a simple request, but it is yet to be done.

One problem is definition. Clearly, full-blown human language must be both semantic and syntactic. It can be cognitive or communicative or both, but the latter is easier to measure. Only spoken language is universal across human societies, but the acoustic-auditory modality is neither necessary nor sufficient. Deaf people read lips and hearing people make signs. Both vocalize paralinguistically, as do many other organisms.

In principle, many functions of language could serve culture. Cognitively, language could have evolved as a labeling or filing system, whether for numbers, ideas, or identities. Such symbol-use need have no social function. Communicatively, language is a useful way to transmit information, especially abstract and arbitrary thoughts. Such symbol-use is necessarily social, as sender and receiver must share a common language for it to work. So far as we know, all humans normally use language both cognitively and communicatively, but these could be decoupled in apes.

In studying culture, the main strength of language is also its weakness. To get beyond behavior (which is directly observable) to knowledge and meaning (which are not), anthropologists like everyone else rely on verbal report from informants. On the other hand, much information about feelings can be inferred from "body language," especially with training. However, speech is a double-edged sword, as informants may bare their souls or lie by commission, omission, or imprecision. Deception by paralanguage seems to be harder, especially with involuntary responses, such as blushing. If your life depended on detecting deception, which would you trust, the content of a word or its spoken inflection? If you chose the latter, then you might not want to trust entirely in verbal report as the sole indicator of culture, in human or chimpanzee.

This dispatch carefully avoids passing judgment on whether or not apes have language. The aim is to show that language and culture are separable. The two are no more necessarily tied by causal co-incidence than are language and bipedality. Communicative language may be a sufficient condition for culture, but it is not a necessary one.

## Culture Is by Definition

Definitions of culture are a dime a dozen, and most are of little use. Most encapsulate an idea or set of ideas, but few are heuristic for pursuing the possibility of nonhuman culture. Especially exasperating are the epigrams beloved of introductory textbook writers: "Culture is what makes us human," "Culture is the human ecological niche," or, "Culture is to human, as water is to fish." Of what empirical use are these?

Equally frustrating are the historical antecedents. Every textbook of introductory anthropology gives Tylor's (1871) seminal definition of culture as "that complex whole which includes knowledge, belief, art, law, morals, custom, and any other capabilities and habits acquired by man as a member of society." Putting aside the inherent sexism, one is left with a vague, all-embracing entity that may include all but digestion and respiration (but then think of antacid tablets and yogic breathing). Something that explains everything explains nothing.

Some have advanced checklists of features, much as Hockett (1960) did for language. Kroeber (1928) tried this for chimpanzee dancing, saying that if it showed innovation, standardization, diffusion, dissemination, durability, and tradition, it would qualify as culture. By analogy, we can recognize a luxury car by ticking off its features, and a BMW will pass and a VW will fail. Does this mean that chimpanzees are 83.3 percent of the way to being cultural if one can tick off five of the six features? Given long generation times in great apes, one would have to wait years, from innovation to tradition, assuming that the other conditions were met along the way. In studying human culture, how many ethnologists in the field have been so patient as to tick off six of six?

More productive in an operational sense may be criteria that approximate essentials, which together capture the gist of culture. Consensually, all seem to agree that culture is *learned* (rather than instinctive), *social* (rather than solitary), *normative* (rather than plastic), and *collective* (rather than idiosyncratic). This minimal combination is a starting point for necessary and sufficient conditions for attributing culture to an organism. Unfortunately for anthropocentrists, the chimpanzees' grooming hand-clasp meets all four criteria (McGrew and Tutin 1978; McGrew et al. 2001).

Another approach is to ask ordinary people what culture (in the rich sense) means to them. When anthropologists do this, and if their informants are patient enough to put up with such a simple-minded question, then the answer is usually some version of: "culture is the way we do things." This elegant phrase contains at least four elements: overt action ("do things"), norms and standards ("the way"), collective consciousness ("we"), and sense of identity (as implied by the whole phrase). So, does this apply to chimpanzees? Overt action is the easiest, as it is seen in both the behavior and the artifacts of elementary technology. Chimpanzees have both tool kits and tool sets (McGrew 1998). Norms and standards are revealed by behavioral diversity at the levels of group, population, and subspecies (Whiten et al. 1999). All known chimpanzee groups, populations, and subspecies scratch themselves, but only M group at Mahale in East Africa does the social scratch, which they do often and predictably (Nakamura et al. 2000). Collective consciousness is pointedly manifest in deadly xenophobia. Parties of chimpanzee males patrol boundaries and kill neighboring rivals, but usually only if three or more aggressors can catch the victim alone (Wrangham 1999). A sense of identity can be inferred when an immigrant female changes her style of doing a common behavioral pattern from that of her community of origin to that of her community of adoption. Their old way of doing things becomes her new way of doing things.

Definitions are useful only if they clarify matters. All else is pedantry. Define culture as you must to tackle the question at hand; just make it clear, fair, and most of all, productive.

## Culture Is Collective

Culture is social (as opposed to solitary), but sociality is only a starting point. Collectivity implies much more. When 51 percent or more of a group behaves in concert, the act becomes a statistical norm, and therefore is typical of the group. Herring shoal, geese flock, bison stampede. Instead, collectivity entails group-oriented action, often with roles, as in an orchestra, which is more than many instruments being played at once. Empirically, roles are social traits that are not intrinsic to actors, but are transferable, able to be donned, as well as shed, like clothing.

Further, to say that something is collective is not to say it is unanimous, but often just the opposite. How often do all humans in a group act, think, or feel as one? Not all humans ski, even in the Alps. Instead the collective is often a subset according to sex, age, kinship, status, and so on in relation to other subsets. In the short term, such collective action is manifest in convention: gentlemen rise, underlings bow, grandparents

56

dote. In the long term, there are institutions: marriage, rite of passage, funeral.

The results of collectivity in culture are emergent properties. By this argument, one cannot break up culture into its components and then recombine those components to reconstitute culture. (Any more than one can reduce an animate organism to its constituent proteins, amino acids, and peptides and then bring it back to life.) Cronk (1999) made the same point about culture in another way by saying that we cannot explain behavior in terms of behavior. Thus, culture defies reductionism and is pervasive.

Finally, it is possible to be a collective creature on many levels at once. One is a European, German, Bavarian, and Münchener at the same time, and each may be indicated differently, by passport, language, patriotism, and taste in beer.

All of the above indubitably applies to human culture, whether in the New Guinea highlands or a Manila barrio. To what extent does any or all of this apply to nonhumans, such as the chimpanzee? At first glance, the task seems impossible. How can we possibly know the mind of an ape, if we have so much trouble comprehending the minds of our fellow humans?

The easiest starting point is the "quack test." The more it looks, sounds, smells, and feels like a duck, the more likely it is to be a duck. For example, grief is apparent in a chimpanzee mother after her infant dies. She is alternately agitated or subdued, distracted or focused. She may be more solicitous to her infant's body than she was to it in life. She seems to be grief-stricken, but this does not mean that she is in mourning, for that is a collective action, not individual sorrow. A sociocultural anthropologist faced with this set of circumstances would adjust her lens accordingly; cultural primatologists need to learn to do so.

One can be guided by function, for evolution ultimately boils down to outcomes in response to natural selection. Dead humans do not pass on genes any more than do dead flatworms (cryogenics apart!). Territorial aggression toward outsiders is natural; a simple rule of "Welcome familiars, but resist strangers" may be enough. Xenophobia is cultural; it is a social phenomenon based on "we" versus "they," and so comes down to collective identity. We can see this when chimpanzees immigrate or congregate. Marshall and colleagues (1999) showed that the long calls of a motley assemblage of captive chimpanzees converged to create a recognizable conventional signal. Regardless of their disparate origins, the apes created a collective dialect.

*Figure 3.4* Two allied males in a quiet moment of contemplation, Mahale K group.

Lest we despair at the task of operationalizing beliefs and attitudes in other species, there are precedents: Tactical deception seemed intractable until Byrne and Whiten (1988) illuminated it. All is inference; the challenge is to find such ingenious ways to increase the probability of more and more accurate inference. It seems likely that social dominance in chimpanzees is a matter of personalities embedded in a collective context (de Waal 1996a). Accordingly, we are likely to understand it only if we act as cultural primatologists, albeit haltingly. We will likely never interview chimpanzee informants, but we can use ethological methods to seek chimpanzee uniqueness, as well as universals.

## Culture Has Escaped from Anthropology

The concept of culture emerged as the core of anthropology in the 1870s, and remained therein for more than a century. Anthropology has been termed the science of culture, and most members of the American Anthropological Association label themselves as cultural anthropologists.

Yet all along, the question of nonhuman culture has lurked in the wings. Morgan (1868), arguably the founder of American ethnology, extolled the technology of the beaver at the same time as he described Iroquois kinship. Kroeber's (1928) consideration of comparative possibilities across species was cited above. Benedict (1935) supported a graduated transition from noncultural lower animals to cultural man. The first primatologist to propose cross-cultural studies was Imanishi (1952), based on early observations of Japanese monkeys. The first explicitly titled book on the subject seems to be *Precultural Primate Behavior*, edited by Menzel (1973a). The first systematic analysis across cultures of chimpanzees covered six wild populations studied for 151 years in total (Whiten et al. 1999).

Cultural primatology is not bounded by anthropology, and the culture concept has diffused to other disciplines. At least three make distinct contributions in very different ways (McGrew 1998): Anthropology asks *what* questions about the constitution of culture, whether these be artifacts in the past or rituals in the present. This is culture as phenomenology. Psychology asks *how* questions about the mechanisms and processes of culture, especially its inventions and their spread. This is culture as information transmission. Zoology asks *why* questions about the survival value and fitness of culture, using the ideas of neo-Darwinian evolutionary theory. This is culture as adaptation. Luckily, cultural primatology calls on all of these points of view.

However reluctant some anthropologists may be to give up exclusive jurisdiction over culture, or to extend the concept to nonhuman species, it is happening anyway. The best strategy for retaining culture in the field of anthropology is to set explicit standards to be met by cultural primatologists and let the chimps fall where they may. Echoing Kroeber, cultural anthropologists should operationalize their criteria and ask primatologists to meet their challenge. The latter are entitled to ask what it would take to satisfy anthropologists and then to go back to the field to seek it among the apes. It is easier to seek holy grails if you know what to look for.

Finally, there is a delicious irony. Some proportion of sociocultural anthropologists find the concept of culture to be outmoded and even obstructive (Kuper 1999). This is hard for nonspecialists to understand, almost as if musical chairs could somehow be played in silence. How strange to think that finally when cultural primatology realizes how much it needs cultural anthropology, the latter may drop its central tenet.

## Culture Is Rich and Complex (But So What?)

Paraphrasing Groucho Marx, a chimpanzee thinking of joining the Culture Club might hesitate to do so, on grounds of suspicion of any club that would have her as a member. Quoting Boy George, the androgynous pop star, if embracing cultural relativism, and so becoming, chameleon-like, "a man without conviction," is the price, it might be too high.

Culture may be a curse, as well as a blessing. Chimpanzees at Bossou have invented "pestle pounding," in which the crown of an oil palm is the mortar smashed by a detached frond as the pestle (Yamakoshi and Sugiyama 1995). The result is a rich, pulpy soup and a good meal, but it likely kills the palm in the process. This is short-term gain but long-term loss, as the palms, like all large organisms, are slow to replace themselves.

Culture may be overrated in several ways. For example, culture is not an explanatory variable. One cannot explain behavioral diversity just by saying that culture made them do it. Cultural determinism is just as silly an idea as genetic determinism. Furthermore, there is often a misplaced value judgment: Culture does not make an organism cooperative any more than nature makes it competitive (Wrangham 1999).

Finally, culture may not be the key to understanding chimpanzee society anyway. As with humans, the more informative level may be one step down, in subculture. Just as it may be simplistic to assume that there is such a thing as (North) American culture, so it may prove for chimpanzees and other species of primates. There is plenty of evidence that shows life as a whole among lower-ranking members of the group to be very different from that of the high-rankers, from sex ratio manipulation to leisure-time pursuits. It may be that cultural primatologists will have to take account of caste or class among their subjects, and seek help from sociologists.

In any event, if nonhumans have culture in any form, then we must be concerned with cultural survival. Just as cultural anthropologists are active advocates on behalf of the traditional societies that they study, so must cultural primatologists do the same. Conservationists may seek to save the species *Pan troglodytes,* but cultural primatologists must seek to preserve cultural diversity. This means going beyond a few famous, long-term study sites like Gombe or Taï. It means safeguarding Tenkere, where the apes make cushions and sandals (Alp 1997), and Tongo, where the apes dig up tubers for moisture (Lanjouw 2002). Both of these populations are unprotected and on the verge of extinction. What a pity it would be to lose them.

## Conclusions

So, are chimpanzees cultural creatures or not? We cannot yet say, to everyone's satisfaction, but the mounting evidence gives a rationale for cultural primatology. If this trend continues, then we must move on from doing beginning ethnography to doing full ethnology. Chimpanzees may use kinship terms (and a guess is that these will be found in their soft grunts, so far undeciphered), and they may have worldviews (for they seem to spend enough time musing). It may be that the ultimate function of culture for community-living apes is social identity. Just as human languages proliferated in areas where there were many distinct human groups, so may it be for our nearest living relations, where cultural identifiers tell who you are and where you come from. It is up to cultural primatologists to find ways to pursue these questions, and so to draw ever-stronger inferences.

Finally, if any of these arguments has merit, then we humans may need to re-think the boundaries of multiculturalism. We may need to be more inclusive in extending our appreciation of cultural diversity beyond anthropocentrism to admit our cousins, the great apes.

It is a measure of the interest shown in the potential of nonhuman culture that after only six years the accompanying chapter (McGrew 2003) needs serious revision. When presented at the Chicago Academy of Sciences conference in 2000, the essay was intended to be provocative and reflective of 10 key issues that were then facing cultural chimpologists. Many of the issues that were featured continue to throb and have engaged a much wider audience (e.g., Laland and Janik 2006). The aim of this brief update is to summarize and to critique some of the new developments, as seen from the viewpoint of a long-standing chimpanzee chaser.

## Chimpanzee Culture? Absurd!

This dispatch sought to convince social scientists of humanistic persuasion that the prospect of the cultured chimpanzees was *not* absurd, but rather that it was a thinkable proposition. However, although recent textbooks in biological anthropology take note of cultural primatology (e.g., Boyd and Silk 2006), there is little indication that sociocultural anthropologists do so. The latest mainstream review of the subject (Perry 2006), although specifically addressed to sociocultural aspects of the topic, for example, social norms, is written (yet again) by a primatologist, with little if any engagement by others who focus on human primates.

There is some irony here, because the nascent field of ethnoprimatology continues to grow. Cormier's (2003) book *Kinship with Monkeys* is a fascinating account of the Guaja foragers of Amazonia who raise orphaned howling monkeys like children after having eaten their parents.

However, chimpanzee ethnography goes marching on, now from over 50 populations, and new study sites that report exciting new data continue to emerge: Ebo (Morgan and Abwe 2006), Fongoli (Pruetz and Bertolani 2006), Gashaka (Schöning, Ellis et al. 2007), and Goualougo

(Sanz et al. 2004). It is now hard even for specialists to keep straight which group of apes does what and where, so that the case for a comprehensive cross-cultural database is even more compelling.

## Chimpanzee Culture? Of Course!

Meanwhile, the ethnographic paradigm (describe behavioral variation in nature and then compare and contrast across populations) has expanded notably beyond *Pan troglodytes*. Hard on the heels of the 2000 conference came Rendell and Whitehead's (2001) treatment of whale and dolphin cultures, which stimulated notable discussion. Van Schaik, Ancrenaz et al.'s (2003) report tackled orangutan culture, making use of Whiten et al.'s (2001) comparative coding format but extending the ethnography into tentative ethnological analyses. Similarly, Perry, Baker et al. (2003) compared behavioral variations across populations of white-faced capuchin monkeys, which took them into new areas, for example, "games." Notably absent are comparable treatments of the other two great-ape species, bonobo and gorilla, although the former was given a tentative two-site comparison by Hohmann and Fruth (2003).

On the nonprimate front, the undoubted star of nonhuman cultural studies is the New Caledonian crow, which makes insect-extraction tools (Hunt and Gray 2003). Multiple populations have been followed, mainly through their artifacts, and ratcheted evolution of extractive technology has been inferred. As with apes, the possibility that social learning is involved has been investigated in parallel, experimental studies of captive birds (Kenward et al. 2005).

## Culture Is Not Behavioral Diversity

Despite clear statements to the contrary (e.g., "Behavioral diversity is neither a necessary nor sufficient condition for culture," McGrew 2003, p. 426), critics continue to assert that cultural primatologists seek variation and then try to rule out alternative genetic or environmental explanations in order to invoke culture by default. This mischievous misdescription of the "ethnographic method" then goes on to show that such trichotomizing is naïve, as of course it is (Laland and Janik 2006; Byrne 2006). Oversimplification of causal mechanisms may have been a problem decades ago (e.g., McGrew and Tutin 1978), but no one seriously asserts now that nature and culture are independent of each other, either in human or nonhuman animals.

The chief and widely cited justification of such finger wagging is Humle and Matsuzawa's (2002) report that variation in the technology of chimpanzees' dipping for army ants is a function of the differential antipredator adaptations of the prey. Earlier reports (e.g., McGrew 1992) had assumed that army ant behavior was a constant, so that population-specific ant-dipping techniques must be culturally determined. Humle and Matsuzawa's demonstration that ape tool length, for example, was a reflection of ant pugnacity was hailed as evidence of environmental determinism, thus rendering any cultural explanation unnecessary. This "refutation" of chimpanzee culture quickly leapt into secondary (Byrne 2006) and tertiary (Boyd and Silk 2006) publications. As it happens, later studies that compared ant dipping at 13 sites across Africa show that variation in techniques and tools used by the apes cannot be explained by the ant taxa present (Schöning, Humle et al. 2008).

## Culture Is beyond Social Learning

One of the most exasperating skirmishes in the culture wars is the repeated assertion that social learning equals culture. Although no one would deny that social learning is a necessary condition of culture, it is equally likely that it is not a sufficient one. To dumb down culture to equivalence with social learning is to reduce a complex phenomenon to a caricature.

If culture as a concept means anything, it is collective. That is, it is characteristic of a set (or subset) of social learning creatures, preferably one that subsists and persists in the real world. Similarly, culture is pervasive, so that the social transmission of customs among the members of the set spills over into all or most aspects of life. From this flow emergent properties such as identity, as manifest in societal behavioral patterns as diverse as xenophobia and ostracism. No one claims that empirical investigation of such tricky topics is easy, but other equally thorny social phenomena, such as tactical deception, have been clarified by ingenious analyses. To show that a guppy will learn from another a route through a maze is nice, but to equate that with culture is tantamount to saying that kicking a football is equivalent to a soccer match.

Meanwhile, clever studies of a variety of organisms, some very far removed from primates, show us yet again that simple dichotomies are usually wrong, and that graduated criteria are more likely. The rubiconic status of teaching as a means of transmitting culture, one that separates humans from other species, seems less and less tenable. Meerkats teach their young to deal with potentially dangerous prey by sequential, ratcheted exposures that take account of the growing competence of the pupils (Thornton and McAuliffe 2006). Even ants use bidirectional feedback in tandem running between teacher and pupil in transmission of crucial information about food finding (Franks and Richardson 2006). Given this, the apparent rarity of teaching by apes is all the more puzzling and so all the more deserving of investigatory scrutiny. When will someone do an experiment in which captive subjects, apes or otherwise, are put in situations in which teaching another is made rewarding to the knowledge holder?

## Tradition Is Not Enough

More and more, one sees the term "tradition" used interchangeably with "culture" as if the two labels were synonymous. Previously (McGrew 2004) I argued that tradition is neither a necessary nor a sufficient condition for culture: behavioral continuity over generations can exist without social learning, and "pop" culture transmitted horizontally exists without intergenerational transmission.

Further, some researchers seem to equate tradition with nonhumans, while culture is reserved for humans, thereby reerecting old speciesist barriers. This is made easier when tradition is defined loosely, for example, "a distinctive behavior pattern shared by two or more individuals in a social unit, which persists over time and that new practitioners acquire in part through socially aided learning" (Fragaszy and Perry 2003b, p. xiii). By these criteria, a successfully completed course of tennis lessons between instructor and pupil would qualify as a tradition. For behavioral patterns to be traditional, surely they have to endure across generations.

A good place to look for full-blown tradition is to track cumulative cultural change, that is, the ratchet effect. Although there is evidence of this in Japanese macaques (Hirata et al. 2001), the ethnographic record over more than 50 years at Koshima is overdue for pointed scrutiny in this regard. Further, there is archeological evidence of ratcheting from the New Caledonian crow (Hunt and Gray 2003). Finally, one wishes for cleverly designed experimental or observational studies of cumulative cultural change, both adaptive and maladaptive, that go beyond mere cultural drift.

## Culture without Language

As research on artificial language acquisition and use by apes ("pongolinguistics") has declined, more sophisticated studies of natural communication have increased (Slocombe and Zuberbuehler 2005a, 2005b). Studies in captivity and in nature suggest that chimpanzees make functionally referential calls of various types (grunts, screams) in varied contexts (foraging, agonism).

Nonprimate mammals besides cetaceans are known to be capable of vocal learning, for example, the gray seal (Shapiro et al. 2004) and the African elephant (Poole et al. 2005). There is even anecdotal evidence of an elephant that mimics human speech, having learned spontaneously to produce Korean words by blowing air from his trunk into his mouth ("Polly pachyderm" 2006). In a wide-ranging review of language faculties, Hauser et al. (2002) concluded that the capacity of recursion (syntax) is the only uniquely human aspect of language, which otherwise shares with nonhumans the sensorimotor and conceptual-intentional systems.

The extent to which variation in acoustic communication across individuals, groups, populations, or regions is cultural or not remains to be seen, but it is worth remembering that vocalizing is not the only mode of sound production. Manual communication, for example, drumming on tree buttresses, varies across populations of wild chimpanzees (Arcadi et al. 2004). Is it too much to hope that these more accessible means of signaling will be explored in experimental studies of captive apes?

## Culture Is by Definition

Definitions of culture continue to proliferate, with no sign yet of consensus. Consequently, some arguments about whether a particular phenomenon is cultural may hinge on a single element; for example, Laland and Janik's (2006) definition of culture founders on its final word. According to them, culture (or tradition) should be group-typical behavioral patterns shared by community members that rely upon socially learned and transmitted information. Almost all of these facets are admirably empirically testable, but how is one to measure "information"?

A similar problem exists with my attempt to boil down culture to a single seven-word sentence: "Culture is the way we do things" (McGrew 2003, p. 433). This epigram contains four (arguably) necessary elements that are listed in order of difficulty of access for study: overt action, norms and standards, collectivity, and identity. However fine this might sound in principle, until it is operationalized in practice, it will be rightly criticized (e.g., Perry 2006).

Definitional challenges also rear their heads in another way, in categorizing phenomena. New variants on old themes show that there are more ways to skin a cat or fish a termite. Goualougo chimpanzees bore holes in the ground before fishing for underground termites (Sanz et al. 2004). Kibale chimpanzees show a dabbing social scratch that is unlike the stroking motion of their Mahale counterparts (Nishida et al. 2004).

## Culture Is Collective

The persistent problem of capturing collectivity is reflected in another recent definition of culture. What is one to do with "Here I define culture as behavioral variation that owes its existence at least in part to social learning processes" (Perry 2006, p. 172)? Again, culture is said to be no more than social learning. The author goes on to define social learning as "changes in behavior that result from attending to the behavior or behavioral products of another." In these terms, any two persons, even strangers, can establish a culture as soon as one learns anything from the other. This minimalist view may be logically sound, but in real-life terms it is barely a shadow of the richness of group-typical customs.

Of all the 10 points raised in the original dispatches, this one has yielded the least progress, but the problem of collectivity is crucial because it must underlie societal norms and is likely to be expressed in group self-regard. If nonhumans have institutions, conventions, roles, and the like, then they must be based on some form of mutual self-awareness. Whether this amounts to shared group consciousness ("we" versus "they") remains to be seen, but many primatologists seem to have strong intuitions about this topic that could be made explicit and operational. Anyone who has been in the thick of a party of chimpanzees on patrol, or in a social hunt, can tell you of the feeling of collective enterprise that is generated. This needs study with ingenious and imaginative methods and measures. I expect that empirical studies of collectivity will show it to be pervasive and encompassing in a variety of social vertebrates and in a variety of guises, at multiple levels, from lineages (clans) to adjacent groups (neighbors).

## Culture Has Escaped from Anthropology

Despite repeated bridge-building attempts (e.g., Perry 2006), cultural anthropologists seem little interested in cultural primatology. Introductory textbooks reveal only cursory treatment, with trite dismissals based on purported basic aspects of human culture that are said to be lacking in nonhuman species, for example, dependent, symbolic, cumulative, or linguistic. In this sense it seems that sociocultural anthropologists continue not to care that the culture concept is "out there" causing a stir in other arenas.

In the original dispatch psychology was said to ask *how* questions, and zoology *why* questions, about culture, and such investigations have continued. Working with groups of captive chimpanzees, Whiten et al. (2005) tackled a basic prerequisite of cultural acquisition, conformity, and showed that it was operating in transmission of problem solution. Horner et al. (2006) showed that transfer of foraging techniques from one individual to another ("transmission chains") is faithfully and similarly replicated in apes and children. It is a bit strange that such long-overdue revelations about the mechanisms of transmission are being done by psychologists rather than anthropologists.

On another front, the entry of archeology into the fray, so that *when* questions can be posed, is under way. After some others' speculative and qualitative false starts, Mercader et al.'s (2002) excavation of a chimpanzee nut-cracking site shows that a scientific archeology of apes in possible. How exciting it will be to see if we can disentangle the archeological records of extinct apes and humans (e.g., Backwell and d'Errico 2001; Goren-Inbar et al. 2002).

## Culture Is Rich and Complex (But So What?)

In the original dispatch I speculated that the destructive extractive technology of pestle pounding of oil palms by the chimpanzees of Bossou (Yamakoshi and Sugiyama 1995) might be a case of killing the goose that lays the golden eggs. Having since visited Bossou and having been shown surviving oil palms, I must temper that surmise until long-term data on survivorship of pounded versus unpounded individuals is assessed.

However, on other fronts, the problem of cultural maladaptation cannot be ignored. As more and more apes are displaced from their natural food sources by human deforestation for agriculture, we should not be surprised if they respond by eating the cultigens that are planted in their stead. Raiding crops may be convenient and tempting for apes in the short run, but it is likely to be disastrous in the long run. Similarly, wild apes that adapt to the presence of human observers at close range by becoming "habituated" may pay a price in terms of vulnerability to deadly pathogens or to being hunted. In all such cases brave and persevering paternalistic protection by field primatologists is no longer optional but is necessary. Field research now carries conservationist obligations.

In the original chapter I called for a rethinking of the boundaries of multiculturalism. By that I meant that we must find new and better ways to enhance and appreciate the overlapping lives of ourselves and other species. Not only must we allow them to survive, or even make sure that they thrive, but also we must try our best to allow them to live out their varied cultural potentials. This is a tall order, but surely it is a good one.

## ACKNOWLEDGMENTS

"Ten Dispatches from the Chimpanzee Culture Wars" by W. C. McGrew was originally published in *Animal Social Complexity*, edited by Frans B. M. de Waal and Peter L. Tyack, published by Harvard University Press, copyright © 2003 by the President and Fellows of Harvard College.

# 4

## GEOGRAPHIC VARIATION IN THE

## BEHAVIOR OF WILD GREAT APES:

## IS IT REALLY CULTURAL?

CAREL P. VAN SCHAIK

Opinions about the importance of social learning of information or innovations among animals in nature vary dramatically. Dugatkin (2000, p. 200) sees culture everywhere: "The vast array of animal behaviors that are touched by the long fingers of culture continues to grow, and my guess is that we have seen only the tip of the iceberg," and "cultural transmission and gene/culture interactions are serious, underestimated forces in evolutionary biology" (ibid., p. 28). De Waal (2001, p. 363) agrees: "The world is chock-full of feathered and furry animals that learn their life's lessons, habits, and songs from one another." Others are not convinced, arguing that social learning is invoked spuriously to explain patterns in nature, in particular among great apes and cetaceans, that are explained more parsimoniously by using developmental models that assume individual acquisition of behavior patterns often attributed to culture (Galef 1992, 2003a Heyes 1993; Tomasello 1994; Laland and Hoppitt 2003; Laland and Janik 2006).

How can such broad variation of opinion arise? Part of the explanation surely lies in the fact that culture is defined in various ways. If culture is meant to involve all socially learned information, including the location of food sources or predators and the identity of food items (Galef and Giraldeau 2001; Danchin et al. 2004), then culture may indeed be ubiquitous. If, on the other hand, it is meant to involve the presence of language-based, locally specific meanings of symbols and institutions (Tuttle 2001), then evidently culture is limited to humans. If, however, culture is taken to refer to the presence in multiple domains of socially transmitted innovations (Whiten and van Schaik 2007), then culture may be common in a moderate number of species, including great apes and cetaceans (and possibly others as yet undocumented).

But varying definitions account for only part of the disagreement. Most of the debate centers on the presence of culture, in the sense of socially transmitted innovations, in wild great apes and cetaceans—taxa in which observational social learning has been documented. Resolving this debate is critical to understanding the trajectory of culture during human evolution; if great-ape behavior is not cultural, we will be forced to rethink the status of the artifacts found among all hominins before *Homo sapiens*. The aim of this chapter, therefore, is to contribute to resolving the debate about culture in nature. Most cases I will discuss concern great apes, especially orangutans, the organisms with which I am most familiar.

There is good experimental evidence for social learning in captivity for animals in general (e.g., Galef and Laland 2005). Most of this social learning relies on very simple mechanisms, such as local enhancement or social facilitation, perfect for the transmission of information but not for that of true innovations, that is, novel behavioral variants that arise rather rarely because of some process of exploration and invention (see Reader and Laland 2003; Ramsey et al. 2007). Experiments on great apes, however, increasingly document sophisticated observational learning by which individual great apes, usually infants or juveniles, copy complex motor acts or their goals of the experts they are exposed to, historically largely humans but increasingly conspecifics (Whiten et al. 2004).

In light of this solid experimental evidence, some may wonder why there is a debate at all since the early criticism rejected culture in wild apes because they were thought to lack imitation (Galef 1992; Tomasello 1994). Is it possible that these imitative abilities represent some by-product of growing up in conditions enriched by interactions with humans and their artifacts—a true spandrel in the sense of Gould and Lewontin (1979)—and would thus not be found in nature? Enculturation is a process in which an animal, so far always a great ape, is "immersed in a system of meaningful human relations that include language, behavior, beliefs, and material culture" (Miles et al. 1996, p. 281). It is true that enculturation may bring out abilities rarely or never seen in conspecifics not reared by humans (e.g., Call and Tomasello 1996). However, imitative social learning is not limited to enculturated apes. Recent work shows that captive great apes born and raised in conspecific groups, adults, as well as immatures, readily learn novel techniques from other group members through social learning techniques that involve imitation (Whiten et al. 2005; Bonnie et al. 2007). It is

71

therefore more parsimonious to argue that enculturation does not produce the ability to learn through imitation but rather creates more complex social inputs than normal rearing does.

I suspect that the real reason that the debate rages on is that the evidence from the wild, at least for great apes and cetaceans, is indirect and,
by the standards used to judge laboratory results, lacking in rigor. The
main technique for demonstrating the presence of culture in nature has
been the use of geographic variation in behavior, also known as the ethnographic method (Wrangham, de Waal et al. 1994), group comparison (Fragaszy and Perry 2003c), or the method of elimination (van Schaik 2003).
Its use has been severely criticized, and I will address the criticism later.

It is especially important to tackle the criticism because to date there
are no easy alternatives to this technique that reach the standards of
demonstration required by the laboratory-oriented critics. First, so far,
no controlled transplantation experiments have been conducted with
great apes (Laland and Hoppitt 2003) because such experiments are lo-
gistically challenging, may produce many additional potentially con-
founding changes, are sometimes ethically questionable, and are usually
illegal. Unplanned experiments can sometimes be used, however. Russon
(2003) studied orphans introduced as a group into areas without a wild
population, and her results strongly suggest an important role for innovation and social learning. Introducing a new behavior pattern is another
possibility, but one difficult to do without introducing animals as well.
However, Biro et al. (2003) introduced nuts new to all but one female in
a chimpanzee community. They noted that the habit spread in a way
consistent with the pattern of social attention.

A second approach consists of longitudinal studies that correlate behavior acquisition patterns with patterns of exposure to experts. For
Sumatran orangutans, for instance, there is the curious case of some ex-
captives that were introduced into the natural population at Ketambe in
the late 1970s (Merrill 2004). One of them, Binjai, was known to be
fond of eating cardboard that she would fish from the garbage pit and of
handling clothes taken from the clothesline. She has several daughters
that remained in the local population. These same behaviors have now
been recorded for them and their offspring (but no records at all for
orangutans of other lineages, despite the occasional opportunity for observation). The artificial nature of the behaviors suggests that this vertical transmission is almost certainly due to social learning; furthermore,
Binjai's offspring and grandoffspring are fully wild orangutans without

any hint of enculturation. Detailed work on capuchins by Susan Perry (this book; see also Perry, Baker et al. 2003) is also consistent with social learning for some behavioral variants. Unfortunately, parent-offspring similarity is also consistent with genetic causes of interindividual similarity and thus might not refute the main alternative of highly prepared individual learning. This method also requires variation in the relevant behavior patterns among the adults within the population, something that is less common in gregarious great apes. It is, however, common in cetaceans, where numerous foraging specializations coexist within the same population and show strong correlations between mother and offspring (Mann and Sargeant 2003).

A third approach is to focus on the actual process of social learning by documenting indicators of social learning that are not easily explained otherwise. Infant Bornean orangutans, for instance, spend almost all their time alone with their mothers. They direct highly selective peering and begging at their mothers for foods that are not yet part of their diet (Jaeggi 2006). As a result, by weaning, infants have diets and repertoires of feeding techniques that are virtually identical to those of their mother (Dunkel 2006). Similarly, Lonsdorf et al. (2004) could attribute sex differences in the trajectory of tool-use proficiency among wild chimpanzees to differential attentiveness. However, it could be argued that this ability is only expressed during a limited sensitive period when the infant is associated with the mother (cf. Biro et al. 2003). Although rapt attention to the use of tools of others by mature animals is commonly seen in some orangutan populations (van Schaik 2004), there are as yet no quantitative data that show that it is highly selective to such contexts. Thus so far, good field evidence exists for purely vertical transmission, but not for the oblique and horizontal forms that produce the more interesting population-wide behavioral variants (but see Noad et al. 2000 for a suggestive case involving whale song).

All these alternative approaches, although incomplete, yield results consistent with the notion that cultural transmission of skills and knowledge is important in wild great apes. Indeed, they suggest that the likelihood that the well-documented social learning seen in captivity is artificial is truly remote. However, this argument is based on the relative plausibility of the cultural explanation versus alternative explanations rather than on solid experimental evidence. As a result, some remain unconvinced by claims for culture based on the geographic technique. Finding the best evidence in favor of culture by using this technique may remove the doubts and allow us all to move on to novel questions about culture in nature.

## The Technique of Geographic Comparisons

If culture is defined as socially transmitted behavioral innovations, then "the behavior of two groups with the same gene pool and with the same type of habitat can differ only by culture" (Kummer 1971, p. 13). Thus behavioral variants differ between two populations because a particular innovation was made in only one of them, then spread to become common in the neighborhood, but was unable to spread beyond it, producing the typical checkerboard pattern of high prevalence at some places and complete absence at others.[1]

Note that the geographic method cannot demonstrate that innovation or social learning actually happened. It uses observable patterns to infer the presence of these processes (Boesch 1996b; van Schaik 2003). So far, it has been used to infer culture in chimpanzees (Goodall 1973; McGrew 1992; Boesch 1996b; Whiten et al. 1999), orangutans (van Schaik, Ancrenaz et al. 2003), capuchins (Panger et al. 2002; Perry, Baker et al. 2003), bonobos (Hohmann and Fruth 2003), and cetaceans (Rendell and Whitehead 2001).

The major alternative to socially transmitted innovations is that the geographically variable behavioral variants are not innovations at all. Instead, they represent behaviors that develop reliably under the given ecological conditions or the given genetic makeup of the population (Galef 1992; Tomasello 1994; Laland and Hoppitt 2003). Their high local prevalence is not due to social transmission but to independent, yet highly predictable and therefore convergent, individual acquisition through prepared pathways.

The geographic technique deals with this potentially serious problem by focusing on those variants with incomplete global prevalence and high local prevalence that are not obviously limited to particular ecological conditions or representative of evolutionarily significant units (ESUs).[2] The core of the criticism of this technique is that this control of ecological or genetic correlates of behavioral variation is not rigorous enough. This criticism gets us into epistemological territory. If one's inclination is not to accept as solid any results but those obtained through rigorous experiments (e.g., Tomasello 1994; Milinski 1997;

Galef 2004), one will reject the results of all geographic comparisons as inconclusive, albeit perhaps suggestive. Indeed, some see no use for correlational data other than as a preliminary phase to experimentation: "The only way to solve the problem is to do a suitable experiment" (Milinski 1997, p. 165).

No one doubts that experiments can demonstrate phenomena beyond reasonable doubt. However, there is no decisive difference between controlling for the effects of variables by physically manipulating them and by doing so statistically. Thus when experiments are fundamentally impossible (such as in paleontology or comparative biology), we can either refrain from developing scientific explanations for phenomena or adopt a second-best approach that chooses between different explanatory models on the basis of experimental tests of their assumptions and nonexperimental statistical tests of their predictions (cf. McGrew 2004). We can then decide whether particular cases are more likely to reflect the outcome of a cultural process (i.e., innovation and social transmission) than of other processes (e.g., convergent individual learning). Although all of us obviously feel more confident about the robustness of experimental findings, in cases where those experiments are not an option, we can still fit the most parsimonious model to the data in hand.

The geographic technique was developed in response to the claim that observational forms of social learning have not been convincingly demonstrated in nature. It thus attempts to infer the presence of culture by eliminating all possible alternative explanations for the observed difference between two populations. Critics (e.g., Laland and Janik 2006) have argued that use of the technique implies a lack of acknowledgment of the complexity of the acquisition process of behavior during development. It is true that a geographic difference in nature may rarely be due to culture alone, given that a complex interaction between genetic makeup, ecological conditions, innovation, and opportunities for social learning usually determines the acquisition of a particular behavior by an individual. However, Kummer (1971) noted decades ago that a genetic/learned dichotomy makes no sense in relation to a particular episode or category of behavior but can logically apply to a "dichotomy of differences," which is exactly the point of this method: a difference in behavior between two populations may perfectly well be caused only by social learning and not by either genetics or individual learning. The technique thus focuses on the extreme cases in order to demonstrate the fact that social learning can in principle account for geographic variation in behavior.

This logic has pointed users of the technique to the kinds of behavioral variants for which they felt that the technique would be most useful: those without clear ecological and genetic correlates. The directive for the geographic technique is therefore clear: maximize the likelihood that one can eliminate ecological and genetic differences between pairs of sites. Even if most currently claimed cases do not quite meet this standard and even if only very few cases survive all attempts at refutation, those that do stand will be the ones that can establish the presence of culture without having to resort to experiments. I will therefore mount a defense of the geographic method later, largely because, as argued earlier, there are no workable alternatives.

The much more difficult problem is what to do with the other cases of geographic variation. Users of the technique have argued, albeit mostly implicitly, that by Kummer's difference logic social learning is also likely to play a role in the acquisition of all the other behavioral variants where ecological and genetic differences cannot be satisfactorily excluded, and perhaps even in those where we know that they exist. That, of course, is a much more controversial claim, and one we will have to examine if we find that the technique is in principle a viable way of demonstrating the presence of culture.

## The Alternative Hypothesis

The alternative to a cultural explanation posits that a particular behavioral variant is in fact developing reliably, given the current environment and given the predominant genetic makeup of the population. How plausible is this alternative? Because the environment plays an essential part in the expression of genes to produce the phenotype (West-Eberhard 2003), we should consider genetic and ecological differences between populations together.

A direct mapping of genes onto behavior is extremely unlikely for the more complex and detailed behavioral patterns usually considered in studies of culture. Virtually all behavioral traits in large-brained animals are based on complex trait genes (QTLs; Glazier et al. 2002), which produces heritability values that are well below 1 and hence intrapopulation variability. Thus on its own, a genetic basis for behavior should not produce the checkerboard pattern of interpopulation variation (defined earlier) seen in cultural species, and a simple genetic determinants model is unlikely to be supported by the data.

However, an interaction between genetic and ecological variation might explain the observed pattern. The species of interest for culture are highly plastic developers. The adaptive significance of plastic development is that it allows the attainment of phenotypes that are closer to the locally optimum phenotype than would be the case without plastic responses to the prevailing conditions during development (cf. West-Eberhard 2003). This process may conceivably make a set of genetically nonuniform individuals that are exposed to virtually identical ecological conditions converge on the same behavior patterns without any social learning among them.

In conclusion, the alternative process that can produce culturelike spatial patterns is not implausible, and this underlines the need to evaluate the geographic technique. However, one may also ask how one can evaluate the alternative hypothesis. Perhaps the best approach is to assess whether some putatively cultural behavior is an innovation. This task requires a powerful operational definition. Reader and Laland (2003) suggested the first observed occurrence in a population as a good operationalization. Ramsey et al. (2007) suggested an additional definition that can be summarized as patchy geographic prevalence due to lack of knowledge rather than unsuitable ecological conditions, genetic variation, or observational artifacts. This definition is not dependent on the presence of solid long-term data but does require solid knowledge of geographic variation and its causes. It can be applied by using comparisons or experiments. Thus van Schaik et al. (2006) found through geographic comparison that 33 of 43 recognized geographic variants could be considered clear innovations, and the remaining 10 probable innovations.

The conclusion drawn by van Schaik et al. (2006) once again rested on parsimony. However, one can also test experimentally whether a particular behavioral variant is an innovation by offering animals that lack the behavior the same ecological conditions as in the area where it is known to occur. If the animals still fail to show the behavioral variant despite being exposed to all the relevant stimuli, the variant is an innovation. If the variant passes the innovation test, its ubiquitous presence in a population is almost certainly due to social learning. Although a negative result can reflect errors in the experimental design, positive results would strongly support the parallel individual-learning explanation.

## In Defense of the Geographic Method
## as Applied to Great Apes

The purpose of this section is to apply the geographic technique in as critical and conservative a way as possible to see if at least some convincing examples remain that compellingly support a cultural interpretation. First, we must make sure that methodological issues are cleared up and that the patterns we wish to interpret are correctly described. Second, we need to demonstrate that at least a few behavioral variants are innovations not linked to ecological or genetic differences. I will use the database on geographic variation in orangutan behavior. The sites in this comparison are shown in Figure 4.1. Table 4.1 shows the results of a second round of comparisons between sites (van Schaik, Ancrenaz, Djojoasmoro et al. in press.), building on a similar first round (van Schaik, Ancrenaz et al. 2003a). In both rounds most site directors

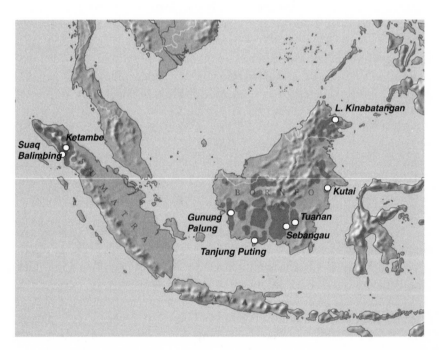

*Figure 4.1* Map of orangutan study sites mentioned in Table 4.1 and the text. Shaded area indicates current orangutan distribution.

and other orangutan observers had jointly discussed the entries in the table and went through multiple iterations of comparisons after new variants had been added. Not all parties were present in both rounds, so some questions marks remain for some sites (considered absences).

## The Problem of False Zeroes

Before we can proceed to evaluate the power of the geographic method, we must eliminate the possibility that we are dealing with artifacts (Table 4.2.a, 1). Some behaviors may seem to produce a checkerboard pattern across populations because their absences in some are artificial; that is, they are due to an inability to recognize the variants or to insufficient observation time. Such recognition problems are not academic. Descriptions are often inadvertently incomplete because details may vary between sites in unexpected ways. Moreover, misunderstandings commonly arise because for most workers the lingua franca of research, English, is their second language. Video clips are the obvious solution to both problems but are not yet available for all behavioral variants. In Table 4.1 I have conservatively included only those variants for which I am confident that all workers referred to the same behavior. Unfortunately, this procedure did not prevent some workers from not recording a particular variant during the first round, yet for whatever reason confidently reporting its absence. A few such artifacts have been identified and removed from the table, hence the disappearance of two putatively cultural variants (play nests and snag crashing) relative to van Schaik, Ancrenaz et al. (2003).

Insufficient observation time is a more insidious problem in case one deals with rarely performed behavioral variants (Table 4.2.a, 2). All sites in Table 4.1 have at least 5,000 hours of observation (most much more than that), the empirically established minimum time to reach a nearly complete repertoire of behavioral variants (van Schaik et al. 2006). However, some behavioral variants are so rare in terms of frequency in some populations that they may produce low prevalence there, if they are recorded at all, whereas they may be less rare in others, thus producing high prevalence there, falsely leading to the conclusion that we are dealing with the checkerboard pattern brought about by the cultural process. The six variants for which we conservatively suspected that this might be possible are entered separately at the end of Table 4.1. I will not consider them in the calculations presented later.

Table 4.1 Geographic variation in orangutan behaviors

| | Site | | | | | | | | | Poss. hidden universal? | Poss. ecological difference? | Why? Perhaps linked to: |
|---|---|---|---|---|---|---|---|---|---|---|---|---|
| | Sua | Ket | GP | TP | Seb | Tua | Ku | LK | | | |
| | Island | | | | | | | | | | | |
| | S | S | B | B | B | B | B | B | | | |
| | Habitat | | | | | | | | | | | |
| | Psw | Dry | Dry | Psw | Psw | Psw | Dry | Psw | | | |
| c1 | Snag riding: Riding on pushed-over snag as it falls, then grabbing on to vegetation before it crashes | A | A | A | C | R | **H** | A | A | No | No |
| c2 | Kiss-squeak with leaves: Using leaves on mouth to amplify sound, then dropping leaf | A | A | C | A | A | **H** | **H** | A | No | No |
| c4 | Leaf wipe: Wiping face with fistful of squashed leaves, then dropping leaves (in kiss-squeak context) | A | A | A | C | A | A | A | A | No | No |
| c9 | Scratch with stick: Using detached stick to scratch body parts | A | A | A | R | R | R | **H** | A | No | No |
| c10 | Autoerotic tool: Using tool for sexual stimulation (female and male) | A | C | A | A | A | A | P | A | No | No |
| c12 | Symmetric scratch: Exaggerated, long, slow, symmetric scratching movements with both arms at same time | C | R | A | A | A | A | A | A | No | No |
| c13 | Twig biting: Systematically passing ends of twigs used for lining of nest past the mouth (sometimes including actual bite) during last phase of nest building | C | A | P | A | **H** | **H** | A | A | No | No |

| Code | Behavior | | | | | | | | | | | Ecological factor |
|------|----------|---|---|---|---|---|---|---|---|---|---|------------------|
| c16 | Leaf gloves/cushions: Using leaf gloves to handle spiny fruits or spiny branch, or as seat cushions in trees with spines | E | C | E | R | E | E | A | R | No | No | |
| c18 | Seed-extraction tool use: Using tool to extract seeds from the protected fruits of *Neesia* sp. | C | E | A | A | E | E | E | A | No | No | |
| c29 | Moss cleaning: Cleaning hands with moss | A | A | A | A | H | A | ? | A | No | No | |
| c30 | Chewing leaves into pulp and then smearing foam over body (selected parts chosen) until fur is all wet | A | A | A | A | H | A | ? | A | No | No | |
| c31 | Male and female use the same nest to spend the entire night (during consortship) | A | A? | R | A | A | A | ? | C | No | No | |
| c33 | Copulation on female's nest | A | H | C | R | A | R | ? | R | No | No | |
| c17 | Tree-hole tool use: Using tool to poke into tree holes to obtain social insects or their products | C | A | A | A | A | A | A | A | Yes | No | Abundance of social insects in tree holes |
| c11 | Raspberry: Spluttering sounds associated with nest building | C | A | A | A | C | A | A | H | Yes | No | Wadging of *Stenoclaena?* |
| c14 | Leaf napkin: Using handful of leaves to wipe latex off chin | A | A | A | A | A | A | C | A | Yes | No | Unusual amount of latex? |
| c21 | Bouquet feeding: Using lips to pick ants from fistful of dry, fresh, or rotting leaves (nests) | C | C | C | C | A | H | A | H | Yes | No | Abundance of weaver ants? |
| c23 | Dead-twig sucking: Breaking hollow (dead) twigs to suck ants from inside | C | A? | A? | C | H | H? | A? | A | Yes | No | Abundance of trees and lianas with ants in twigs? |

**Table 4.1** (*continued*)

| Site | Sua | Ket | GP | TP | Seb | Tua | Ku | LK | Poss. hidden universal? | Poss. ecological difference? | Why? |
|---|---|---|---|---|---|---|---|---|---|---|---|
| Island | S | S | B | B | B | B | B | B | | | |
| Habitat | Psw | Dry | Dry | Psw | Psw | Psw | Dry | Psw | | | Perhaps linked to: |
| c24 Slow loris eating: Capturing and eating slow loris hiding in dense vegetation | H | H | A | A | A | A | A | A | No | Yes | Loris abundance? |
| c25 Nest smack: Smacking sounds associated with nest building | A | A | A | A | R | C | ? | A | No | Yes | Feeding on waxy leaves? |
| c26 Carrying leaves to different tree in which subject starts to build nest using carried leaves as lining, pillow, or cover (Tuanan: *Campnosperma*) | A | A | A | R | A | H | ? | A | No | Yes | High mosquito abundance? |
| c27 Branch cushion: Covering big branch(es) with few leaves or leafy branches, then resting on it | A | H | A | H | H | C | ? | A | No | Yes | Need to save energy? |
| c28 Throat scrape: Deep throat sound made by female toward offspring | A | A | A | R | A | H | ? | A | No | Yes | Need to save energy? |
| c32 Tooth cleaning: Chewing and spitting out leaves after eating (sticky?) fruit | A | A | P | A | A | A | ? | C | No | Yes | Especially sticky fruit |
| c3 Kiss-squeak with hands: Using closed fists, open fists (like trumpet), or flat hands on mouth (or fingers in mouth) to amplify sound | H | C | P | P | C | C | H | C | Yes | No | |

| Code | Behavior | | | | | | | | | | | Ecological reason |
|------|----------|---|---|---|---|---|---|---|---|-----|-----|------|
| c6 | Bunk nests: Building a nest a short distance above the nest used for resting (during rain) | A | A | A | P | R | R | A | H | Yes | No | (Heavy rain at night everywhere) |
| c7 | Sun cover: Building cover on nest during bright sunshine (rather than rain) | A | H | A | A | R | A | C | C | Yes | No | (Bright sun during day everywhere) |
| c8 | Hide under nest: Seeking shelter under nest from rain | A | R | R | A | R | A | C | P | Yes | No | (Rain during day everywhere) |
| c15 | Branch as swatter: Using detached leafy branches to ward off bees/wasps attacking subject (who is raiding their nest) | H | H | R | A | A | A | H | H | Yes | Yes | Variable abundance of stinging bees |
| c19 | Branch scoop: Drinking water from deep tree hole using leafy branch (water dripping from leaves) | H | A | A | A | H | A | A | A | Yes | Yes | Variable need to get water from deep tree holes |

*Note:* Sites are ordered from west to east. Sites are Sua = Suaq Balimbing; Ket = Ketambe; GP = Gunung Palung; TP = Tanjung Puting; Seb = Sebangau; Tua = Tuanan; Ku = Kutai, Mentoko area; LK = Lower Kinabtangan. Islands: S = Sumatra; B = Borneo. Habitat: Dry = dryland forest, usually mixed riverine and lowland dipterocarp; Psw = peat-swamp forest. Numbers in first column refer to numbers in van Schaik, Ancrenaz et al. (2003); new variants numbered from c24 on. C = customary, shown by all or most relevant individuals at a site; H = habitual, shown by multiple individuals; P = present with unknown prevalence, usually meaning rare; R = rare, i.e., low prevalence; A = absent; E = absent for clear ecological reason. Bold entries refer to variants thought to have spread, and to be maintained, by social learning.

**Table 4.2** The two main problems of the geographic technique and their possible solutions

| Problem | Solution |
|---|---|
| a. False positives (incorrectly inferring culture) | |
| 1. Variant not recognized at one or more sites | Better communication between sites; videotapes |
| 2. Very low frequency creates low incidence, artificially producing variation similar to that due to culture | Longer observation time |
| 3. Subtle ecological differences between sites explain absences | 1. Better attention to natural history <br> 2. Experimental production of ecological differences in same population <br> 3. Demonstrate that variant is innovation |
| 4. Genetic differences between populations explain differences | 1. Compare genetic variation across and within taxonomic units with variation in repertoire of variants <br> 2. Demonstrate that variant is innovation |
| b. False negatives (incorrectly inferring absence of culture) | |
| 1. Miss existence of geographic variation because of insufficiently precise description (e.g., feeding techniques; nest-building techniques) | Better descriptions; videotapes; mutual site visits |
| 2. Variant is an innovation, despite clear ecological or genetic correlation | Try to provoke variant in experimental conditions (captive, wild): study variation within populations |
| 3. Universal behavior that is nonetheless cultural | ? (probably very rare outside humans) |

*Eliminating Ecological Effects*

Subtle ecological differences between sites may undermine the cultural interpretation (Table 4.2.a, 3). In Table 4.1 I indicated all 11 cases where ecological differences may reasonably be posited as being responsible for differences between sites. Not only does this leave 13 cases where ecological causes for behavioral variation are highly unlikely, but I will also try to show that in most of the remaining 11, the ecological influence is not compelling or is acting through an effect on the likelihood of innovation. I will now examine several cases in more detail.

Ecological influences are extremely unlikely for various variants. Orangutans that are harassed by conspecifics or distressed by the presence of predators (including humans) often emit kiss-squeaks, loud kissing sounds. Orangutans may embellish these kiss-squeaks by placing the hand on the mouth in most or all sites. However, in just a few sites the kiss-squeaks are placed on leaves or bunches of leaves stripped just beforehand (c2, c4), and where this happens, their prevalence is high. It is difficult to see how ecological differences between sites could have affected such variation. Chimpanzees have numerous communication variants that have no link to ecology (Whiten et al. 2001). The best known of these is perhaps handclasp grooming (see also Nakamura 2002).

Some ecological influence might seem to be present for variant c17, tool use on tree holes. Among the sites represented in this sample, it is found only among the swamp orangutans at Suaq Balimbing. This use is largely for honey from stingless bees, but also to some extent to extract termite larvae and ants (eggs, pupae, and adults). A detailed comparison with Ketambe, a dryland forest only 70 kilometers from Suaq, was made to examine the possible role of ecological differences (Fox et al. 2004). The incidence of tree holes (the percentage of trees with them) is very similar between Suaq and Ketambe, but those at Suaq were roughly twice as likely to be occupied by social insects. The most commonly taken resource, honey from stingless bees, was also more abundant at Suaq: 9 percent of occupied tree holes versus 5 percent at Ketambe. Per tree hole, animals at Suaq had about 3.2 times the number of opportunities to extract honey. Thus if we limit our conclusion to the predominant context, honey extraction, tool use is more likely at Suaq but is clearly not impossible at Ketambe.[3] Under an ecological model we should perhaps have expected a higher proportion of tool users at Suaq than at Ketambe, but not the near-ubiquitous prevalence at Suaq (van Schaik 2003) versus complete absence at Ketambe. Instead, the most parsimonious hypothesis (Fox et al. 2004) is that the tree-hole tool was more easily invented at Suaq than elsewhere and was subsequently maintained by social learning. This hypothesis also explains why Suaq animals, when they enter the hill habitat, where we assume that tree-hole conditions are very similar to those of dryland forests elsewhere, readily use tools on tree holes (van Schaik 2003, 2004).

Even though the behavioral variant is customary, variation among the females at Suaq in the frequency of using tree-hole tools is dramatic (van Schaik, Fox et al. 2003). The females at Suaq can be divided into clusters on the basis of range use and association (Singleton and van Schaik 2002). The northern females at the site are far less gregarious than the central and southern females. This difference is stable over time and also includes adolescent females, who tend to settle in or near their natal range. However, female home ranges show high overlap. By focusing on the overlap zone, we can eliminate all ecological effects. In this zone females of both clusters are equally likely to forage for insects and hence have equal opportunity to engage in tree-hole tool use, yet the northern females do so seven times less often (van Schaik, Fox et al. 2003). Their low rate of tool use suggests that they have more difficulty recognizing opportunities for profitable extraction. The difference, then, is primarily one of experience, which is probably built up during the formative years and may depend critically on the frequency of opportunities to engage in the task from start to finish during this period. The central and southern females are far more gregarious and thus are much more likely to associate with experienced foragers and to develop a sharp search image for opportunities for tool-supported extractive use of tree holes. This pattern is also most consistent with the absence of an effect of ecological differences on differences in tool use among individuals.

Another well-studied variant (c18) is the extraction of seeds from the large, woody *Neesia* fruits (Figure 4.2; van Schaik and Knott 2001; van Schaik 2004; van Schaik and S. A. Wich, unpublished). Because of the high energy content of the seeds, we expect them to be in the optimum diet everywhere (van Schaik and Knott 2001). Although the fruits may vary somewhat in size among sites, they also do so from year to year at a single site (van Schaik 2004), and this allows one to argue that ecologically the task is basically similar everywhere. Nonetheless, the solutions, if present, vary. In northwestern Sumatra tools are used for this purpose in three large coastal swamps (Kluet, which contains Suaq; Singkil; and Tripa) and, as we established recently, one upland swamp (Batang Toru).

*Figure 4.2* *Neesia* fruits and their uses: (a) mature, nondehisced fruits *(right)*, recently dehisced fruits *(center, below)*, and old dehisced fruits *(upper left)*; (b) nondehisced fruit, with one valve broken off by flanged male orangutan at Suaq Balimbing; (c) dehisced fruit, with one valve broken off, at Batu-Batu; (d) dehisced fruit with tool sticking out, found on forest floor at Suaq Balimbing; (e) same (but fruit several months old), found in Batang Toru, at Aek Ura-Ura; (f) same, found in Tripa swamp. Photos by Perry van Duijnhoven, except for (e) (Carel van Schaik).

In contrast, seeds are made available by breaking off one valve of the dehisced fruit at one nearby swamp site across an impassable river (Batu-Batu). Finally, *Neesia* fruits are ignored by orangutans at three other known sites, all far from the coast (Figure 4.3; see also van Schaik 2006).

This geographic pattern is most consistent with a cultural explanation (genetic influences will be discussed later). However, the total exposure of orangutans to the extraction problem has until recently varied significantly. The three large coastal swamps had large orangutan populations of well over 500, as well as large stands of *Neesia* trees. Batang Toru was within dispersal distance of an extensive nearby swamp population of comparable size. Batu-Batu was a smaller, more isolated swamp, but with many trees. However, the sites without any extraction at all are areas where trees are rare, and in two cases they also have low orangutan densities. Thus exposure to exploration opportunities may influence the likelihood of invention or the subsequent maintenance through social

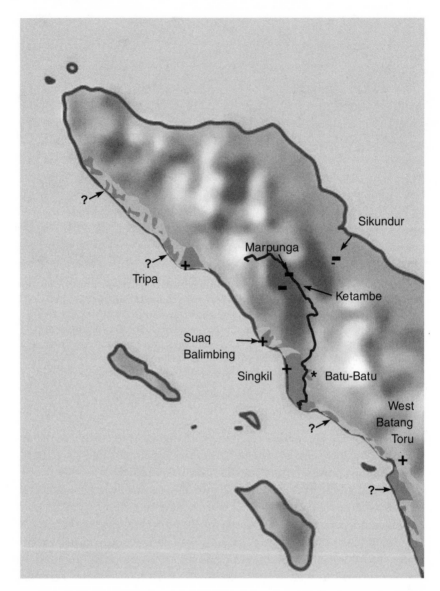

*Figure 4.3* Geography of *Neesia* use by orangutans in northwestern Sumatra. Plus signs refer to tool-based extraction; the star refers to extraction by breaking off part of the dehisced fruit; minus signs indicate that the fruits are ignored by orangutans. Unknown responses are due to forest conversion (south) or inability to visit areas (north). Dark shading indicates the extent of swamp forests around 1980, and light shading indicates the original distribution of swamp forest, on the basis of the presence of peat soils. Peat soil distribution, after Laumonier (1997). Map prepared by Perry van Duijnhoven.

learning, with the cognitively richest (and most efficient; van Schaik and Knott 2001) tool-based solution in the populations with the most abundant opportunities for invention.

Subtle ecological effects may also have played a role in the presence of several other variants (see Table 4.1). Thus several variants may reflect a strong tendency to save energy, imposed by food shortage. One example is branch pillows, in which animals break or bend one or two twigs onto a sturdy branch used for resting. However, the frequency of producing branch pillows at Tuanan does not show any obvious correlation with food abundance (M. van Noordwijk and van Schaik unpubl.). Moreover, although food shortages are encountered at all Bornean study sites from time to time (Marshall, Ancrenaz et al. in press), these variants are not recorded for all the sites. Thus the link with ecology, if present, is too tenuous to account for the geographic variation of this variant.

Likewise, orangutans at some sites make sounds during the last phase of nest building while piling small leafy twigs, usually ventral side up, on one side of the nest to serve as a pillow, often accompanied by twig biting (nonfunctional biting at the end of the already-picked twig as if to remove it from the tree). In some sites no sound is made at all; where it is made, it varies from smacking sounds to raspberry sounds. Where sounds occur, they have universal prevalence, but their origin is unclear. At Suaq the sounds are similar to those made when expelling the last remaining fibers after expectorating wadges created by chewing on a long piece of terminal stem of the climbing fern *Stenoclaena palustris*. At Tuanan they are similar to those made when chewing on the waxy leaves of the liana *Leucomphalos callicarpus*. Both of these vegetable items are commonly eaten at the sites concerned. In principle, all individuals in the local population could have converged on exactly these sounds through some associative learning process, but why they would do so is mysterious. At least as plausible is the alternative possibility that the urge to copy behaviors of role models led immatures to copy the behaviors, even those that seem to be functionally irrelevant.

The case of feeding on slow loris *(Nycticebus coucang)* shows similarities to that of tree-hole tool use. Slow loris occur at all study sites involved in this comparison but may be more common among the Sumatran sites because of the higher productivity and thus higher abundance of fruits and insects (Marshall, Ancrenaz et al. in press), major foods of slow loris. Thus an individual may more easily develop a search image for slow loris in the Sumatran sites. Because not all individuals are known to capture slow loris in these sites, the argument for social learning is weaker in this case.

In conclusion, we can with confidence eliminate a role of ecology in causing differences between sites in at least some behavioral variants, and the evidence points to these variants being innovations (e.g., kiss-squeak variants). In other cases ecological effects certainly exist, but the pattern of variation across sites is such that their influence seems most likely on eliciting particular innovations, subsequently maintained by social transmission, rather than predictable presence whenever the ecological conditions are suitable (e.g., tree-hole tool use, *Neesia* tool use). In a few others (e.g., slow loris feeding, nesting sounds) there is not enough information to decide. Overall, then, a cultural interpretation of the orangutan geographic pattern remains the most compelling one for many of the variants. Whiten et al. (2001) drew the same conclusion for chimpanzees.

### Eliminating Genetic Effects

Genetic differences between islands and populations may undermine a cultural interpretation (Table 4.2.a, 4). Unfortunately, detailed genetic information is currently absent. However, it is not clear how an analysis of genetic effects would proceed. If two units (individuals, populations, subspecies) show genetic differences—which they inevitably will—then how can we determine that these genetic differences are in any way related to the behavioral differences in the absence of information on the actual genes involved in the traits that differ? The most obvious solution to this problem is to find a system with variation in behavior, on the one hand, and in genetics, ecology, and association, on the other hand, and find the best-fitting predictor of the behavioral similarities between individuals and populations. We are currently taking this approach (see discussion), but results are not yet available. That leaves us for now with cruder tests that compare between taxonomic units or nearby populations. They will be used to distinguish between a simple genetic determination model (where genes determine traits regardless of ecological conditions, similar to species differences in behavior), a gene-environment interaction model, and finally the culture model.

The island forms of orangutans have recently been elevated to species level (Groves 2001), although estimates of their genetic distance vary widely (Muir et al. 2000), and they do not show complete mtDNA lineage sorting (e.g., Kanthaswamy et al. 2006) and thus are denied separate ESU status by this criterion. Nonetheless, one might expect island differences in behavior consistent with this taxonomic difference. Many differences exist that are not considered cultural (e.g., Delgado and van Schaik 2000), but Table 4.1 also indicates that 16 of the 24 behavioral variants (including only those that are not possible artifacts of limited observation time or attention, i.e. hidden universals) are found on only one of the islands, 12 on Borneo and 4 on Sumatra.[4] Inevitably this must reflect to some extent that we have data on six Bornean sites and only two Sumatran ones (varying geographic prevalences prevent a simple statistical test). However, far more important is that none of the 12 variants limited to Borneo is found at all six sites—a pattern inconsistent with a simple genetic determination model.

A more detailed examination of possible genetic effects is possible in Kalimantan, where four sites are within the geographic range of a single subspecies, *Pongo pygmaeus wurmbii:* Gunung Palung, Tanjung Puting, Sebangau, and Tuanan. Not a single variant is consistently present at all four sites (Table 4.1), a pattern inconsistent with a simple genetic determination model. Indeed, even those that were recognized as lacking any ecological influence still show a patchy presence across these sites, a finding inconsistent with a genes-ecology-interaction model.

Finally, we can examine nearby sites with similar habitats and not separated by obvious dispersal barriers. Under any genetic model they should be largely similar, with subtle differences due to minor ecological effects.[5] Under the culture model they may be rather different because of two effects. First, animals of different populations may adopt behavioral variants because others in the population perform them, which will increase behavioral differences between populations (Boyd and Richerson 1985). To date, evidence for such conformity is lacking in orangutans (van Schaik 2004) but stronger in chimpanzees (Boesch 2003), where experimental work in captivity also suggests its presence (Whiten et al. 2005). Second, innovations may get lost when suitable habitat conditions are separated by areas of habitat unsuitable for the expression of innovations wider than the most common dispersal distance.

Unfortunately, no clean comparisons are possible in orangutans (Table 4.3). Suaq and Ketambe have rather different habitats, whereas two impassable rivers (known to be barriers to gene flow elsewhere on Borneo; Goossens et al. 2005) separate ecologically similar Sebangau and Tuanan. These pairs share only about a third of the putatively cultural variants found in at least one of the pair of sites, but so do the chimpanzee sites shown in the same table, where all comparisons are between sites with broadly similar habitat and (historically) without major dispersal barriers. Under a strict genetic determination model of geographic variation in behavior, we had expected a high percentage of shared variants, and given that all but one of the pairs have similar habitats, far more than half the variants should also be shared under the genes-ecology-interaction model. Even the two adjacent chimpanzee communities at Mahale share less than half their variants. If this result is not a sampling artifact, it points to some constraints on the adoption of variants to which an individual is exposed, such as the conformity postulated earlier (cf. Boesch 2003).

The finest level is that of individual variation within sites. As mentioned earlier, there is some evidence of variation in various putative cultural traits among females within sites at Suaq (tree-hole tool use) and Ketambe (cardboard chewing, clothes lifting), both consistent with a cultural interpretation, although not necessarily inconsistent with a mixed genetics-ecology model in which response thresholds are under some genetic control. Turning to repertoire-level comparisons, we can add recently discovered variation in Tuanan, central Kalimantan. On the basis of the innovation matrix presented by Van Schaik, Ancrenaz, Djojoasmoro et al. (in press), it can be seen that members of the same cluster show greater overlap in innovation repertoires than members of different clusters (Figure 4.4), although the power is so small that this pattern does not reach statistical significance when examined in relation to association time.

Another way to approach the influence of genetics is to focus on the distribution of individual variants (cf. van Schaik 2006). Consider again the geographic pattern in *Neesia* tool use (Figure 4.3). A simple genetic determinism model is unlikely to hold in this case. First, the Tripa, Kluet, and Singkil animals formed until recently one contiguous population with those at Ketambe, Marpunga, and Sikundur (where the fruits are ignored), which allowed genes to diffuse from one place to the other. Second, between Batu-Batu and Batang Toru the rivers are small and form no major dispersal barriers. The orangutans may therefore be genetically rather similar, but because swamps are intermittently present, habitat barriers may prevent the spread of innovations.

An equivalent case in chimpanzees is nut cracking. It shows a highly disjunct distribution in two different subspecies (Boesch et al. 1994; McGrew et al. 1997; Morgan and Abwe 2006). Within the western subspecies *(Pan troglodytes verus)* a dispersal barrier forms the boundary of

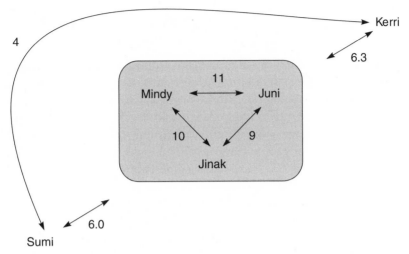

*Figure 4.4*  Mean number of innovations shared per dyad among adult female orangutans studied at Tuanan. The females are arranged in relation to home range overlap; the three central females form a clear cluster with high range overlap and high association.

the behavior's distribution. Moreover, within the limited region there is no obvious pattern in which species of nuts are cracked (Biro et al. 2003). This whole pattern is more suggestive of differential innovation and extinction than of the action of simple genetic predispositions (cf. Wrangham 2006).

Overall, the data are inconsistent with a model that posits a simple genetic determination of behavioral variants. It is not always easy to reject the model in which ecological conditions and genetic predispositions interact in the absence of any social learning. Although such a model might explain major fitness-enhancing skills such as seed extraction with tools, because there may be selection on the underlying abilities, it is inherently implausible for the majority of variants that do not represent major local adaptations. It seems unrealistic to expect a genetic basis for fitness-neutral variants, such as placing kiss-squeaks on leaves rather than on a hand, or making smacking rather than raspberry sounds (or no sounds at all) when adding a pillow to a nest, at small spatial scales (see Whiten et al. 2001 for many similar examples among chimpanzees). A more plausible interpretation is that these behaviors are acquired simply because they arose by chance through innovation, and animals copy others.

93

## Discussion

### *Beyond Reasonable Doubt?*

If nothing else, this evaluation of the geographic method for orangutans supports the proposition that many geographically variable behavior patterns are most parsimoniously explained as innovations that arise with varying probability in a population and are then spread and maintained with varying probability by social learning. Ecological and genetic effects on behavioral variants are present and may even be of major importance in the great-ape species studied here, but they act by affecting the likelihood that particular types of innovation arise instead of bringing about the reliable and convergent development of these behavior patterns in numerous individuals.

This conclusion is consistent with the main body of data on geographic variation, as well as findings on reintroduced ex-captive orangutans (Russon 2003), patterns of variation within populations (van Schaik, Fox, et al. 2003), longitudinal data on infant development (Jaeggi 2006; Dunkel 2006), patterns in repertoire size across populations (van Schaik, Ancrenaz et al. 2003), distance effects on similarities in repertoire (ibid.), studies of innovation in the wild and the features of highly prevalent behavioral variants (van Schaik et al. 2006), and the emerging body of experimental data on social learning in captivity (Whiten this book). This conclusion is also consistent with patterns not reviewed here. Thus van Schaik, Ancrenaz et al. (2003) showed that for both chimpanzees and orangutans the time spent in association predicted the size of the cultural repertoire, as expected when the density of opportunities for social learning affects the number of variants maintained in equilibrium in a population (see also van Schaik 2003). Likewise, van Schaik et al. (2006) showed that putative cultural variants were more salient than those that remained rare, in that they were performed more frequently and were more likely to concern subsistence skills or communication signals. Similar conclusions can be drawn for chimpanzees (Whiten this book) and capuchins (Perry this book).

Some lingering doubts remain over possible subtle genetic effects, however. As mentioned earlier, we plan to use pairwise similarity tests between individuals to compare the effects of genetic distance, ecological similarity (overlap in plant species), and association time, creating variation in these variables by using sites on either side of an impassable river (M. Bastian, C. van Schaik, and M. Krützen, unpubished). More subtle between-site comparisons at the repertoire level can also still be developed, where we use barriers for genes and behaviors (dispersal barriers) or for behaviors only (habitat barriers) in a carefully chosen array of sites. One can predict different patterns on the basis of whether expression of the behaviors depends on local ecological conditions: those that do not should be more homogeneous in space than those that do, in the absence of any genetic barriers.

I already mentioned an altogether different approach that, although so far untried, may be promising. The status of behavioral variants differs under the culture model and the alternative model: they are either innovations or regular behavior patterns. One can experimentally offer suitable conditions in a wide range of captive or rehabilitant animals. If all animals readily adopt the variant, ecological variation was arguably responsible for the variation in nature; if some readily do but others do not, genetic factors were responsible; finally, if none (or only a small minority) of the animals adopt the variant, that is, if the variant is an innovation, the variation was cultural.

### Other Uses of the Geographic Technique?

The questions everyone would like to answer are how important culture is in nature, and which behavioral domains it pervades. To answer these questions, we need estimates of the size of the cultural repertoire ands its contents. At present we have no techniques to estimate behavioral repertoire size. Future work, discussed later, may yield an effective approach, but for now we may ask whether the geographic technique can provide some preliminary answers to these other questions as well.

For estimates of repertoire, the sum total of both false positives and false negatives must be minimized (cf. van Schaik et al. 2006 on innovations). The geographic technique was designed to demonstrate the presence of culture and therefore tries to minimize the number of false positives (Table 4.2.a), although many of the more tentative assessments of cultural variants may still be false positives. However, the geographic technique may also produce false negatives, that is, deny cultural status to several innovations that are not recognized as such.

Table 4.2.b lists such false negatives. It is impossible right now to assess their importance, but the methodological artifacts (Table 4.2.b, 1) should be removable, and geographically uniform behavioral patterns (3) will only rarely be cultural, at least in great apes. The main unknown, therefore, is the proportion of behavioral variants that are innovations despite having clearly recognizable ecological or genetic correlates. I concluded earlier that with culture demonstrated, many such cases are to be expected, but their proportion remains fundamentally unknown (cf. Laland and Janik 2006).

Estimates of cultural repertoires derived through the geographic technique will therefore be seriously biased. However, this bias need not preclude comparisons of repertoire size and composition because the same kind of innovations will be missed in all populations within a species. Thus if the geographic technique is applied cautiously, it may be used to compare cultural repertoires across populations, even if some of the actual assessments used in the comparison are wrong. Van Schaik, Ancrenaz et al. (2003) did this and found patterns consistent with predictions based on social learning and diffusion, but not easily explained by other models. Obviously, however, the biases preclude firm statements about the contents of great-ape cultures, and future work will have to develop assessments of the full content of great-ape cultures.

### The Future

As suggested by the discussion of techniques to estimate content and size of cultural repertoires, the question that most urgently awaits resolution is the nature of the interaction between ecology, genetics, and the cultural process. The most plausible form of this interaction is that ecology and genetics affect the three key individual-level parameters of the cultural process: innovation, acquisition through social transmission, and retention. Within species their most important effect is probably on innovation. The probability of innovation depends on the cognitive leap required, which in turn may depend in part on ecological opportunities and genetic predispositions. The probability of social adoption depends on opportunities for social learning (van Schaik 2003), the salience of the variant (van Schaik et al. 2006), and perhaps social conformity (Whiten et al. 2005). The quality of the feedback generated by performing the innovation (Galef 1995; Galef and Whiskin 2001) and perhaps the prevalence in the local population may affect the likelihood of retention.

The most plausible scenario for the origin of checkerboard patterns is that all three parameters vary significantly, and we observe a particular variant at a particular site when it happens to be in its "on" phase (invented recently enough not to have gone extinct again; cf. van Schaik and Pradhan 2003). Under this scenario we should expect many variants to have rather short life expectancies in a population, especially those that are not salient and provide poor individual feedback. Not surprisingly, there is little information on the longevity of cultural variants in the wild. However, some cultural practices are at least many centuries old (nut cracking in capuchins: Urbani 1998; chimpanzees: Mercader et al. 2002), whereas for others the apparent origin was witnessed (e.g., pestle pounding in chimpanzees: Yamakoshi and Sugiyama 1995) or recent changes in prevalence were recorded (Matsusaka et al. 2006), and still others appear to be short-lived "fashions" (Goodall 1986; Perry, Baker et al. 2003). The highly patchy checkerboard patterning shown by most variants also suggests dynamic changes with frequent independent origins rather than spread from a single origin (cf. Whiten et al. 2001). A combination of modeling and empirical studies will help reveal the importance of culture in nature and thus the force of cultural evolution in nonhuman primates and early humans.

## Notes

1. Prevalence is defined as the percentage of individuals that show the variant, regardless of the frequency with which they show it (cf. Ramsey et al. 2007; van Schaik et al. 2006).

2. ESUs are defined as units that are reciprocally monophyletic, that is, show complete lineage sorting for mtDNA, and are significantly differentiated at the nuclear level, that is, have significant $F_{st}$ values (Moritz 1994).

3. As to the other tree-hole inhabitants, orangutans at Ketambe had a chance seven times higher of extracting ants but did not invent this kind of tool use (those at Suaq had a chance roughly seven times higher to extract termites).

4. Additional but nonsystematic information on one more Sumatran site reduces the island effect indicated in Table 4.1, but the general conclusion remains the same (see van Schaik et al. in press.

5. An interesting prediction is that variants restricted to males should be more heterogeneous than variants restricted to females, given the greater average dispersal distance of males. Unfortunately, in orangutans too few variants are limited to one sex to allow testing, but perhaps testing is possible in chimpanzees.

ACKNOWLEDGMENTS

I am especially grateful to the numerous orangutan fieldworkers who contributed to the data presented in Table 4.1 (to be published in more detail in van Schaik et al. in press). The Wildlife Conservation Society, the L.S.B. Leakey Foundation, the National Geographic Society, and the A.H. Schultz Foundation have over the years supported my orangutan work. Current work is in collaboration with Universitas Nasional (Jakarta), and I thank the BOS Foundation for making station facilities available, and the Indonesian Institute of Sciences (LIPI) and the Ministry of Forestry for permissions. I benefited from discussions with Gustl Anzenberger, Meredith Bastian, Judith Burkart, Lynda Dunkel, Jeff Galef, Barbara Hellriegel, Christine Hrubesch, Adrian Jaeggi, Kevin Laland, Stephan Lehner, Tatang Mitra Setia, Kisar Odom, Susan Perry, Anne Russon, Signe Preuschoft, Simon Reader, Suci Utami, Maria van Noordwijk, Erin Vogel, Serge Wich, and especially Michael Krützen.

# 5

## THE IDENTIFICATION AND DIFFERENTIATION
## OF CULTURE IN CHIMPANZEES
## AND OTHER ANIMALS:
## FROM NATURAL HISTORY
## TO DIFFUSION EXPERIMENTS

ANDREW WHITEN

In the robustly argued article that gave its name to the present book, Galef (1992) assessed "The question of animal culture" and, finding the evidence wanting for such purported (and hitherto-celebrated) cases of animal culture as food washing in Japanese macaques and termite fishing in chimpanzees, concluded that "both analyses of field reports and laboratory studies of social-learning phenomena suggest that animal traditions rest on processes quite different from those supporting culture in humans. Animal traditions are, therefore, analogs rather than homologs of human culture. *Consequently, it can be misleading to speak of an evolution of culture in animals* (e.g. Bonner 1980); this usage suggests homology when there is evidence only of analogy" (p. 92: italics added).

At a more specific, methodological level, Galef stressed the limitations of "unobtrusive observation" in establishing that putative traditional differences between different groups of conspecifics are acquired through imitation, or indeed through any form of social learning. This skepticism was shared by others, who suggested that it is logically impossible, without experimentation, to refute the hypothesis that the behavioral variation at issue is actually due to unrecognized differences in the environment that steer individual learning in different directions (Tomasello 1994).

Did such warnings damp the efforts of field ethologists to map "animal cultures" in the wild? It seems not. To the contrary, the present century has seen an unprecedented flurry of reports that suggest ever more complex cultural repertoires in an expanding range of animal taxa, reviewed in Whiten and van Schaik (2007) and this book. From one perspective, this

99

should be celebrated, for it marks a very special phase of achievement in the behavioral sciences: results of the necessary decades of fieldwork are being collated, sometimes in large-scale international collaborations, to reveal the first bird's-eye views of putative cultural variation in a growing number of species. We are in an exciting era of new discoveries. However, there is a countervailing perspective, that such celebration may be premature. The crucial question at the heart of this book is (as I see it) the extent to which scientific progress has truly been made since Galef's and Tomasello's critiques. In short, has the vaunted recent evidence for animal cultures been overhyped, or not? Of course, the answer will depend on just what we mean by "animal cultures."

## What Is Culture?
### The Status of Definitions

Two distinct steps appear necessary to decide whether (some) animals evidence any psychological or behavioral phenomenon, such as culture. Step 1 is defining one's terms ("culture," in the present case), and step 2 is assessing the scientific evidence for what one defined in step 1. Step 2 is surely the only truly scientific one, where we can have an expectation that eventually, given publication of replicated data that have been refereed as having adequate objectivity, the scientific community will converge on the truth of the matter about whether species X does or does not display culture, as defined. In short, step 2 aspires to establish the empirical fact of the matter. By contrast, there is no empirical fact of the matter about how "culture" (or any such phenomenon) should be defined in step 1. We can make no observations nor complete any experiments that will resolve this. At best we can hope to establish a vocabulary of mutually understood terms. This may seem self-evident, but sometimes writers seem to imply that they know better than do others what the "true" meaning of a term like "culture" really is. Doing so can surely reflect no more than conceit. Instead, we must each clarify the definition of the crucial terms we are applying when we write our essays, recognize that others may offer different definitions, and acknowledge that neither can be empirically more "correct" than others. Of course, some authors' definitions may be more mutually intelligible than others and/or share a closer affinity with everyday usage, where the term is already in ordinary dictionaries, so some definitions might be judged "better" in those respects, but this is still a different business from the scientific job done in step 2.

100

In summary, given any particular definition of culture (step 1), we should be able to use the tools of science to empirically verify how well it applies to a particular species or population of animals (step 2). As Laland and Hoppitt (2003, p. 151) noted, "Culture is as rare or as common as it is defined to be." With this perspective in mind, I can tackle some key definitions.

### Defining "Tradition"

Authors fall into two broad camps, with those in one camp recognizing no distinction between "tradition" and "culture" and those in a second arguing that it is important to do so, typically by requiring more demanding criteria for any traditions that are to be elevated to some higher "cultural" level (Whiten et al. 2003; Whiten 2005, Box 1). Interestingly, the writings of the editors of this volume appear to provide strong examples of each different persuasion. The first approach, in which culture is simply treated as a synonym for tradition, appears to have got into its stride during early progress in studying the case of birdsong, so that titles like "The cultural transmission of bird song" and "Cultural evolution in bird song" became routine (e.g., Marler and Tamura 1964; see Catchpole and Slater 1995 for a review). However, other animal traditions were assimilated to this approach, so that in a popular ethology textbook like that of Slater (1985) an illustration titled "A classic case of cultural transmission: The opening of milk bottles by titmice" (in a section simply headed "Culture") would raise few eyebrows (although Galef's, presumably, would have begun to twitch).

Defining "tradition" is thus vital, both because this will immediately define culture for some researchers, as noted earlier, and also because even where "culture" is taken to require more, this typically amounts to adding additional criteria on top of the existence of tradition. At the start of the most recent volume devoted to the subject, thus giving it some authority, is the definition of tradition as "a distinctive behaviour pattern shared by two or more individuals in a social unit, which persists over time and that new practitioners acquire in part through socially aided learning" (Fragaszy and Perry 2003b, p. xiii). Variants on this populate the animal behavior literature, as when authors such as McGrew (2004) expect a tradition additionally to show intergenerational transfer. However, on my reading, there is less contention among researchers about the definition of "tradition" than about that of "culture." I suspect that most would accept the definition of Fragaszy and Perry, as I am happy to do here.

101

### Defining "Culture"

For those who are content to equate culture and tradition, the definition of Fragaszy and Perry quoted earlier does the job. If an animal can be shown to display a tradition, that is culture. Thus, for example, Laland and Hoppitt (2003, p. 152) state that "there is better evidence for culture in fish than in primates" because there are experimental translocation data that demonstrate a tradition of spatial behavior in fish, whereas there are no equivalent translocation data concerning the putative traditions of wild chimpanzees.

Others require additional criteria for culture, usually with a rationale of the kind that (1) the scale and scope of human culture make it an inevitable benchmark for talk of culture in any other species; and (2) there is considerably more to human culture than the existence of a tradition such as we see in the dialects of songbirds or the route preferences of fish.

The result of this is usually that some more "humanlike" requirement is added, but different writers vary in what they pick on. In his 1992 article Galef focused on two particular transmission mechanisms, imitation and teaching, on the assumption that the existence of these in nonhuman species, as well as in humans, could potentially mean that we are looking at homologous (phylogenetically connected) cultural phenomena. Remaining unconvinced that either mechanism had been convincingly demonstrated, Galef found it unproblematic to accept the existence of traditions in animals, but not culture in the sense in which he defined it.

I have since suggested that transmission mechanisms represent only one of three main dimensions with respect to which one might distinguish grades of culture (Whiten et al. 2003; Whiten 2005). The two others concern the behavioral content of traditions and the spatiotemporal patterning of traditions at the population level. With respect to any of these three dimensions, one can pick on particular manifestations that would restrict culture to humans. With respect to content, for example, one might note the uniqueness of human language, religion, or many other cultural contents. With respect to spatiotemporal patterning one might pick out the cumulative elaboration of culture over time (Tomasello 1999a). However, restricting the definition of culture to humans is obviously of limited use in providing a framework for comparisons among nonhuman species or for understanding the evolutionary precursors of human culture. Instead, I have proposed that within each of these three broad classes of comparison, some nonhuman species evidence various intermediate aspects of culture, the nature of which nevertheless goes beyond the mere existence of a tradition (Whiten et al. 2003; Whiten 2005). This replaces the black-or-white question, "Does species X have culture or not?"—which I see as not particularly illuminating—with a richer exercise in comparative biology in which one may need to be prepared to discover a more mosaic-like evolutionary distribution of culturally relevant characteristics. Later I use this framework to review the evidence for chimpanzees and some selected comparisons. Byrne et al. (2004) likewise argue that it will be fruitful to go beyond a culture/no-culture dichotomy, but they offer an alternative framework.

"So," a reader might respond, "you are evading the interesting and straightforward question whether a species such as the chimpanzee has culture, and instead raising the prospect of a messier, bit-of-culture-here, bit-of-culture-there kind of answer?" Well, yes, to the extent that the whole idea is to dissect culture into possibly numerous elements that can be independently compared across species; we can then address empirically the question whether some of these cluster in ways that suggest co-evolution of larger complexes, possibly up to and including one we simply want to call "culture."

That having been said, it might nevertheless be helpful to highlight one of the three dimensions outlined earlier as offering a primary criterion for talking of a species as cultural. I suggest that this should be the dimension that concerns the patterning of traditions in time and space. If one discovered an alien being that exhibited very numerous traditions, with each community expressing a multitude of such traditions that made it unique, we would probably say that this species was cultural even if we knew nothing of the precise transmission mechanisms at its disposal and irrespective of the actual behavioral content of the traditions exhibited. Of course, in the biological world it may be that cultures defined in this way will necessarily depend on certain relatively sophisticated transmission mechanisms and/or will exhibit traditions with diverse and sophisticated content, but those are, in principle, separate empirical questions.

Whiten and van Schaik (2007) have noted that this perspective can be pictured in terms of a "culture pyramid" (Figure 5.1). The basal layer represents social information transfer, which increasingly appears to be widespread in vertebrates (Laland and Hoppitt 2003) and at least present among some invertebrates (Worden and Papaj 2005), although its true

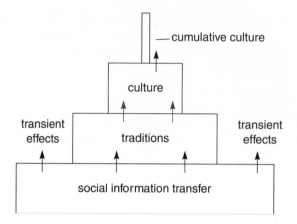

*Figure 5.1*  A culture pyramid. The base of the pyramid represents social information transfer, widespread in the animal kingdom (Danchin et al., 2004). The second level represents the longer-term subset of social learning consequences that are traditions. The third level distinguishes the even smaller set of cultural phenomena defined by multiple traditions, and the fourth level denotes the complexities that result from cumulative cultural evolution. After Whiten and van Schaik (2007).

distribution is still far from clear. However, much of this learning may serve short-term functions like targeting currently productive foraging sites (Danchin et al. 2004), so that only a subset of such socially learned behavior becomes sustained in the form of traditions, which constitute the second level of the pyramid. Where a species has communities that are distinguished by multiple diverse traditions, we have culture as defined by Whiten and van Schaik, denoted by the smaller third level. The fourth level is the even smaller subset of cultures that ever became cumulative. I now focus on the culture (third) level, dissecting it from a comparative perspective, with chimpanzees providing our case-study data.

### The Comparative and Evolutionary Dissection of Culture
#### *Spatiotemporal, Population-Level Patterning of Traditions*

The well-studied behavior of our closest living relatives, chimpanzees, has historically provided several of the earliest references to apparent cultural differences among wild animals (Goodall 1973; McGrew and Tutin 1978). In particular, as more and more information was published, authors began to tabulate differences between study sites in an expanding number of behavior patterns that were suggested to represent traditions (Goodall 1986; McGrew 1992; Boesch 1996b; Boesch and Tomasello 1998). In reviewing the literature on social learning and traditions for a journal's special issue on primate cognition (Whiten 2000), I was both impressed and frustrated. The frustration arose for several reasons, but principally because (1) there was often a lack of consistency in how the behavior patterns were labeled and/or defined (or they were not defined); and (2) reliance on what happened to have been published in the primate literature offered little confidence that the picture was at all definitive.

Fortunately, the leaders of the principal field sites were willing to pioneer a different approach designed to overcome these restrictions, which required them to pool the decades of information they had amassed in a systematic two-step analysis (Whiten et al. 1999). In the first step, contributors proposed candidate cultural variants that they suspected to exist on the basis of their working knowledge of chimpanzee behavior at their own and other field sites. This generated 65 candidate cultural variants that were consensually labeled and defined. In step 2, each of these was coded for each site by the site leader and coworkers, with the particular aim of discriminating between behavior patterns that were either common (at either of two levels, denoted customary or habitual) or absent without any apparent straightforward environmental or ecological explanation (for example, absence of relevant raw materials at the site). "Cultural variants" (equivalent to "traditions" as defined in this chapter) were then picked out as those that were common in at least one community of chimpanzees but absent without environmental explanation in at least one other. Genetic explanations were not considered likely either because differences were recorded between geographically adjacent populations, or the behavior patterns involved complexities such as tool use, the kinds of behavior that are known to be learnable, and in particular socially learned, in chimpanzees (Whiten et al. 1999, 2001; de Waal 1999).

Two major discoveries of this study were, first, that as many as 39 putative traditions were identified, spanning foraging techniques, tool use, courtship gambits, and social behavior, and second, that each chimpanzee community exhibited a suite of these that made it culturally unique. These complexities contrasted markedly with prior studies of animal traditions, the majority of which identified no more than a single case of tradition, such as birdsong dialect. Thus while we may acknowledge the widespread existence of such traditions among animals, the existence of multiple traditions becomes one sensible basis for differentiating a more refined concept of "culture" because, as already suggested earlier, this is what we are refer-ring to when we discriminate any two human cultures on the basis of the multiple traditions that make each unique. Chimpanzees have nowhere near the levels of such cultural complexity that humans attain (Levinson 2006; Nielsen in press), but they appeared to peg out an interesting status against which other species might be systematically compared.

This is indeed what has happened, as van Schaik, Ancrenaz, and col-leagues (2003) precisely replicated our approach for orangutans and found a strikingly similar picture of multiple and diverse traditions (see van Schaik this book for an overview). Both the chimpanzee and orang-utan projects have since entered a second round, with the chimpanzee consortium constituted since 2003 as the Collaborative Chimpanzee Cultures Project (CCCP2), the current updating program still in progress (but see McGrew 2004 and Whiten 2005 for additional cultural variants described in recent publications), and the orangutan consortium already offering an extended and refined analysis (see van Schaik this book). Field researchers who are studying other primates and cetaceans have in recent years also published evidence consistent with the notion of animal "cultures" as defined by multiple and diverse kinds of traditions (Rendell and Whitehead 2001; Panger et al. 2002; Perry, Baker et al. 2003; see Whiten and van Schaik 2007, Whitehead this book, Perry this book, and Mann and Sargeant this book for comparative reviews and appraisals). Indeed, the makeup of the present book indicates a tacit acknowledg-ment that this body of work is in the present century shaping the scien-tific landscape for the concept of animal culture that Whiten et al. (2003) and Whiten and van Schaik (2007) advocated, insofar as over half the 10 empirical chapters (2, 4–7, 9) fit this mold.

However, inherent limitations in these naturalistic approaches have also repeatedly been aired (Galef 2004; Laland and Hoppitt 2003; Fragaszy and Perry 2003a; Laland and Janik 2006; and see peer commentaries on Ren-dell and Whitehead 2001 and Perry, Baker et al. 2003). Although at a more detailed level they break down into such questions as "Have genetic expla-nations satisfactorily been rejected?" and "Have all ecological constraints been recognized?" the concerns can perhaps be boiled down to a core issue: how safe is the inference that the claimed traditions are truly the result of social learning? In this book authors give their own perspectives on how progress can best be made on this issue. Here I focus on two approaches in our own work, one that extends to experimental studies with captive apes and another that revisits the data from the wild.

106

## The Two-Cultures Experiment

From a purely scientific standpoint, the ideal experiment would arguably be a translocation, in which a chimpanzee (or chimpanzees) proficient in one putative tradition would be shifted to a distant community that currently lacks the behavior. For example, a chimpanzee from West Africa, skilled in cracking nuts by using natural hammer and anvil materials, would be translocated farther east to where members of the same subspecies do not crack nuts (Boesch et al. 1994). However, no chimpanzee researcher has thought this ethical. In practice, field experimentation has been limited to the pioneering work of Matsuzawa and colleagues, who introduced novel nuts to an existing nut-cracking community in Bossou, Guinea (Matsuzawa 1994). Over the years cracking the new nuts spread, which the researchers interpreted as evidence of a new or modified tradition (Biro et al. 2003). However, in the wild it was not possible to run a control condition to assess whether discovery at the individual level could account for this. Thus although the results are consistent with social learning, they are far from clinching the matter.

Given such constraints in the field, I have initiated a parallel research program with captive chimpanzees, called the Two Cultures project. This has developed into a collaborative program with Frans de Waal at the Yerkes Center of Emory University, more recently extended to a similar facility at Bastrop, Texas, in which we have used "diffusion experiments" of a kind pioneered by researchers on other species (Lefebvre 1986; Laland and Plotkin 1990; see Mesoudi, Whiten, and Dunbar 2006, Mesoudi, Whiten, and Laland 2006, and Horner et al. 2006 for reviews) to answer the question, "Can chimpanzees sustain traditions through social learning?" In such experiments a model is trained to perform some new task, and we then study whether the skill diffuses across other individuals. Social learning is assessed by comparison with at least one control group, such as one with no model. This group-centered approach differs from the approach that has characterized the substantial number of previous experimental studies of social learning in apes, which have typically focused on a single observer watching a single model, all too often a human (Whiten et al. 2004 review the 31 such studies published in the preceding 15 years, and Tomasello and Call 1997 discuss earlier ones). In retrospect, it may seem surprising, given that social learning experiments have been pursued largely because of their presumed relevance to the issue of culture, that it has taken so long for any such study to be published concerning nonhuman primates.

Rather than compare only a no-model group with one provided with an initial model, we have developed a more powerful "three-group, two-action" (3G2A) design. Here, in addition to a no-model control group, each of two groups is seeded with a model, one pretrained to execute a task using one method, the other using a different method. If these spread differentially in the two groups, we then have particularly powerful evidence of a capacity to sustain different traditions in the species studied. Testing for such a contrast was first advocated by Galef and Allen (1995) in a study that investigated dietary choice by rats, but, surprisingly, appears not to have been exploited again until our present work.

To date, six such 3G2A experiments have been run at Yerkes and Bastrop, involving five different action patterns. All have eventuated in statistically significant differential spread of the traditions initiated, with one exception that provided evidence of social learning but not of consistent traditions (Hopper et al. 2007). These studies have been of two main types: open diffusion and transmission chains (or diffusion chains). In open diffusion one individual is temporarily removed from the group, trained to become an expert in one technique, and then reintroduced. The same happens for the alternative technique in a second group, not visible to the first. It is then "open" for others to observe the model and/or attempt the task, and we are interested in who watches whom and the manner in which any successful task solutions come about. A good example is our first study (Whiten et al. 2005), which used a task nicknamed the "panpipes," a naturalistic problem insofar as chimpanzees needed to use a tool to free food items stuck behind a blockage in the higher of two pipes. One model was trained to achieve this by lifting the blockage so the food could roll toward the chimpanzee, the other by inserting the tool through a small flap to push the blockage backward so the food could then fall into the lower pipe and roll forward to the chimpanzee (Figure 5.2). The two techniques spread preferentially in the groups in which they had been seeded, with the poke technique being virtually universal in one and the lift technique predominating in the other. However, some corruption occurred, particularly with several chimpanzees discovering how to use the poke method (Figure 5.2). Although this corruption might at first sight suggest that the traditions would not be sustained, in fact, retests two months (Whiten et al. 2005) and eight months later (unpublished data) showed that the differential pattern was

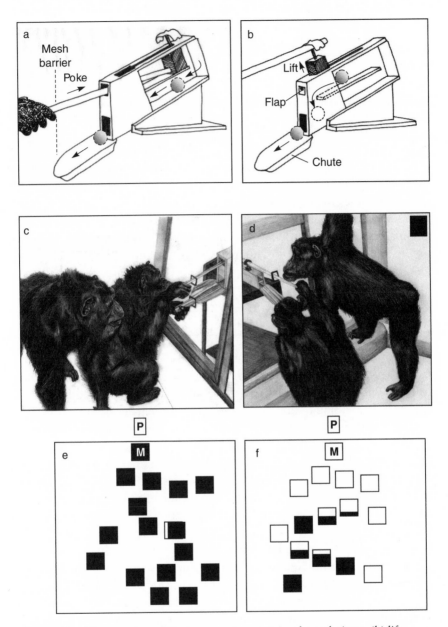

*Figure 5.2* The panpipes diffusion experiment: (a) poke technique; (b) lift technique; (c, d) poke and lift performed and observed; (e) spread of the poke technique; (f) spread of the lift technique (each square represents one chimpanzee's preferred method in first 30 successes: black = poke; white = lift; M = model; P = panpipes; distance of each chimpanzee to model represents order of acquisition). Drawings (c) and (d) courtesy of Amy Whiten.

maintained and indeed was slightly more robust than in the first phase of spread, amounting to a stable mixed tradition, incorporating what could be called "subcultures" within one of the groups.

However, given that several individuals typically watched each chimpanzee working on the task, our ability to tease out who learned from whom is limited. It is here that the second type of study, the transmission chain, makes a unique contribution. In this, still working within the general 3G2A design, we allow only one naïve individual to watch each of the two initial models. If that second individual becomes proficient at the task, it becomes the model for a further naïve individual, with the original model absent. This process then continues along a chain of further chimpanzees. Running such a chain in a fashion likely to reveal the true nature of social learning in chimpanzees is a complex and challenging task because to be meaningful it must respect such factors as relationships of affinity and rank between successive pairings, and a century of primate social learning studies has passed (Whiten and Ham 1992) without executing any such study until now.

In our recent study we used a relatively simple task that we think ought to be suitable for comparative purposes among a wide range of mammals and birds. It is a variation on the earlier idea of an "artificial fruit" (Whiten et al. 1996), in this case allowing access to food inside a box by either sliding to one side a doorway on the front or instead lifting an additional smaller door set into this sliding element, optionally using an additional tool to then help extract the food. Working sensitively in relation to the social dynamics of pairings, we were able to run two linear chains involving five and six transitions, respectively, with some side branches where pairings had to be aborted, for example, when the current expert would not perform while the partner we chose was isolated with her or him. The key finding was that each technique was passed faithfully along the respective chains, such that the final chimpanzee was using the same method (either slide or lift door) as had the trained expert who was the first in that chain (Horner et al. 2006).

These chains are modest in length, but as Horner et al. (2006) note, six such transmission events between mothers and daughters in the wild would represent around 80 years, much longer than the longest existing field studies. Together, the open-diffusion and transmission-chain results demonstrate unequivocally that chimpanzees have the capacity to sustain differential traditions across multiple "cultural generations," and that such traditions will tend to spread spontaneously in the group context—

principles that have been contentious in the research on wild chimpanzees. Of course, such experiments do not tell us whether each putative case in the wild is truly a tradition: experimental confirmation of this would require an experiment for each. What it does do is provide a clear positive answer to the long-standing criticism that the studies from the wild have not been able to provide an unequivocal demonstration that traditions will spread with fidelity among chimpanzees through social learning. In my view, experimental results of this kind provide important background information about the social learning propensities of the species under study and thus play a key role in helping specify the assumptions most appropriate to making parsimonious interpretations of findings from the wild.

Results from each of the six experiments concluded so far constitute firsts of various kinds for the nonhuman primate literature. In addition to the two experiments previously summarized, a third extends the panpipes study to Bastrop chimpanzees, such that at the time of writing, 76 chimpanzees have been tested with this task (Hopper et al. 2007), a fourth yielded evidence of high-fidelity copying of an arbitrary local convention (the "token" study; see Bonnie et al. 2007 and de Waal and Bonnie this book), and further experiments demonstrate the spread of two new traditions across larger numbers of chimpanzees at Bastrop (Whiten et al. 2007). Together, these provide an even more important first in my view. They constitute the originally proposed two-cultures experiment, for if, as advocated earlier, animal "culture" is defined by the existence of multiple traditions, we now have experimental confirmation of this at Yerkes. One Yerkes group is dominated by panpipe pokers, token-to-bucket conventions, and door lifters; the other by panpipe lifters, token-to-pipe conventions, and door sliders. In fact, the existence of similar complexities at Bastrop means that we really have a four-cultures experiment coming to fruition. With the implications of these studies in mind, let us return to the wild.

### Identifying Social Learning in the Wild

Despite its superior ability to identify social learning, our program of diffusion experiments would make little sense without the prior and ongoing studies of wild chimpanzees. Field research provides the only sensible first step in any behavioral study, including the subject of culture. I see the role of the pair of initial surveys (Whiten et al. 1999, 2001) as providing our best estimate of the scope of traditions in the wild. Of course, some of this picture may well require revision in the future—that is the nature of science—but achieving what we have to date is essential as a foundation for (1) the kinds of experiments previously described, teasing out answers to secondary questions best addressed in this way; (2) more extended and refined approaches to field data (the ongoing CCCP2 project); and (3) parallel comparative studies with other species (see, for example, van Schaik this volume). For this reason, although constructive criticism is healthy in the development of any field of science, I tend to view some of the critiques of the field analyses that have come forward in recent years (Galef 2004; Laland and Janik 2006) as a little harsh and premature. In the extended essay of the pair (Whiten et al. 2001) we were careful to describe the charts as mapping "putative" cultural variants, acknowledging the circumstantial nature of the evidence, and we offered substantial critical discussion of the arguments for and against the classifications of behavior we published at that time. In short, we stand by the value of these initial analyses while recognizing the inherent difficulties in establishing the truth of the matter concerning animal culture in the wild.

Several methodological refinements to the evidence from the wild have been provided by parallel work of van Schaik and colleagues, reviewed in this volume. Since the CCCP2 project is currently still in train, I will discuss only one modest contribution here, a preliminary attempt to more explicitly document the nature of the circumstantial evidence available. Table 5.1 lists 36 of the 39 putative traditions, omitting three variants of "nuthammer" from the original charts because these simply covary and are perhaps more properly collapsed together to constitute a single variant. For the remaining 36, three columns in the table are designed to indicate the strength of evidence for social learning on a 0–3 scale. The first of these is concerned with assessing genetic alternatives and is based directly on geographic information listed in the two prior columns. The second and third concern, respectively, assessment of environmental determinants and evidence of chimpanzees closely observing the behavior patterns at stake. The ratings shown for these have been agreed by W. C. McGrew and myself, and they should be regarded only as interim illustrations of the approach, pending more comprehensive work in the CCCP2 project.

*Assessing genetic explanations for variations.* Before turning to the table, I note de Waal's (1999, p. 635) comment on our 1999 article, that "genes determine general abilities, such as tool use, but it is hard to imagine that they instruct apes how exactly to fish for ants or whether or not to make cushy seats out of vegetation." Indeed, at present we know of no examples

**Table 5.1 Ranking three types of circumstantial evidence for social learning**

| Behavior patterns from Whiten et al. 2001, Table 3 | Cmy | Nst Abs | NG | NE | Ob |
|---|---|---|---|---|---|
| Index-hit (squash ectoparasite on arm) | Taï | Bos | 2 | 3 | 3 |
| Leaf-clip, fingers (rip single leaf with fingers) | Taï | Bos | 2 | 3 | 3 |
| Leaf-clip, mouth (rip parts off leaf with mouth) | MM | Gom | 2 | 3 | 3 |
| Stem pull-through (pull stems noisily) | Kib | Bud | 2 | 3 | 3 |
| Rain dance (slow display at start of rain) | Taï | Bos | 2 | 3 | 3 |
| Termite-fish using leaf midrib | MK | MM | 3 | 2 | 3 |
| Termite-fish using nonleaf materials | MK | MM | 3 | 2 | 3 |
| Shrub-bend (squash stems underfoot) | MM | MK | 3 | 3 | 2 |
| Leaf-inspect (inspect ectoparasite on hand) | Bud | Kib | 2 | 3 | 3 |
| Pestle-pound (mash palm crown with petiole) | Bos | Taï | 2 | 2 | 3 |
| Nut-hammer, stone hammer on stone anvil | Taï | Sass. | 2 | 3 | 3 |
| Aimed-throw (throw object directionally) | MM | MK | 3 | 2 | 2 |
| Food-pound onto wood (smash food) | Gom | MM | 2 | 2 | 3 |
| Bee-probe (disable bees, flick with probe) | Taï | Bos | 2 | 2 | 3 |
| Marrow-pick (pick bone marrow out) | Taï | Bos | 2 | 2 | 3 |
| Food-pound onto non-wood surface (e.g., stone) | Gom | MM | 2 | 2 | 3 |
| Ant-dip-wipe (manually wipe ants off wand) | Gom | MM | 2 | 2 | 3 |
| Fluid-dip (use of probe to extract fluids) | Taï | Bos | 2 | 2 | 3 |
| Self-tickle (tickle self using objects) | Gom | MM | 2 | 3 | 2 |
| Leaf-squash (squash ectoparasite on leaf) | Gom | Kib | 1 | 3 | 3 |
| Knuckle-knock (knock to attract attention) | Gom | Kib | 1 | 3 | 3 |
| Branch-slap (slap branch, for attention) | Bud | Kib | 2 | 3 | 2 |
| Leaf-napkin (leaves used to clean body) | Gom | MK | 2 | 3 | 2 |
| Leaf-strip (rip leaves off stem, as threat) | Kib | Bud | 2 | 3 | 2 |
| Leaf-dab (leaf dabbed on wound, examined) | Kib | Bud | 2 | 3 | 2 |
| Hand-clasp (clasp arms overhead, groom) | MM | Gom | 2 | 3S | 1 |
| Nut-hammer, wood | Taï | Bos | 2 | 1 | 3 |
| Club (strike forcefully with stick) | Gom | MK | 2 | 2 | 2 |
| Lever open (stick used to enlarge entrance) | Gom | MM | 2 | 2 | 2 |
| Expel/stir (stick expels or stirs insects) | Taï | Bos | 2 | 2 | 2 |
| Leaf-groom (intense "grooming" of leaves) | Kib | Lo | 0 | 3 | 3 |
| Ant-fish (probe used to extract ants) | MM | Kib | 1 | 2 | 3 |
| Ant-dip-single (one-handed dip stick on ants) | Taï | Lo | 0 | 2 | 3 |
| Fly-whisk (leafy stick used to fan flies) | Taï | Bos | 2 | 2 | 1 |
| Seat-vegetation (large leaves as seat) | Taï | Lo | 0 | 2 | 1 |
| Branch din (bend, release saplings to warn) | Lo | Taï | 0 | 2 | 1 |

*Note:* For each of 36 putative chimpanzee traditions listed in Whiten et al. (2001), ratings on a three-point scale are given for three sources of circumstantial evidence consistent with the hypothesis that the behavior is socially learned. Cmy = Customary: A site where the

**Table 5.1** (*continued*)

behavior is customary, i.e., common in all able-bodied individuals of at least one sex. Nst abs=Nearest absent: the nearest site to the Cmy one listed in the previous column at which the behavior is absent. Bos=Bossou; Lo=Lope; MM=Mahale M; MK=Mahale K; Go=Gombe; Kib=Kibale; Bud=Budongo. For further details of sites and locations, see Whiten et al. 2001. NG=Nongenetic: Significance of regional evidence that variation is not genetic; 0=Cmy and Nst abs communities are subspecies (e.g., *troglodytes* versus *verus*); 1=C and N columns are within subspecies (e.g., Gombe versus Kibale); 2=C and N columns are the closest studied sites within subspecies (e.g., Mahale versus Gombe); 3=C and N columns are neighboring communities (e.g., Mahale M versus Mahale K). NE=Nonenvironmental: Strength of evidence that absence is not simply environmentally constrained: 0=variations explicable through obvious environmental differences; 1=may be explicable through environmental differences; 2=unlikely to be explicable through environmental differences; 3=there are *no* apparent, relevant environmental differences; 3S=behavior is social, no apparent relevant environmental constraints. Ob=Observation: Strength of evidence that chimpanzees spend time observing the behavior in others; 0=Observation of behavior never seen; 1=observation has been seen, but no focused attention has been noticeable; 2=Attention is obviously repeatedly directed at activity; 3=Intense and repeated visual interest in activity.

of genetic differences that cause behavioral differences among common chimpanzees, whereas there exists a mass of studies that demonstrate that behavior such as tool use is highly learnable (Tomasello and Call 1997), as well as that chimpanzees can socially learn foraging and other techniques (Whiten et al. 2004, 2005). This evidence contrasts markedly with that from studies with tool-using birds, which have shown great competence emerging without benefit of social learning (Tebbich et al. 2001; Kenward et al. 2005). From this perspective, it might seem a reasonable working assumption to discount genetic explanations for chimpanzee behavioral variations across Africa. However, it could be that we know of no genetic causes of behavioral differences because nobody has properly investigated this, so it may be wise to attend to the issue more explicitly.

For each behavior pattern listed in Table 5.1, there is a column that lists a site where the behavior is customary and another that lists the closest site at which it is absent without any apparent straightforward environmental explanation. The next column then ranks the likely genetic difference between the two communities on the basis of geographic separation. The strongest case for rejection is where the communities are neighbors, because in that case there will be frequent interbreeding between them. There are only 4 cases of this because in the 1999 and 2001 analyses only two communities studied were neighbors. Since then other examples such as that concerning fine details of the way in which handclasp grooming is conducted have been published for these two communities (McGrew 2004; Whiten 2005 for reviews), and studies of other neighboring communities are identifying additional behavioral variations: Boesch (2003) lists 6 further cases for the North and South communities in the Taï Forest. At the other extreme, only 4 cases rely on comparisons between subspecies. Of the rest, the great majority (25) represent contrasts between the closest sites within the same subspecies, including Gombe and Mahale, just 170 kilometers apart. Accordingly, and particularly given the nature of many of the behavior patterns at stake, notably tool use, it appears unlikely that genetic differences will underlie the behavioral variations. The example that most deserves scrutiny on this account is leaf-groom, a still relatively mysterious behavior in which chimpanzees break off from grooming each other to briefly groom leaves on attached vegetation. This is absent at both West African sites (subspecies *verus*) but common at others in East Africa (subsp. *schweinfurthii*). Moreover, at least two techniques for killing ectoparasites use leaves in the east (leaf-squash and leaf-inspect), whereas this is done without leaves in the west. Although this might represent an interesting case of leaf-based cultural evolution in the east, we cannot rule out the possibility that *schweinfurthii* has a certain genetically based predilection for leaves in these contexts that generates the several leaf-based behaviors catalogued. It is difficult to see how this can be resolved in the wild, but clues might be gleaned from any captive situations shared by members of these subspecies, given access to appropriate leaves.

*Assessing environmental explanations for variations.* Two preliminary concerns of critics should be mentioned before inspecting the table. First, Laland and Janik (2006) note that Whiten et al. (1999) identified only 3 among their 65 candidate cultural variations that should be set aside as explicable for local environmental reasons (as when nests occur at ground level in predator-free areas), and they worry that this is a surprisingly low number, perhaps indicating that potential environmental explanations were little considered. To the contrary, phase 1 of our procedure invited participants to propose likely candidate traditions for filtering in the subsequent coding exercise of phase 2, so they would have already aimed to set aside cases where an environmental explanation would be likely. In fact, this latter category is likely to be huge, most notably because of local variations in dietary selections. Thus the 65 and 39 behavior patterns scrutinized in phases 1 and 2 already constitute a tiny subset of the variation in chimpanzee behavior across Africa.

Second, a study by Humle and Matsuzawa (2002) has shown that chimpanzees at Bossou use versions of the two techniques classed as cultural variants in our surveys and adjust their use according to environmental circumstances. Long tools are used together with a bimanual wiping technique for the more aggressive of two species of ants and preferentially at burrow entrances. Critiques such as that of Laland and Janik (2006) have accordingly suggested that this may mean that such factors are the explanations for cross-site differences, not social learning. Humle and Matsuzawa's studies do indeed provide excellent data of the kind needed to take "animal cultures" research to increasingly rigorous levels (and our broad survey work is the foundation on which these new, more refined studies are being built). However, it should be noted that the authors themselves do not interpret their results as discounting social learning. In a recent review of the work, Humle (2006, p. 469) says that "variations in prey emphasis and technique used suggest that social learning might still explain most of the between-site variation observed in ant-dipping behaviour and prey emphasis" and concludes that "ant-dipping . . . still exhibits strong cultural patterns and appears to remain a good example of culture in chimpanzees."

In Table 5.1 ratings on a 0–3 scale are given to indicate judgments about the plausibility of environmental explanations in each case. As expected from the phase 2 criteria for inclusion, which required fieldworkers to judge that such environmental explanations were implausible, scores of 0 or 1 are absent, with a single exception that thus queries a potential earlier miscoding. However, inherent uncertainties in this process are indicated by the fact that about half the codings are at level 2, half at level 3. The evidential bases for these codings are varied. For level 3 they include cases where the behavior is purely social, such as handclasp, or where the materials required appear available at sites where the behavior pattern is absent, such as leaves required for actions like leaf-clip, or nuts and hammer materials for nut cracking. Level 2 codings are given where some environmental explanation is easier to conceive but is still judged relatively implausible. Examples include the prospect that access to the crowns of palm trees for pestle pounding might vary in difficulty, that differences in prey habits might affect the utility of using tools for their extraction, like bee-probe, and that differences in the need for a resource might affect patterns like fluid-dip. The distribution of codes 1–3 in this column expresses quite well the status of our knowledge on this important issue. On the one hand, without experimentation these judgments are always subject to some uncertainty about how well environmental factors can be excluded. On the other hand, it is the view of the CCCP participants that such a chart represents our best estimate of the scope of cultural variation among chimpanzees. Accordingly, the best scientific course is the combination of (1) publishing charts such as this one (and those contributed by parallel projects summarized elsewhere in this volume) and (2) making increasingly explicit the evidential base for the interpretations inherent in them.

*Clarity of observational information acquisition.* In the case of some tasks, such as fine tool use, chimpanzees can be observed to spend extended periods intently observing expert practitioners. This was the primary fact that led Goodall long ago (1973) to infer cultural transmission in chimpanzees; it likely underlies the judgments of many fieldworkers in our surveys but has never been explicitly included in these publications and is thus information hidden from readers and commentators not familiar with chimpanzees. Of course, such observation does not guarantee social learning, but that this is its function is supported by a recent report by Lonsdorf et al. (2004) that showed that infant female chimpanzees, who spend significantly more time observing their mothers termite fishing than do their male peers, master the skill over a year earlier and match their mother's tool-length preference. Accordingly, it would be helpful in the future to have quantitative measures of attentional behavior, such as lengths of observational bouts in relation to the stage of skill acquisition. Of course, some activities can be observed and learned about at a distance, so high scores are not necessarily to be expected for all behavior patterns listed in Table 5.1.

In summary, the broad picture provided by the chimpanzee fieldwork community and the diffusion experiments outlined earlier now converge on a consistent central conclusion, that (1) chimpanzees can sustain multiple traditions that include a diversity of types of behavior; and (2) different groups may therefore develop unique cultural profiles constituted by a suite of such traditions. These are referred to in the first rows of Table 5.2, which summarizes my overall comparative scheme. In rows 3 and 4 this table lists the two other aspects of population-level patterning of traditions discussed in the analysis of Whiten et al. (2003). One of these addresses the question whether there exists cumulative culture (recall Figure 5.1), where the cultural

**Table 5.2 A comparative dissection of "culture"**

| | Humans | Chimpanzees (*Pan troglodytes*) |
|---|---|---|
| *1. Population-level patterning* | | |
| Multiple, diverse traditions? | Yes: countless (arguably each word in everyday language is a tradition or "meme"). | Yes: Whiten et al. (2001) identified 39 traditions of diverse kinds. |
| Populations differentiated by multiple traditions? | Yes: cultures (e.g., Inuit, Masai) are distinguished by numerous traditions. | Yes: Whiten et al. (2001) described 9–24 traditions per community, each set unique. |
| Cumulative evolution of traditions? | Yes: Extremely elaborate, progressive (Tomasello 1999a). | Minimal at best, and disputed (Boesch and Tomasello 1998). |
| Clustering of traditions through core ideas? | Yes (e.g., LeVine 1984), but evidence debated (Boyd et al. 1997). | Unknown: speculatively, some clusterings may be consistent with this (Whiten et al. 2003). |
| *2. Transmission mechanisms* | | |
| Teaching? | Yes: but may be rare in hunter-gatherers and thus rare until very recently everywhere. | Minimal "scaffolding" (Whiten 1999) in limited contexts at best; disputed. |
| Sophisticated observational learning processes? | Yes: including imitation, emulation, learning from verbal instruction (Tomasello 1999). | Yes: including emulation and imitation of sequential actions (Whiten et al. 2004). |

| | | |
|---|---|---|
| Selective acquisition? | Yes: Sensitivity to e.g. model's goals (Nielsen 2006). | Yes: Sensitivity to, e.g., causal effectiveness of actions (Horner and Whiten 2005). |
| Convention and conformity? | Yes: Some traditions are merely arbitrary conventions; strong conformity to norms common. | Recent but controversial evidence of conventions (Bonnie et al. 2007) and conformity (Whiten et al. 2005). |
| Recognition of transmission process? | Yes: ranging from "do-as-I-do" "Simon says" games to propaganda. | Yes, only great apes and dolphins have been able to learn "do-as-I-do" rule. |
| **3. Content of traditions** | | |
| Social | Includes, e.g., language, gestural conventions, moral norms, ceremonies, and institutions. | Includes use of tools (e.g., leaf-clip in courtship), grooming conventions, possibly dialects. |
| Nonsocial | Includes, e.g., tool-construction methods, hunting, trapping, clothing, medicine, shelters. | Includes nontool foraging techniques, as well as tools used for foraging, comfort, and hygiene. |

*Note:* Eleven aspects of culture are discriminated within three broad categories, and their manifestations are indicated for two species (see Whiten et al. 2003, 2004, and the present text for detail), but any species, including those discussed in the present volume, can be compared within this scheme. Each aspect represents one way in which "culture" goes beyond the mere existence of a tradition. The classification is anthropocentric to the extent that each aspect is fully represented in human culture; in principle, other categories that go beyond the mere existence of tradition and are more fully manifested in species other than humans could be added, but none appear to have been suggested in the current literature.

achievements of one generation have been built by elaborating on earlier ones. The other concerns the question whether traditions form clusters characterized by a common conceptual core, as some cultural anthropologists have argued is characteristic of human cultural life (so this could arguably form an alternative fourth layer in the culture pyramid of Figure 5.1). For neither of these did Whiten et al. (2003) find more than elementary and tentative evidence that would justify more extended discussion here. McGrew (2004) has since argued that more evidence for cumulative culture among nonhuman primates exists than is recognized in the literature. Notwithstanding such debates, it seems to me that a vast gulf separates even the strongest cases for cumulation in chimpanzees and the elaborate achievements of millennia (and particularly the most recent centuries) of human cumulative cultural evolution.

## Mechanisms of Social Learning

Whiten et al. (2003) considered five different aspects of social learning mechanisms, with respect to which they suggested that it could be instructive to compare the capacities and outcomes for different species (Table 5.2). The latter four of these, which required the most extensive treatment, were the subject of a recent review more comprehensive than space permits here (Whiten et al. 2004; see also Whiten 2005; Whiten and van Schaik (2007). Accordingly, I treat these elements here in relatively abridged fashion, highlighting just one particularly interesting recent finding from our work.

### Teaching

To my knowledge, teaching has not been described for East African chimpanzees. Claims of teaching have been restricted to the specific context of nut cracking in Taï chimpanzees, which appears to be the technological skill it takes chimpanzees the longest to master (Boesch 2001). However, the majority of this is limited to maternal tolerance of hammer and nut scrounging, supplemented by limited active provision of these. This is far from the elaborate forms that teaching can take in the human case, whether by nonverbal demonstration or verbal instruction. However, there is a big question mark over whether there is much of such explicit teaching in hunter-gatherer life, and thus whether teaching was in fact significant during much of the evolutionary past of *Homo* (or even *H. sapiens*) (Whiten et al. 2003).

Nevertheless, teaching may be of considerable interest for comparative research. It may be more critical for species for which hunting is a more obligate way of life in both sexes than is the case for chimpanzees (Caro and Hauser 1992; Thornton and McAuliffe 2006). As always, fruitful comparison will depend on how the concept is defined. Teaching was recently claimed in ants (Franks and Richardson 2006), but this was limited to one ant that was moving toward a foraging goal waiting while a naïve follower caught up with it, an act mirrored much of the time by primate mothers, including chimpanzees, who will intermittently wait while their naïve infant catches up with them en route to some food source known to the mother.

### Sophisticated Observational Learning Processes

Twenty-two experimental studies of chimpanzee social learning conducted in the preceding 15 years were reviewed in Whiten et al. (2004), and for reasons of space, readers must be advised to consult that article for a detailed analysis of "how apes ape" relevant to this section and the three that follow, for which only outlines are given. Among the conclusions of this review that I highlight here are that (1) there is evidence for imitation (Custance et al. 1995; Hopper et al. (2007), the criterion for culture advocated by Galef (1992) alongside teaching; (2) this evidence extends to copying of sequential structure in actions (Whiten 1998), with some tentative evidence of copying hierarchical structure (Whiten et al. 2004); and (3) chimpanzees have at their disposal a "toolbox" of social learning mechanisms that includes imitation and emulation, deployed according to context-dependent need (Horner and Whiten 2005).

An additional recent finding relates to a long-standing debate about whether chimpanzees do employ imitation or, as Tomasello (1996) argued, are essentially emulators (learning about the environmental results of actions rather than actions themselves), unless they experience "humanizing" rearing conditions described as "enculturation." Since our diffusion experiments are themselves mute on the precise nature of the social learning processes involved, we have since used "ghost conditions" to present a naïve set of chimpanzees with what is normally just the environmental result of what they do with the panpipes: either the blockage alone or the tool-plus-blockage configuration was repeatedly moved so as to release food, using a simple fishing-line setup. Interestingly, no chimpanzee learned from these experiences (Hopper et al. (2007), although they typically learned once they later saw an ape model

121

(unpublished data). This suggests that the chimpanzee-as-emulator theory has significant limitations. If we recall Thorndike's definition of imitation as "doing an act from seeing it done" (1898, p. 50), it does indeed appear that at least for a challenging task like the panpipes, chimpanzees gain crucial information from seeing an ape actually do the action.

### Selective Acquisition

This does not mean that apes mindlessly ape (mimic) what they see, however. Horner and Whiten (2005) showed that an observed element would be incorporated into the copy depending on whether it could be perceived to be causally relevant (in a transparent version of the task) or not (when the model worked on an opaque version). This could be regarded as an additional aspect of the "sophisticated observational learning processes" outlined earlier (like all such classifications, there is ample scope for either splitting or lumping within it), but it is separately highlighted here because the efficiency it suggests may be important to a species for which it is adaptive to acquire the multiple diverse traditions referred to here as culture.

### Convention and Conformity

Important swathes of human culture are not inherently functional, as a foraging technique is, but instead are merely local, arbitrary conventions. The local meanings of gestures and words are examples—they work only because each community treats them in the same way. Tomasello et al. (1997) found that chimpanzees failed to learn novel food-getting gestures a model was trained to perform, which raised the prospect that being able to recognize such arbitrary conventions may be a distinctively human capacity. Bonnie et al. (2007; see de Waal and Bonnie this book), by contrast, in an open-diffusion study, found that alternative, experimentally seeded, arbitrary actions spread differentially and with high fidelity. These were object-directed actions rather than pure bodily gestures, but the results are consistent with the interpretation of certain object-directed courtship gestures as local traditions among wild chimpanzees (Boesch 2003).

The issue of conformity is related insofar as this shares the characteristic of adopting an act "just because others do it" (see de Waal and Bonnie this book). However, like some other terms discussed earlier, conformity may be defined in different ways. Most straightforwardly, it refers to preferring to adopt the behavior of the majority where two or more options exist (Laland 2004). In a recent study we unexpectedly obtained evidence for conformity of a kind previously much studied within human social psychology, where individuals conform to the option a majority of companions perform even when they are familiar with the other option. In this case individuals in our panpipes study who discovered the technique common in the other group nevertheless showed a significant tendency to converge on the norm in their own group later (Whiten et al. 2005). In the study of Bonnie et al. it was found that individuals ignored one single nonconforming individual and focused on their group norm instead. These recent findings echo an earlier result obtained in dyadic testing, where copying of the sequential structure of a complex action emerged over time despite the fact that subjects had been successful in solving the artificial fruit task involved in other ways (Whiten 1998).

### Recognition of the Copying Process

Curiously, it has been possible to train only great apes and dolphins to copy actions on demand (the "do-as-I-do" paradigm; Custance et al. 1995). Whiten (2000) suggested that this means that apes are able to recognize when they are imitating, as do human children. Attempts to train monkeys to "do as I do" have not been successful (Whiten et al. 2004). The implications of this difference for the acquisition of traditions remain unclear, but it does imply a higher level of cognition in apes in relation to the process of social learning, shared with the most cultural of species, humans.

## Cultural Contents

Whatever the patterning of traditions in time and space and whatever the transmission mechanisms involved, the actual contents of traditions vary greatly across taxa, and this represents the third major section of my proposed scheme for comparative analysis. In Table 5.2 examples are given for humans and chimpanzees under two subheadings concerning social and nonsocial content, but there could be many alternative ways of classifying such content, and such taxonomies could in principle be very complex insofar as they would need to deal with the whole gamut of human cultural behavior, as well as that of other species, such as grooming conventions.

All 11 distinctions made in the scheme illustrated in Table 5.2 are illustrated here only for chimpanzees, the prime focus of the present chapter, and humans, who, as the most cultural of species, provide an essential anchor point for any such endeavor (although this primary anthropocentric perspective should not be allowed to limit the scope of the distinctions a comparative analysis deals with; Fragaszy and Perry 2003a). However, the concept embodied in the table is intended to provide a framework in relation to which it should be possible to compare any species that have been shown to display traditions. Whiten and van Schaik (2007) have recently done this in relation to the first two rows of Table 5.2, reviewing evidence for chimpanzees, orangutans, capuchins, dolphins, killer whales, and New Caledonian crows.

## Concluding Remarks

The conclusions I wish to emphasize in closing are of two kinds: methodological and substantive. The methodological message is that, as in ethology generally, an eclectic approach enriches our understanding of the phenomena of interest. Here I have drawn principally on both observational field research and experiments with captive animals. Each of these has made major contributions to elucidating complementary aspects of chimpanzee culture that cannot be achieved in any other way, and the same will be true for all species. Moreover, each approach can inform the other, because the fieldwork guides appropriately naturalistic experimentation and the experimental results provide knowledge that shapes the most appropriate inferences in the wild.

The substantive, empirical conclusion is that the scheme for comparative analysis that I have offered (summarized in Table 5.2) does discriminate aspects of culture that are empirically tractable, as confirmed by the chimpanzee data I have discussed (although the question of cultural cores admittedly remains challenging in this respect). At present the effort applied to chimpanzee research in this area is sufficiently disproportionate that many of the issues raised in Table 5.2 are not currently answerable for other species in the way they are for chimpanzees, but the questions they pose provide ample scope for future studies in the increasingly exciting subject of "animal culture."

ACKNOWLEDGMENTS

This chapter was written during the tenure of a Leverhulme Major Research Fellowship. The recent empirical studies summarized here were supported by the Biotechnology and Biological Research Council and the Economic and Social Research Council. I am grateful to Marietta Dindo, Lydia Hopper, and Kevin Laland for comments on the manuscript.

# 6

# HOW MIGHT WE STUDY CULTURE?

## A PERSPECTIVE FROM THE OCEAN

HAL WHITEHEAD

As a setting for behavior and evolution, the ocean is different. Compared with the habitats of most of the other animals that have earned the "cultural" label, the ocean is three-dimensional, vast, and connected. Its inhabitants have no contest with gravity and so have relatively low costs of transport but must subsist within trophic webs based on microscopic autotrophs that are extremely patchy in both space and time. These and other differences, including a general difficulty of study, mean that ocean scientists are usually more holistic, collaborative, technologically dependent, and empirically driven than those who study terrestrial habitats and organisms. In the ocean, designed, replicated experiments are quite rare, being difficult to achieve and often perceived as having limited value in systems that naturally vary enormously over large scales of both space and time.

Given these major contrasts, it is not unexpected that our perspective of culture in the ocean is different from the terrestrial models that have informed the work of most of the other authors of the chapters of this book (but see the chapter by Sargeant and Mann, who work in a near-shore system that is somewhat intermediate in character between land and the pelagic ocean). This is both because culture really is different in the ocean—for instance, there is more vocal culture and fewer material tools, as might be expected in a fluid medium that transmits sound particularly well—and because ocean scientists have approached the phenomenon with their different ways of thinking and operating.

Among scientists who have considered the cultures of the ocean, I am perhaps the most pelagic. My study sites (such as the South Pacific or the Sargasso Sea) are thousands of kilometers across, mostly far from shore; my study animal, the sperm whale *(Physeter macrocephalus),* is huge and wide ranging and (except a few terminally sick calves) cannot be kept in captivity; and I have never conducted a successful experiment. I collect simple data opportunistically—principally photographic records of individual identity together with place and time—and then analyze them, using fairly sophisticated statistical techniques when necessary. Here is my perspective on culture and how we might study it, informed by studies of whales and dolphins (the cetaceans), and their principal habitat, the ocean, but hopefully relevant more widely.

## There Is Culture in the Ocean

Although there had been some previous speculation about the role of culture in whales and dolphins, particularly by the pioneering cetologist Ken Norris (e.g., Norris and Schilt 1988), until 2001 (when there was a well-attended workshop on culture at the biennial marine mammal conference and the publication of a review of culture in whales and dolphins with commentaries; Rendell and Whitehead 2001), finding culture was not an objective of studies of cetology. Rather, patterns of behavior were discovered that could be explained in no other way than by culture, or whatever one wants to call the process by which social learning homogenizes behavior within sections of a population.

The first case in which cetologists stumbled into culture, and in some ways still the most compelling, is the song of the humpback whale *(Megaptera novaeangliae)*. Male humpbacks sing long, elaborate songs on their winter breeding grounds (Payne and McVay 1971). In any ocean basin at any time, virtually all the humpbacks sing the same song, but the communal song evolves gradually over periods of months and thus substantially over periods of years (Payne 1999). Songs in different ocean basins have different content and evolve independently but conform to the same set of rules (Payne 1999). An exception occurred off the east coast of Australia in 1997. The population dropped the Southwest Pacific song that it had been singing and adopted a quite different song, the song of the Indian Ocean, used by humpbacks off the west coast of Australia (Noad et al. 2000). There is no conceivable mechanism for such patterns other than animals listening to one another's songs and adjusting their own accordingly to produce stereotypical group behavior—culture. And the sudden switch off eastern Australia was the first known cultural revolution in nonhumans.

A second example of cetologists stumbling into culture comes from the study of sperm whales of the South Pacific by me and my colleagues. In

1992–1993 we made a survey of the females and immatures across this huge ocean, trying to map the structure of the sperm whale populations. We expected to find genetic and other differences among sperm whales at two levels: social groups and geographic areas that might relate to oceanographic features. Group-specific features were prominent (White-head et al. 1998); groups have a partially matrilineal structure, so that a female is grouped with close female relatives and their dependent young, as well as unrelated females and their young. But in place of a geographic structure we found a clear clustering of groups based on their vocal repertoire—there were clans with distinct dialects, clans that spanned thousands of kilometers of ocean and overlapped (Rendell and White-head 2003). Where the clans did overlap, groups from different clans possessed different movement strategies, habitat use, and feeding success (Whitehead and Rendell 2004). This picture could not be explained by ecological differences and individual learning, because the clans used the same areas quite differently, or by genetic factors, because genetically un-related animals were performing distinct socially integrated and clan-specific behavior. There had to be social learning, and so culture.

These two examples illustrate a quite consistent feature of long-term studies of cetacean behavior, the stumbling across patterns that can only be explained, or can most easily be explained, by social learning. Other ex-amples are considered by Rendell and Whitehead (2001) and its commen-taries. One of the frustrating aspects of the study of culture in cetaceans, as well as in other species, is that culture is hard to pin down in situations in which it affects how an animal interacts with its environment, for exam-ple, through foraging behavior. The patterns observed could be due to culture but also to individual learning in different environments. The re-markably diverse foraging strategies of sets of bottlenose dolphins (*Tur-siops* sp.) in Shark Bay, Australia (Mann and Sargeant 2003), are likely to be the results of social learning and so perhaps culture, but there are doubts (Sargeant and Mann this book). We need better methods for sepa-rating the causes of behavioral diversity, and I will suggest some.

## What Is Culture?

In the humpback and sperm whale examples I have called such patterns "culture," but I could have used "tradition" or "socially learned patterns of group-specific behavior" or something else. Why invoke the contro-versial, value-laden "c" word? There are several reasons.

Fundamentally, humpback songs and sperm whale clans are culture as English-speaking people generally use the term. Most people's connotation of culture is that it is learned behavior characteristic of a particular set of individuals, "the way we do things" (McGrew 2003, p. 433). Our own preferred formal version of this is the following: "Culture is information or behavior—shared by a group[1] of animals—which is acquired from conspecifics through some form of social learning" (Rendell and Whitehead 2001, p. 310).

In this definition the behavior that is culture has two principal elements: it is socially learned and shared. The type of social learning is left open for good reasons (discussed later), and the sharing makes culture more than just social learning. The definition implies that a cultural type can be common to all members of a population or species (these then become the "group"), and cultural evolution may promote conformism (Henrich and Boyd 1998), but in practice complete homogeneity may make it hard to identify a cultural type as culture (and impossible using the methodology I discuss later).

Our definition is similar to definitions used by others who study the culture of nonhumans in the wild (e.g., McGrew 2003) and to those used by cultural theorists (e.g., Richerson and Boyd 2005), as well as by some anthropologists and sociologists (e.g., Lenski et al. 1995). But it does clash with two other major classes of definition.

It rejects the belief of some psychologists (e.g., Galef 1992; Tomasello 1994) that among social learning processes, culture should be restricted to imitation and teaching. Although the mechanism by which information or behavior is transmitted between individuals is important, we and others have argued that it should not be used to restrict the definition of culture on several grounds (Whiten and Ham 1992; de Waal 2001; Boesch 2001; Rendell and Whitehead 2001; McGrew 2003). These include the lack of clarity in definitions of imitation and other types of social learning, the uncertainty about which social learning mechanisms underlie human culture, and that types of social learning other than imitation and teaching can lead to characteristic group-specific behavior. Restricting culture to phenomena produced by imitation and teaching also leads to unresolvable dichotomies for field scientists, including cetologists. How can we know how a young sperm whale acquired her movement strategies?

I also reject definitions of culture that are explicitly or implicitly (because of our inability to study certain phenomena in other species) restricted to humans. An example of the former is "what makes us human" (quoted in McGrew 2003) and of the latter "shared values" (Ripoll and Vauclair 2001). These are of no use to a cetologist and preclude using the enormously powerful comparative method, even when study is directed at humans.

The definition of culture that I use encompasses a large number of behavioral activities from a wide variety of organisms. So be it. Narrowing the definition of culture will necessarily exclude features of human culture and/or make inoperable the study of nonhuman culture (Rendell and Whitehead 2001). It is not the presence of culture in a species that is particularly interesting; it is its extent, variety, and effects (see Laland and Janik 2006).

For me, effects are most engaging. A culturally determined behavior is interesting in its own right, but it is much more important when it affects ecology, population biology, life history, genetic evolution, or conservation biology. When culture is important for a species, affecting a substantial portion of behavior, it can become an important driver of many aspects of biology. And as the cultural theorists have clearly shown, cultural evolution proceeds differently from genetic evolution (Boyd and Richerson 1985). For humans, "culture is essential" (Richerson and Boyd 2005): we cannot understand humans and our place on earth without including our culture and its evolution. A culturally driven species is much more likely to show rapid behavioral change, maladaptive behavior, group-specific behavior, niche construction, and ecological domination (Richerson and Boyd 2005).

What role does culture have in the biology of nonhumans? Could culture have been key in promoting the evolution of large brains and intelligence in a range of species (Wyles et al. 1983; Wilson 1985; Reader and Laland 2002; van Schaik 2006)? The evidence is growing that the behavior, evolutionary trajectories, and environmental roles of whales and dolphins are affected by culture. But how strongly? These are the important questions, the ones that need addressing most pressingly (Laland and Janik 2006). Disputes about whether imitation has been proven are distractions.

## How Should We Study Culture in Wild Animals?

In what follows, I will outline some protocols for the study of culture in the wild that are based on my experience at sea and working with data collected at sea. They are far from comprehensive. I omit experiments, which can be revealing even in the field (Matsuzawa 1994) and potentially even at sea. Another class of data that I will not touch on are the "one-off" observations that can be enormously revealing, for instance, whether immigrants adopt the characteristic behavior of their new groups, or whether behavioral norms suddenly change (Noad et al. 2000; Sapolsky and Share 2004). In contrast, the approach that I advocate will be based on the techniques of multivariate data analysis—techniques that have become vital in our studies of offshore cetaceans—and will address the importance of culture in determining an animal's behavior and its effect on other areas of biology. Subsections will consider quantitative techniques of addressing the following questions:

- What is the contribution of social learning in producing the diversity of a type of behavior among individuals, compared with the contributions of genetics, ecology, or other factors?
- Is socially learned behavior group-specific and thus culture?
- Could patterns suggestive of culture result instead from preferential association between individuals with similar behavior?
- Should these methods be amended when one is studying communal behavior?
- Can we construct a "cultural index" to compare the importance of culture in different populations or species?
- What are the attributes of culture that we might measure to work toward a "taxonomy of culture" (Slater 2001)?
- How does culture affect other areas of the biology of a species, ecologically or evolutionarily?

The techniques I consider are in some respects new, but they are built on methods that have been successful in other contexts and on our initial investigations of culture in sperm whales (e.g., Whitehead et al. 1998; Rendell et al. in prep.). These ideas are at an early stage of development, and if they prove useful, methodological guidelines will be refined. They are founded on the challenges of studying cetacean cultures but should be relevant for other species.

## How Much of the Variation in a Behavioral Pattern Is Determined by Social Learning?

How should the interindividual variation in a behavioral activity (such as a movement pattern, foraging method, or type of grooming method) be portioned into causation by differences in genes, ecology, and social learning (Laland and Janik 2006)? This is challenging but not necessarily impossible. Here is an approach that is outlined by Legendre and Legendre (1998, p. 559) for congruent ecological problems. It is based on similarity matrices, each indexed by the individuals in the study. These are matrices with the same labels (the names or numbers of the individuals) on the rows and columns. Each entry gives a similarity between the row individual and the column individual. The diagonal elements, the self-similarities, are not used. In this exposition I assume that the matrices are symmetric, that is, that the similarity of A with B is the same as that of B with A. This means that only the lower or upper triangle of the matrix needs to be shown. As an example, some artificial data arranged in this way are shown in Table 6.1.

The first similarity matrix measures *genetic relatedness* (e.g., Table 6.1A). It indicates the probabilities that the individuals share a gene through common descent and thus acquired the same behavior genetically. This could be derived from knowledge of genealogy in a very well-studied population but will more normally be the product of molecular genetics, nowadays often produced using microsatellites, with relatedness then being calculated using software such as KINSHIP (Queller et al. 1993).

To indicate the probability that two individuals acquired the same behavior through developmental processes (including individual learning) in similar ecological conditions, we need a matrix of *ecological similarity* (e.g., Table 6.1B)—how similar are the habitats of each pair of individuals? This could be achieved for fairly small-scale studies (e.g., comparing behavioral patterns of bottlenose dolphins in an inshore environment) by using estimates of range overlap from geographic information systems, and over large scales (e.g., comparison of chimpanzee populations across Africa) by ecological similarity measures (Krebs 1989, pp. 293–309).

We also need a measure of the opportunity two individuals may have to acquire the same behavior through social learning, a *social similarity* matrix (e.g., Table 6.1C). Once again, we can use methods developed for another agenda, in this case association indices (Cairns and Schwager 1987). Association indices are estimates of the proportion of time that individuals are associated. Several have been proposed (e.g., simple ratio, half-weight, twice weight) that attempt to correct for different types of bias in data collection (see Cairns and Schwager 1987, Ginsberg and

Table 6.1 Artificially constructed examples of similarity matrices for 12 individuals that could be used to investigate the role of social learning in determining behavior

A: Genetic relatedness

|     | #1 | #2 | #3 | #4 | #5 | #6 | #7 | #8 | #9 | #10 | #11 | #12 |
|-----|-----|-----|-----|-----|-----|-----|-----|-----|-----|-----|-----|-----|
| #1  |     |     |     |     |     |     |     |     |     |     |     |     |
| #2  | 0.26 |     |     |     |     |     |     |     |     |     |     |     |
| #3  | 0.01 | 0.34 |     |     |     |     |     |     |     |     |     |     |
| #4  | 0.04 | −0.01 | 0.48 |     |     |     |     |     |     |     |     |     |
| #5  | 0.07 | 0.01 | 0.02 | 0.13 |     |     |     |     |     |     |     |     |
| #6  | 0.50 | 0.12 | −0.26 | 0.49 | 0.04 |     |     |     |     |     |     |     |
| #7  | 0.06 | −0.02 | −0.01 | 0.00 | −0.02 | 0.10 |     |     |     |     |     |     |
| #8  | 0.51 | 0.24 | −0.04 | 0.25 | −0.01 | 0.51 | −0.02 |     |     |     |     |     |
| #9  | −0.02 | −0.03 | 0.03 | 0.01 | 0.14 | 0.00 | 0.63 | 0.40 |     |     |     |     |
| #10 | −0.09 | −0.01 | 0.00 | 0.11 | 0.47 | 0.02 | 0.00 | 0.32 | 0.20 |     |     |     |
| #11 | 0.40 | 0.55 | −0.02 | −0.09 | 0.00 | 0.01 | −0.01 | 0.01 | 0.52 | −0.01 |     |     |
| #12 | −0.04 | −0.01 | 0.26 | 0.00 | 0.50 | −0.09 | 0.02 | −0.02 | 0.00 | 0.01 | 0.14 |     |

B: Ecological similarity (e.g., proportional range overlap)

|     | #1 | #2 | #3 | #4 | #5 | #6 | #7 | #8 | #9 | #10 | #11 | #12 |
|-----|-----|-----|-----|-----|-----|-----|-----|-----|-----|-----|-----|-----|
| #1  |     |     |     |     |     |     |     |     |     |     |     |     |
| #2  | 0.91 |     |     |     |     |     |     |     |     |     |     |     |
| #3  | 0.87 | 0.56 |     |     |     |     |     |     |     |     |     |     |
| #4  | 0.49 | 0.37 | 0.29 |     |     |     |     |     |     |     |     |     |
| #5  | 0.18 | 0.32 | 0.25 | 0.53 |     |     |     |     |     |     |     |     |
| #6  | 0.03 | 0.12 | 0.26 | 0.79 | 0.60 |     |     |     |     |     |     |     |
| #7  | 0.06 | 0.02 | 0.04 | 0.54 | 0.76 | 0.77 |     |     |     |     |     |     |
| #8  | 0.21 | 0.12 | 0.11 | 0.25 | 0.36 | 0.41 | 0.81 |     |     |     |     |     |
| #9  | 0.24 | 0.13 | 0.08 | 0.21 | 0.34 | 0.55 | 0.75 | 0.98 |     |     |     |     |
| #10 | 0.03 | 0.06 | 0.11 | 0.11 | 0.87 | 0.36 | 0.85 | 0.56 | 0.88 |     |     |     |
| #11 | 0.00 | 0.05 | 0.04 | 0.09 | 0.23 | 0.15 | 0.55 | 0.31 | 0.41 | 0.52 |     |     |
| #12 | 0.04 | 0.01 | 0.06 | 0.08 | 0.33 | 0.10 | 0.11 | 0.22 | 0.52 | 0.65 | 0.81 |     |

C: Social similarity (association indices)

|     | #1 | #2 | #3 | #4 | #5 | #6 | #7 | #8 | #9 | #10 |
|-----|-----|-----|-----|-----|-----|-----|-----|-----|-----|-----|
| #1  |     |     |     |     |     |     |     |     |     |     |
| #2  | 0.71 |     |     |     |     |     |     |     |     |     |
| #3  | 0.44 | 0.36 |     |     |     |     |     |     |     |     |
| #4  | 0.29 | 0.47 | 0.59 |     |     |     |     |     |     |     |
| #5  | 0.18 | 0.12 | 0.23 | 0.11 |     |     |     |     |     |     |
| #6  | 0.43 | 0.52 | 0.56 | 0.49 | 0.30 |     |     |     |     |     |
| #7  | 0.16 | 0.12 | 0.34 | 0.24 | 0.46 | 0.47 |     |     |     |     |
| #8  | 0.33 | 0.44 | 0.23 | 0.42 | 0.13 | 0.32 | 0.66 |     |     |     |
| #9  | 0.11 | 0.02 | 0.13 | 0.44 | 0.30 | 0.33 | 0.14 | 0.71 |     |     |
| #10 | 0.00 | 0.00 | 0.21 | 0.31 | 0.47 | 0.56 | 0.35 | 0.26 | 0.11 |     |

**Table 6.1** (*continued*)

*C: Social similarity (association indices)*

|  | #1 | #2 | #3 | #4 | #5 | #6 | #7 | #8 | #9 | #10 | #11 | #12 |
|---|---|---|---|---|---|---|---|---|---|---|---|---|
| #11 | 0.11 | 0.00 | 0.00 | 0.00 | 0.27 | 0.35 | 0.35 | 0.23 | 0.11 | 0.02 | | |
| #12 | 0.24 | 0.31 | 0.46 | 0.58 | 0.00 | 0.01 | 0.01 | 0.12 | 0.55 | 0.05 | 0.01 | |

*D: Behavioral similarity (continuous)*

|  | #1 | #2 | #3 | #4 | #5 | #6 | #7 | #8 | #9 | #10 | #11 |
|---|---|---|---|---|---|---|---|---|---|---|---|
| #1 | | | | | | | | | | | |
| #2 | 1.00 | | | | | | | | | | |
| #3 | 0.81 | 0.51 | | | | | | | | | |
| #4 | 0.57 | 0.47 | 0.89 | | | | | | | | |
| #5 | 0.54 | 0.38 | 0.60 | 0.26 | | | | | | | |
| #6 | 0.60 | 0.69 | 0.91 | 0.27 | 0.63 | | | | | | |
| #7 | 0.08 | 0.34 | 0.39 | 0.66 | 0.56 | 0.88 | | | | | |
| #8 | 0.45 | 0.85 | 0.50 | 0.44 | 0.71 | 0.73 | 0.53 | | | | |
| #9 | 0.47 | 0.28 | 0.39 | 0.74 | 0.59 | 1.00 | 0.39 | 1.00 | | | |
| #10 | 0.03 | 0.00 | 0.33 | 0.78 | 1.00 | 0.68 | 0.64 | 0.53 | 0.65 | | |
| #11 | 0.36 | 0.41 | 0.72 | 0.12 | 0.38 | 0.99 | 0.62 | 0.38 | 0.64 | 0.58 | |
| #12 | 0.25 | 0.49 | 0.58 | 0.69 | 0.36 | 0.61 | 0.33 | 0.38 | 0.99 | 0.16 | 0.70 |

*E: Behavioral similarity (1:0)*

|  | #1 | #2 | #3 | #4 | #5 | #6 | #7 | #8 | #9 | #10 | #11 |
|---|---|---|---|---|---|---|---|---|---|---|---|
| #1 | | | | | | | | | | | |
| #2 | 1 | | | | | | | | | | |
| #3 | 1 | 0 | | | | | | | | | |
| #4 | 1 | 0 | 1 | | | | | | | | |
| #5 | 0 | 0 | 0 | 0 | | | | | | | |
| #6 | 0 | 0 | 0 | 0 | 0 | | | | | | |
| #7 | 0 | 0 | 0 | 1 | 0 | 0 | | | | | |
| #8 | 0 | 0 | 0 | 0 | 1 | 0 | 1 | | | | |
| #9 | 0 | 0 | 0 | 0 | 1 | 0 | 0 | 1 | | | |
| #10 | 0 | 0 | 0 | 0 | 1 | 0 | 0 | 1 | 1 | | |
| #11 | 0 | 0 | 0 | 0 | 0 | 0 | 1 | 0 | 0 | 0 | |
| #12 | 0 | 0 | 1 | 0 | 0 | 0 | 0 | 0 | 1 | 0 | 0 |

*Note:* A: matrix of genetic relatedness, such as might be derived from molecular genetic analysis of microsatellites (although relatedness values are theoretically positive, values calculated from molecular data by using popular techniques may contain small negative values). B: ecological similarity, which could be the proportion of range overlap calculated using geographic information systems data. C: social similarity, an association index that estimates the proportion of time two individuals are associated. D: continuous measure of behavioral similarity (e.g., proportion of call types shared). E: 1:0 measure of behavioral similarity (e.g., same/different dialect).

Young 1992, Whitehead 2008, and Whitehead and Dufault 1999 for guidance on choosing association indices). In this context, probably more important than the choice of index is the definition of association. Association should be defined in such a way that associated animals could learn from one another. For instance, if we know or strongly suspect that social learners and their models are generally in particular age-sex classes (e.g., animals learn most from adult females and before their sexual maturity), then the social similarity measure (the association index) can be restricted to data collected when the members of the dyad were both in appropriate age-sex classes (assigning zero social similarity if they never were in the characteristic learner-model classes at the same time). Boogert et al. (2008) and Morrell et al. (in press) have found by using experiments that individuals of some species learn more effectively from familiar individuals than from strangers, but not disproportionately more from more familiar familiars. In such a case the measure of association might be binary (1 = familiar; 0 = stranger).

Finally, there is the dependent variable, *behavioral similarity*. Diet similarity could be measured by a niche overlap index (Krebs 1989, pp. 379–380), while dialect similarity could be measured by the proportion of call types shared (as in Yurk et al. 2002). In some cases behavioral similarity might be a 1:0 measure, for instance, the individuals use the same type of grooming handclasp or they do not. Examples of continuous and 1:0 behavioral similarity data are shown in Tables 6.1D and 6.1E.

Having constructed these matrices, I suggest making a multiple regression of the nondiagonal elements of the behavioral similarity matrix (those actually shown in Tables 6.1D and 1E) on the corresponding elements of the relatedness, ecological similarity, and social similarity matrices, thus examining quantitatively how behavioral similarity is dependent on the other variables:

Behavioral similarity = constant + genetic relatedness + ecological similarity + social similarity + error

The standard partial regression coefficients (called beta-weights in the social sciences) of each of the independent variables (genetic, ecological, and social similarities) indicate the strengths of their effects in explaining behavioral similarity (Sokal and Rohlf 1994, p. 614). These measures are "standard" because the effects of the different variances in the different variables have been removed, so, for instance, doubling all the ecological similarities would not affect the standard partial regression coefficients (while it would halve the normal regression coefficient). They are "partial" because they measure the effects of one independent variable (such as ecological similarity), given the presence of the others (genetic similarity and social similarity). These are comparable measures, ranging from −1 (perfect inverse linear relationship) through 0 (no relationship) to +1 (perfect linear relationship) of the significance of genetic variation, ecology, and social learning in determining behavior. The jackknife method (see Efron and Gong 1983) can be used to estimate the precision of these standard partial regression coefficients.[2]

In such analyses the $P$-values from the multiple regression output, automatically presented by most statistical analysis software packages, should not be used because of dependence (the behavioral similarities between A and B, B and C, and A and C will not be independent). Instead, if we are working within the somewhat-discredited hypothesis-testing framework (see, for instance, Johnson 1999), then we should use permutation tests, variants of the Mantel test (Smouse et al. 1986). For example, to test whether the contribution of social learning in explaining behavioral similarity is significantly greater than zero, we can permute individual identities only in the social similarity matrix. To do this, take the individual identities (#1, . . ., #12 in Table 6.1), and randomly permute them (giving, say, #7, #1, #3, #6, #2, #5, #12, #9, #11, #4, #8, #10). These are then used to rearrange the data in Table 6.1C, so that the new ecological similarity between #1 and #6 is that originally between #2 and #4, and so on. The regression analysis is repeated with this new Table 6.1C to give a new standard partial regression coefficient $b_S(*1)$. This randomization process is performed many times, 1,000 being fairly standard, to give a set of partial regression coefficients that use permuted identities: $b_S(*1), b_S(*2), . . ., b_S(*1,000)$. If the real standard partial regression coefficient of the social similarity matrix, $b_S$, is greater than that of more than 95 percent of the matrices with permuted identities $\{b_S(*1), b_S(*2), . . ., b_S(*1,000)\}$, we might reject, at $P<0.05$, the null hypothesis that social learning does not affect the behavior.[3]

The results of using these techniques on the data in Table 6.1 are shown in Table 6.2. Both the continuous behavioral similarity (Table 6.1D) and the 1:0 behavioral similarity (Table 6.1E) were regressed on the genetic (Table 6.1A), ecological (Table 6.1B), and social (Table 6.1C) similarities. In the case of the continuous measure of behavioral similarity (Table 6.1D), social similarity has the largest effect, and one that is

Table 6.2 Significance of genetics, ecology, and social learning in determining behavior as expressed by standard partial regression coefficients, using the data in Table 6.1

| Independent variable | Continuous behavioral similarity (D) | 1:0 behavioral similarity (E) |
|---|---|---|
| Genetic similarity, genes (A) | 0.005 (SE=0.172); $P=0.474$ | 0.012 (SE=0.159); $P=0.456$ |
| Ecological similarity, developmental processes such as individual learning in different ecological circumstances (B) | 0.209 (SE=0.133); $P=0.019$ | 0.377 (SE=0.204); $P=0.001$ |
| Social similarity, social learning (C) | 0.541 (SE=0.107); $P=0.000$ | 0.251 (SE=0.212); $P=0.015$ |

Note: A–E refer to the five similarity matrices in Table 6.1. Also shown are estimated standard errors (SE) from the jackknife procedure and the results of permutation tests for the significance of a particular independent variable (1,000 permutations of identities in the corresponding independent similarity matrix; null hypothesis that the independent variable has no effect on behavioral similarity, alternative hypothesis that it has a positive effect).

significantly greater than zero, whereas the 1:0 measure (Table 6.1E) is more determined by ecological differences.

This proposed technique has a number of attractive features. It does not pre-position genetics and ecology as the null hypothesis and culture as the alternative, as in many studies of nonhumans, or the reverse, as in studies of humans. It gives a comparable estimate of the significance of each factor in explaining behavioral variation, with a measure of confidence. The method could be simplified in some circumstances. If ecological factors are deemed irrelevant for some behavior, such as the pattern of vocalizations used in a social setting, then there might be only two independent variables. The method could also be extended in various ways. Other potential dependent variables could be included, such as gender similarity, age difference, or more than one measure of ecological similarity (e.g., range overlap plus another measure of habitat similarity) or social similarity (based on two or more definitions of association).

However, the method does possess limitations. It assumes that the measures in the independent-variable matrices are measured without error and are linearly related to the values that are relevant to the animals' behavior (so that if the social similarity between two individuals increases by a certain amount, their expected behavioral similarity increases by the same amount, whatever the original social similarity). If the independent measures used in the analysis do not truly represent the factors that influence behavior, or there are measurement errors, the influence of the factor on behavior will tend to be underestimated. Errors in measuring behavioral similarity (the dependent variable) will tend to reduce all the standard partial regression coefficients. Of the four matrices, I suspect that ecological similarity will often cause the greatest difficulty. With sufficient microsatellites, genetic relatedness can be estimated with reasonable precision, and, in my experience, association indices calculated in different ways tend to be well correlated, but a measure of ecological similarity might omit the crucial ecological factor that influences individual learning or other development of behavior (Laland and Hoppitt 2003). This needs to be guarded against as far as possible, perhaps by trying various measures of ecological similarity.

It could be that although we have chosen and measured appropriate variables, they are not linearly related. If this is the case, but the relationship is monotonic (e.g., an increase of a certain amount in social similarity leads to an increase in behavioral similarity, but the amount varies with the original social similarity), then ranking elements within some or all matrices will help.[4]

Sometimes two of the independent measures may be so closely related that their effects cannot be separated. For instance, ecological similarity based on range overlap might be highly correlated with social similarity from an association index, a situation that perhaps might result from little preference among available social partners. In such instances the regression analysis will either fail or will indicate strong colinearity, so that little confidence can be placed on the standard partial regression coefficients. Although there are methods for reducing the effects of colinearity (see, for instance, Kleinbaum et al. 1988, pp. 206–218), its presence is symptomatic of a lack of ability to distinguish between the effects of the independent variables. In such cases we do not have the data to separate social learning from individual learning or other developmental effects in different ecological conditions.

## There Might Be Social Learning, but Is It Culture?

In the previous section, I did not use the word "culture." The proposed regression method examines the influence of social learning, but the definition of culture that I use includes another ingredient in addition to social learning; it is a characteristic feature of groups. There are instances in which social learning might be present but not, according to this definition, culture. A study population could have a continuous distribution across its habitat, but there might be little overlap of the ranges of individuals. Animals might learn from neighbors, so that there is a correlation between social similarity and behavioral similarity, but there is no division into behaviorally homogeneous groups. A second example of social learning without culture comes from the ocean. Sayigh et al. (1990) showed that the individually specific signature whistles of female bottlenose dolphins were particularly unlike their mothers' signature whistles. This has to be the result of a form of social learning, because genetic, environmental, and maternal effects would all promote similarities between mother and daughter. Such a learning process would likely produce a statistically significant negative standard partial correlation coefficient for social similarity if the method previously described were used to study it. However, it is not culture, because there is no resulting group of animals with similar signature whistles.

Once the study population is divided into groups,[5] we can use a matrix correlation and Mantel test (see Schnell et al. 1985) to examine whether there is behavioral homogeneity within groups. As an example, in Table 6.3 the individuals of Table 6.1 are assigned to groups, a "1" indicating that two individuals are members of the same group, and "0" that they are not. With a group-similarity matrix, we correlate the off-diagonal elements of this matrix (those actually shown in Table 6.3) with those of the behavioral similarity matrices (e.g., Table 6.1D or Table 6.1E). These are called matrix correlations, and we can test whether they are significantly different from zero by using Mantel tests, permutation tests in which the identities of the individuals are scrambled in one of the matrices (the same process as described earlier for testing standard partial regression coefficients). The results of this procedure are that the group designations of Table 6.3 are positively and significantly related to the continuous behavioral similarities of Table 6.1D (matrix correlation$=0.305$, $P=0.006$, two-sided test), indicating culture. Because we do not have strong evidence for social learning in the case of the

138

Table 6.3 Group delineation matrix indicating which individuals are in the same social group

|     | #1 | #2 | #3 | #4 | #5 | #6 | #7 | #8 | #9 | #10 | #11 | #12 |
|-----|----|----|----|----|----|----|----|----|----|-----|-----|-----|
| #1  |    |    |    |    |    |    |    |    |    |     |     |     |
| #2  | 1  |    |    |    |    |    |    |    |    |     |     |     |
| #3  | 1  | 1  |    |    |    |    |    |    |    |     |     |     |
| #4  | 1  | 1  | 1  |    |    |    |    |    |    |     |     |     |
| #5  | 0  | 0  | 0  | 0  |    |    |    |    |    |     |     |     |
| #6  | 0  | 0  | 0  | 0  | 1  |    |    |    |    |     |     |     |
| #7  | 0  | 0  | 0  | 0  | 1  | 1  |    |    |    |     |     |     |
| #8  | 0  | 0  | 0  | 0  | 0  | 0  | 0  |    |    |     |     |     |
| #9  | 0  | 0  | 0  | 0  | 0  | 0  | 0  | 1  |    |     |     |     |
| #10 | 0  | 0  | 0  | 0  | 1  | 1  | 1  | 0  | 0  |     |     |     |
| #11 | 0  | 0  | 0  | 0  | 1  | 1  | 1  | 0  | 0  | 1   |     |     |
| #12 | 1  | 1  | 1  | 1  | 0  | 0  | 0  | 0  | 0  | 0   | 0   |     |

*Note:* The social groups are {#1, #2, #3, #4, #12}, {#5, #6, #7, #10, #11}, and {#8, #9}. This delineation was obtained from the social similarity matrix of Table 6.1C by using Newman's (2007) eigenvector modularity method and has a modularity of 0.11.

1:0 behavioral similarity of Table 6.1E, adding the test for group-specific behavior in a quest for culture is probably not appropriate.

The preceding assumes that the study population has been divided into groups, but how should this be done? Sometimes, as with many primate groups or killer whale *(Orcinus orca)* pods, the delineation is well established or obvious, but in other cases there is no a priori designation. Usually it should be based on social relationships, so we can use a social similarity matrix such as Table 6.1C to produce groups. There are several ways to do this, including using cluster analysis (Whitehead and Dufault 1999; Wittemyer et al. 2005), Bayesian methods (Durban and Parsons submitted), and network analysis techniques such as Newman's (2006) new eigenvector modularity method. The latter is relatively easy to implement (certainly in comparison with the Bayesian methods) and has the attractive feature that it gives a measure of how well the population is segmented by the groups, that is, modularity. Modularity is the difference between the proportion of the total association within groups and the expected proportion, given the tendency to form groups of the different individuals (Newman 2004). For randomly assigned groups, modularity equals 0.0, and if there are no associations between members of different groups, it equals 1.0. Newman (2004) suggests that if modularity is greater than about 0.3, then the divisions between groups are "good." The group identities of Table 6.3 were obtained from the association data in Table 6.1C by eigenvector modularity. However, the modularity was only 0.11, which indicates a rather uncertain separation of groups.

I have suggested the derivation of groups from a social similarity matrix, such as one that gives association indices. It may be tempting, but I think generally wrong in the search for culture, to derive groups from the behavioral similarity matrix. This is circular because we are saying that there is behavioral similarity within groups that are defined on the basis of similarity of the same behavior. However, once culture has been attributed using the methods outlined previously, defining groups by using a social similarity matrix or some other method unrelated to the cultural behavior being examined, we may wish to use similarity among groups in this behavior to define higher levels of social structure, as with the clans of killer and sperm whales (Ford and Fisher 1982; Rendell and Whitehead 2003).

## Social Similarity and Behavioral Similarity?

The procedure that I have outlined for investigating the role of culture in behavior assumes that similarities in behavior that correlate with social associations are caused by social learning, once environment and genetics are controlled for. There is a possibility that the relationship could operate in the opposite direction, that social similarity could be a function of behavioral similarity. Perhaps individuals tend to associate with other individuals whose foraging techniques, for example, are similar. Would not this pattern suggest social learning if we used the regression procedure? This depends on what determines foraging technique in the first place. If it is not social learning, then it must be genes, environment, and/or something else (such as gender or ontogeny). If these factors are well captured by the other independent factors (e.g., genetic similarity, ecological similarity, gender similarity, age difference) in the regression, then the standard partial regression coefficient between social similarity and behavioral similarity will be small, and social learning will not be supported, even though social and behavioral similarities are well correlated. This consideration emphasizes the importance of including all potentially important factors in the regression and using measures that are relevant to the animals themselves.

A second way of addressing the direction of any relationship of social similarity and behavioral similarity might be to consider sensitive periods for social learning. If we suppose, as suggested earlier, that the social similarity measure indicates the dyadic relationship during a time period when the members of the pair were both in appropriate age-sex classes for social learning (e.g., juvenile and adult female), then, if behavioral similarity is measured later, after any social learning has taken place, we control for the direction of the social-behavioral relationship.

## Communal Behavior and Culture

A rather neglected side of the animal culture debate is communal behavior: behavior that is conducted by a whole group of animals. This class of behavior is perhaps particularly important for cetaceans. In some cases, such as synchrony or foraging formations, it cannot be separated into individual behavior; the behavioral activity is described by the relationship between acts of different individuals. In others, such as movement patterns of sperm whale units (Whitehead and Rendell 2004) or dialects of killer whale pods (Ford 1991), although individual animals could behave differently, they do not. All members of a unit move together, and all members of a pod use the same dialect.

Communal behavior among members of a permanent social group does not imply culture or even social learning. Social insects can perform some remarkable communal behavior, but this seems to be based on quite simple individual actions determined by interactions between genes and environmental and social cues (Sumpter 2006). The key factors are that the groups, or at least some of them, have distinctively different behaviors. Then we can start to parse these differences into genetic, ecological, social learning, and other causes, perhaps using the regression technique described earlier.

This could be done at the level of the individual, as in the example of Tables 6.1 and 6.2. For communal behavior, if social similarity is analyzed at the level of the individual, it generally will be very high among members of the same group and very low between individuals of different groups, and the behavioral similarity for communal behavior will be maximal between members of the same group and low between some different groups (but not necessarily all, because different groups could have similar communal behavior). This means that the correlation between the behavioral similarity and social similarity matrices, and the standard partial regression coefficient of social similarity, will necessarily be high unless environmental similarity or genetic similarity, on their own or together, also explain much of the behavioral similarity. A problem here is that even sympatric groups will not always be in exactly the same place. So environmental similarity, if defined sufficiently finely, will correlate well with group membership and thus behavioral similarity. Consequently, the social and environmental similarity factors will be co-linear and cannot be disentangled (see Laland and Hoppitt 2003). For behavioral types that are unlikely to be influenced by fine-scale environment, such as conventional social signals, this may not be too much of an issue, but with foraging behavior it often will be.

An alternative is to perform the analysis at the level of the group. Behavioral similarity and ecological similarity between groups can be derived much as with the individual analysis, and genetic similarity can be calculated as the mean relatedness between members of the different groups. But social similarity needs to be modified, because groups will not usually learn directly from one another. Instead, if social learning is present, it will principally consist of learning within groups. Thus social similarity among groups due to social learning should be correlated with shared ancestry—groups that had common members in the recent past or were part of the same ancestral group will tend to behave similarly. From this follows an approach that we have employed to look at sperm whale culture: using mitochondrial DNA as a marker of ancestral links among social groups (Whitehead et al. 1998; Rendell et al. in prep.). This is based on the well-founded assumptions that sperm whale society is matrilineally based, so a matrilineally transmitted gene traces the social ancestry, and that mtDNA does not code for the behavior that we are interested in. Instead, microsatellites are used to estimate relatedness and the probabilities that genes that code for behavior are shared. The full analysis of the sperm whale data, using both mtDNA as a marker for ancestral social links and microsatellites to estimate the probability that coding genes are shared, has not yet been completed using the methodology that I have proposed here. However, dialect similarity is correlated with mtDNA similarity but not with microsatellite-based estimates of relatedness or geographic distance, an indicator of environmental similarity, and this leads to a conclusion of social learning and culture (Whitehead et al. 1998; Whitehead 2003a; Rendell et al. in prep.). I can say "culture" because in this case, and generally with communal behavior, we do not need to worry about the second part of the protocol, because the behavior is, by definition, group-specific.

## How Much of the Behavioral Variation of a Population or Species Is Culturally Determined?

I have suggested ways in which the contribution of culture to a particular behavioral activity might be assessed. Can we extend this to give an estimate of the contribution of culture to an entire behavioral repertoire? Here are some ideas.

One could derive some overall measure of behavioral similarity between pairs of individuals, so that the measure of behavioral similarity (e.g., Table 6.1D), rather than referring to a particular activity, integrates over the whole behavioral repertoire. Perhaps the behavioral similarity between A and B could indicate the probability that at a randomly chosen time the type of behavior exhibited by A is essentially the same as that which B exhibits when in the same general behavioral state. To give a very simplistic example, if A spends 20 percent of her time foraging with a sponge on her beak, 40 percent foraging by chasing fish onto beaches, 30 percent resting with her head out of the water, and 10 percent resting with her tail out of the water, while for B these proportions are 0, 50, 25, and 25 percent, then the A-B behavioral similarity is 0.6 ($[0.2 \times 0.0] + [0.4 \times 1.0] + [0.3 \times 0.5] + [0.1 \times 0.5]$).

However, such a methodology would downplay brief but significant activities such as alarm calls, as well as ignore large-scale behavioral patterns such as the diurnal arrangement of behavioral states (e.g., when to rest). Another problem is that the relevant environmental similarity and social similarity matrices may be different for different behavioral activities. For instance, for an alarm call, the relevant environment might describe the types of predators and possible defense strategies (are there trees to climb or crevices to enter?), whereas the distributions and abundances of possible prey would probably be more appropriate for foraging behavior.

Because of these difficulties, performing separate analyses for different behavioral activities and then integrating the results may be preferable. One might perform several analyses for different behavioral activities, such as those shown in Tables 6.1 and 6.2, and then calculate, for each activity, the ratio of the standard partial regression coefficient for social similarity to that for environmental variation; these could then be integrated by using the mean (perhaps weighted) of these values. However, the result of such an integrated analysis will generally depend heavily on which behavioral activities are selected for analysis, how well the similarity matrices reflect the true similarities in the different factors (especially the environmental similarity), and how the results for the different activities are weighted.

These are all challenging issues, and it seems unlikely that such methods will produce robust measures that can be used to compare quantitatively the significance of culture in the behavioral repertoires of different species. However, the methods may be more feasible intraspecifically, so that if the same measures and methods are used on several populations, the relative levels of cultural influence on behavior might be compared. This would assist with the testing of hypotheses such as van Schaik's (2006) prediction that more cultural lineages should have greater intelligence.

With regard to interspecific comparisons, Reader and Laland (2002) used published reports of the incidence of socially learned behavior as an operational index of the significance of culture in different primate species. This type of approach is vulnerable to a number of biases. Some of these, such as research effort, can be corrected for, at least partially, in various ways, but there will always be the lingering concern that even expert witnesses may attribute social learning when it is not occurring or miss social learning when it is. Nonetheless, such methods seem the most feasible current ones of comparing cultures across species.

## Types of Culture

Although there may be no good operationally useful index of the significance of culture that can be used to compare different species, there are many reasons for comparing cultures (Slater 2001). Culture in nonhumans is studied through behavioral variation, and different forms of culture correspond to different types of behavioral variation. Here are some measures of behavioral variation that could be used to describe cultures and potentially to compare behavioral activities within a species or culture in different populations of a species or among species:

- How large, in numbers and geographic extent, are the social groups that use a characteristic cultural behavior (see, for instance, Rendell and Whitehead 2003)?

- What proportion of a social group consistently shows a behavioral type? Is the behavior restricted to, or favored by, a class of animal (e.g., females; Smolker et al. 1997)?
- How fast does the probability that an individual shows a behavioral type change with time, or, if a trait is quantitative, what is the pattern of autocorrelation? Is temporal change in group-specific behavior gradual (evolution; e.g., Deecke et al. 2000) or sudden (revolution; Noad et al. 2000)?
- If the behavior changes with time, does it "ratchet" (increase in complexity; Tomasello 1994)?

The answers to such questions will inform about transmission mechanisms, as well as whether there are processes that stabilize culture, such as conformism. However, their role is perhaps most important in considering the ecological and evolutionary effects of culture. Spatial and numerical scale and ubiquity, as well as the rates and directions of cultural change, are all vital factors if we are to understand how cultural variants affect patterns of resource use, intraspecific competition, and the evolutionary changes of the cultural type itself, as well as its potential effects on genes.

In addition to these attributes of a particular behavioral type, we can and should examine how they, and the social groups that possess them, interact. Do social groups, populations, or subpopulations possess culturally distinctive behavior of several types (i.e., multifaceted cultures such as those of chimpanzees and sperm whales; Whiten et al. 1999; Whitehead and Rendell 2004), or is there only one distinctive cultural behavior? Second, how do the sets of animals with different cultures interact? There is a range of possibilities, including no interaction if cultural variants are delineated by firm barriers to movement, limited interaction if cultures are specific to territorial groups (as in chimpanzees), and multicultural populations in which groups with distinctively different cultures use the same habitat and often encounter one another (as in killer and sperm whales; Rendell and Whitehead 2001). The last possibility has particular implications for the evolution of cultural variants, as well as genes, through cultural group selection if the sympatric groups compete for resources, and the culturally determined behavior influences success (Whitehead 2005). To examine these issues, we need to investigate cultural variation over numbers of social groups, behavioral types, and substantial scales of space and time.

Writing from the perspective of behavioral variation, I have omitted one of the most significant attributes of a culturally determined behavior, its transmission mode. Between whom, and how, does social learning take place? Is the transmission vertical, between parent and offspring, or horizontal, among members of the wider community of the same generation? Is the learning imitation, emulation, teaching, or simple exposure? To address these topics needs individually or dyadically oriented studies. However, population studies of behavioral patterns can give some insight. For instance, cultural attributes that are stable over generations are much more likely to be vertically transmitted, whereas passing "fads" must be horizontally transmitted. If a culture is going to ratchet, it probably needs to be transmitted by imitation or teaching (Boyd and Richerson 1985, p. 35).

## Does Culture Affect Other Areas of Biology?

There are many areas of biology that culture could potentially influence. The effects might be on either ecological or evolutionary scales.

Culture can potentially change the way in which individuals interact with their environment and thus can affect feeding success or survival rates. Generally, because of cultural selection, we would expect cultures to be adaptive, increasing survival and reproductive success, but different cultures might do so to different degrees, and cultural evolution can lead to maladaptive behavior (Richerson and Boyd 2005). Thus we might want to examine how groups with different cultures differ in their feeding success, survival, or reproductive success. We have done the first and third of these with sperm whales and have found that groups of different cultural clans differed in their feeding success in ways that were environmentally dependent (Whitehead and Rendell 2004), and had consistently different reproductive success (Marcoux et al. 2007).

In addition to such ecological-scale interactions, culture can affect a wide range of biological attributes evolutionarily. Culture has been proposed as a driver of brain complexity in a variety of species (Wyles et al. 1983; Wilson 1985; Reader and Laland 2002; van Schaik 2006). In cultural societies there will be selection for individuals who possess brains that can most effectively learn from social partners and process the socially acquired information. Brains are expensive organs. Maintaining them and optimizing their use may drive the evolution of other attributes, such as lengthened life histories (see Lefebvre et al. 2006). Alternatively, or perhaps additionally, the evolution of life-history attributes could be driven directly by culture: long juvenile periods could be adaptive for species in which social learning is very important, and menopause might prolong the lives of older females who have particularly important information that is useful for their grouped relatives (Whitehead and Mann 2000). Menopause, in which females characteristically have long postreproductive lives, is known only for humans, killer whales, and short-finned pilot whales *(Globicephala macrorhynchus)*. The first two of these are particularly cultural species, while the poorly known short-finned pilot whales are good candidates for culture (Rendell and Whitehead 2001). Culturally acquired behavior might also lower mortality through, for instance, culturally determined communal defense against predators or the smoothing of environmental variation from communal knowledge of a wide range of resource types (Whitehead 2003a). Other products of such processes might be the apparently greater ranges and ecological successes of the more cultural species (Rendell and Whitehead 2001).

Finally, group-specific cultural selection might have reduced genetic diversity in those genes that are being transmitted in parallel with selective cultural traits through the process of cultural hitchhiking (Whitehead 1998, 2005; Whitehead et al. 2002; see also Laland and Janik 2006, Box 2). In humans this might explain the extraordinarily low diversity of Y-chromosome genes if the current population is descended from just one or a few tribes that possessed a selectively advantageous patrilineally transmitted cultural advantage that was passed on in parallel with the Y-chromosome genes (Whitehead et al. 2002). Four species of whale that are known to have matrilineal social systems (females generally stay grouped with their mothers) all have exceptionally low matrilineally transmitted mitochondrial DNA: the killer whale, the sperm whale, the short-finned pilot whale, and the long-finned pilot whale *(Globicephala melas)*. I have proposed that in these species selection favored particular groups with matrilineally transmitted cultural advantages, so that the mitochondrial haplotypes present in those groups now make up the diversity of the entire species (Whitehead 1998, 2005).

These are all speculations, interesting and reasonable, but speculations. How can we investigate them? The most obvious manner is using the comparative method, comparing cultural and other attributes among species. A major problem is defining a "cultural index" that can be assigned in a reasonably unbiased way across species. As noted earlier, I do not think that the proportion of behavior that is culturally determined, a nearly ideal measure in many respects, could be estimated in any currently feasible manner that would be satisfactory for interspecific comparisons. Instead, we may need to fall back on relatively simple 1:0 measures such as whether there is any evidence for culturally determined vocalizations (or tool use, or extractive foraging methods, or other behaviors; see Reader and Laland 2002). It is obviously important to use only species where research effort has been sufficient that the answers to such questions are reasonably authoritative (or to use research effort as a factor in the analysis). Another approach, more indirect, is to use attributes that seem to be well correlated with culture, as in my argument that matrilineal social systems seem to provide a good substrate for the development of stable group-specific cultures and thus cultural hitchhiking (Whitehead 1998, 2003a).

An approach often used in examining the evolution of human culture is to try to trace the changes in culture and other attributes in evolutionary time. For instance, Richerson and Boyd (2005) looked at the development of the human brain (from paleontological results) and of human culture (from archeological data). Unfortunately, nonhumans rarely leave archeological records of past cultures, and, as far as we know, this is never the case in the ocean, where tool use is exceedingly rare, and water movement is a great leveler and homogenizer. However, we might be able to make some rather weak inferences about culture by using this type of approach. For instance, increases in brain size (see Marino et al. 2004), perhaps linked to additional cultural capacity, might be shown, in evolutionary time, to be correlated with increased habitat or diet diversity through paleontological studies.

## Why Is There a "Question of Animal Culture"?

In many respects it is frustrating that there is a question of animal culture. Both in the ocean and on land it is clear that animals learn from one another, and in some cases this leads to behavior that is characteristic of groups. The use of the word "culture" raises skepticism and sometimes antagonism that would not be present if we just used "tradition" (although see Galef 2003a). Partly this stems from an irrational desire to keep culture for humans, but there is a better reason to be careful with the "c" word: culture changes almost everything. Once culture starts determining a substantial part of behavior, social structures, ecological relationships, and evolution (both cultural and genetic) are subject to a set of forces very different from those experienced by organisms for which trait inheritance is almost entirely genetic.

This is a strong reason not to give in to the skeptics and retreat into "traditions," as suggested by some (e.g., Mann 2001). If animals have what we call culture in humans, we should say so both for accuracy and so that its implications, both biologically and in other realms, such as ethics (Fox 2001), can be assessed.

The "question of culture" implies a hypothesis-testing framework: a null hypothesis that a particular form of behavior is determined only by genetics and developmental factors plus individual learning, and an alternative that culture plays a part. This traditional manner of doing science has been much criticized (e.g., Johnson 1999) and has been replaced in many areas by techniques such as model selection and Bayesian estimation that, rather than assuming "yes/no" dichotomies for factors, estimate their relative significance. The issue of culture in nonhumans should join the trend. "Is there culture?" should be replaced by "How important is culture?" "How stable is culture?" and "What kinds of culture?" I hope that the techniques outlined in this chapter may assist in this conceptual and methodological transformation.

## Summary

The attributes of the ocean, including its size, dimensionality, connectedness, tropic structure, and patterns of variability, have led to rather different methods and emphases of research than those employed on potentially cultural terrestrial species. This oceanic perspective deemphasizes controlled experiments. Instead, I suggest that culture may be best studied by using the methods of data analysis applied to patterns of behavior. In particular, similarities between individuals or groups in their behavior can be regressed upon genetic, ecological, and social similarities, allowing the influences of genetic determination, individual learning, and other developmental processes in different environments and social learning to be compared in a quantitative, nonhierarchical fashion.

A second feature of culture, the group-specific nature of behavior, can be assessed by using matrix correlations and Mantel tests of behavioral similarity against group membership. We can, and should, investigate the attributes of culture: its extent, temporal stability, and reliability. Of great significance is how culture may affect other areas of biology, ecologically or evolutionarily. Ecological-scale interactions may be examined by correlating cultural and other attributes across social groups, while evolutionary-scale correlations often use the comparative method across species, a method that contains an important and largely unresolved challenge: how can we compare the extent of culture among different species in a robust manner? Animals have culture. We should study its extent, importance in controlling behavior, effects on other areas of biology, how culture evolves, and how cultural capacity evolved.

## Notes

1. I have replaced "population or subpopulation" in our original definition by "group" for simplicity. Groups, as I use the term here, are social entities whose members interact behaviorally with one another, but they may contain from a few individuals each to an entire population.

2. To use the jackknife, first remove individual A from the analysis and recalculate the standard partial regression coefficients. Suppose that the standard partial regression coefficient for social similarity without individual A is $b_S(-A)$. This is repeated for all $n$ individuals, and the jackknife standard error of the estimated standard partial regression coefficient for social similarity ($b_S$) is the standard error (over all individuals X) of $n \cdot b_S - (n-1)b_S(-X)$.

3. This is a one-sided test, so that the alternative is that social similarity is positively related to behavioral similarity. There are cases when a two-sided test might be more appropriate. One such is considered later in this chapter.

4. In some cases, following the ideas of Hemelrijk (1990), it may make sense to rank elements of the social similarity matrix within rows, so that the likelihood that A learns from B is the rank of B among the associates of A. This implies that the probability that A learns from B depends not on the absolute proportion of time the pair are associated, but rather on the amount of time A is with B compared with how much A is with other members of the study population.

5. A key feature of such an assignment is that the groups are transitive, so that if A and B are members of the same group and B and C are members of the same group, then A and C are necessarily also from the same group.

ACKNOWLEDGMENTS

My ideas about culture have been enormously influenced by an extremely constructive collaboration with Luke Rendell. This chapter is built on our work together.

Discussions with Robert Latta and Michael Krützen helped spark the analytical approach developed here. David Lusseau has guided me into the powerful but complex field of network analysis. Many thanks to Marie Auger-Méthé, Shane Gero, Kevin Laland, Leah Nemiroff, and Luke Rendell for their perceptive and useful comments on drafts.

## 7

# FROM SOCIAL LEARNING TO CULTURE:

# INTRAPOPULATION VARIATION IN

# BOTTLENOSE DOLPHINS

BROOKE L. SARGEANT AND JANET MANN

## Introduction

A colleague studying sharks in Australia was once asked by a tourist, "We *know* we can commune with dolphins, but can you commune with sharks?" It is a common perception that the cognitive abilities and social complexity of bottlenose dolphins (*Tursiops* spp.) rival those of any other nonhuman animal. If prompted, laypersons are likely to report that dolphins have language, communicate with people, teach their offspring to hunt, baby-sit each other's offspring, rescue stranded people from the sea, and defend people from sharks. Although such statements are both anthropomorphic and hyperbolic, some are, perhaps surprisingly, partially rooted in real data.

Bottlenose dolphins have been called our cognitive cousins because they exhibit an impressive array of cognitive skills rarely observed outside the great apes. These include passing the controversial, but widely applied, Gallup test of mirror self-recognition (Reiss and Marino 2001), program-level imitation (Tayler and Saayman 1973; Roitblat 1988, 1991), language-like skills (reviewed in Herman 2002a), vocal learning (common in birds, but relatively rare in mammals; Janik and Slater 1997), mental representation (e.g., cross-modal representation of echoic and visual information; Harley et al. 2003), metacognition (Smith et al. 2003), exceptional memory (Mercado et al. 1998), behavioral innovation (Herman 2002a), and tool use with marine sponges (Smolker et al. 1997; Mann and Sargeant 2003; Krützen et al. 2005). Although such abilities are not restricted to hominoidea and bottlenose dolphins (tool use, for example, has been documented in more than 30 avian species [Tebbich et al. 2001; Lefebvre et al. 2002; Kenward et al. 2005], primates [e.g., van Schaik et al. 1999, 2003b; Whiten et al.

1999; Fragaszy et al. 2004], elephants [Hart et al. 2001], and otters [Hall and Schaller 1964], among others), they do push the boundaries for known capabilities of nonhuman animals and blur the constantly shifting line distinguishing *Homo sapiens* from all other animals.

In this chapter, we will discuss whether dolphins have culture, which is a hotly debated behavioral trait often considered a hallmark of being "human." We will also investigate the interplay between social learning and individual specialization in foraging, a relationship important in understanding our study population of bottlenose dolphins in Shark Bay, Western Australia, where matrilineal traditions help create a patchwork of foraging tactics.

## Defining Culture

An immediate challenge to determining which species have culture is providing a definition of culture that is biologically meaningful, does not inherently prevent or require attributing culture to particular species, and is widely acceptable to a diverse community of scientists (Laland and Hoppitt 2003). Although recognizing some simple requirements for a satisfactory definition is easy, finding a definition that meets those requirements is not (e.g., Heyes and Galef 1996; Rendell and Whitehead 2001 and peer commentaries). We agree with others that a definition encompassing social learning broadly is more useful than one restricted to certain social learning mechanisms (e.g., teaching, imitation) (e.g., Laland and Hoppitt 2003), and such a definition would capture many human cultural behaviors arising from diverse mechanisms of social learning (e.g., Henrich and Boyd 1998), while permitting comparisons across taxa. Of commonly used definitions that center on social learning, many consider culture as any information or behavior that develops through social learning (e.g., Boyd and Richerson 1996), or further stipulate that it be shared by a population, "subpopulation," or "groups" (e.g., Rendell and Whitehead 2001; Laland and Hoppitt 2003). The former approach would designate any socially learned behavior as "culture," and the latter does not clarify what constitutes a subpopulation or group. Thus, these definitions potentially label all socially learned behaviors (even those shared by two individuals) as "culture" because, technically, socially learned behaviors must be shared by at least two individuals *and* two individuals arguably compose a subgroup.

Consider a case in which an individual learns its mother's hunting behavior, but then no longer associates with her after weaning. Consequently, mother and offspring are the only two individuals in the population engaging in the hunting technique. Would these two individuals comprise a culture? We suggest that most behavioral scientists would be disinclined to label two individuals that share a single socially learned behavior as a distinct "culture," regardless of whether those individuals were humans or nonhumans. The issue of the minimum group size required for culture has rarely surfaced in discussions of animal cultures because intrapopulation variation in socially learned behaviors is typically quite low in populations studied. As we shall discuss, many putative cultural behaviors of bottlenose dolphins, particularly foraging behaviors, are not common to an entire population, but rather to subgroups within populations. Thus, the degree to which a socially learned behavior is shared exists on a continuum, ranging from only two individuals to an entire population. Such intrapopulation variation presents challenges to many current definitions of culture and suggests that more consideration is necessary to determine whether there is any reason to demarcate steps along the continuum. For example, information about group identity or social affiliation may be most likely to be shared by large proportions of a population and may differ *qualitatively* from behaviors with other functions. In fact, some argue that, consistent with use of the term "culture" in studies of human behavior, the term should be reserved for socially learned behaviors that perform social functions, such as promoting shared values or group identity (Premack and Hauser 2001; Henrich and McElreath 2003; Castro and Toro 2004; Tomasello et al. 2005).

We consider socially learned behaviors as points along a continuum (Figure 7.1), in which a higher tendency for the behavior to provide shared benefits for individuals results in more individuals engaging in the behavior. Because socially learned behaviors that provide social identity, communicative, or affiliative functions are generally considered essential to human culture and are more likely to be widespread among individuals, we view these as qualitatively different from behaviors with other functions and therefore consider them cultures. Although social functions are harder to identify in nonhuman animals, this definition does not exclude nonhuman animals from having culture, but remains consistent with the term's traditional use. Therefore, we will discuss the evidence for socially learned behaviors in bottlenose dolphins, but will reserve the term "culture" for those behaviors that likely function to promote group identity, affiliation, and/or social cohesion.

Foraging behaviors are often designated as cultural because of the role social learning frequently plays in their development. Socially learned foraging behaviors may be particularly common because foraging skills are closely tied to fitness, and social learning can increase the efficiency of behavioral development (Laland 2004). The usefulness of learned foraging behaviors, however, depends on exposure to appropriate environmental contexts and should be widely shared only when individuals forage cooperatively, forage in groups, or forage in similar habitats that enable them to exploit similar resources. In the former cases, affiliative or cohesive functions are likely to be involved.

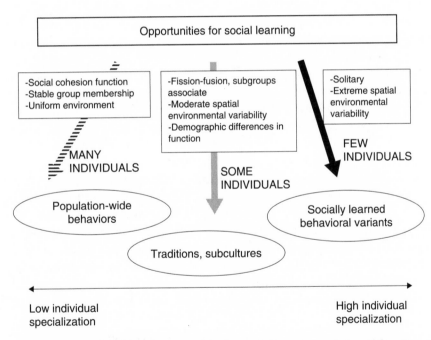

*Figure 7.1* Model of how opportunities for social learning and shared function determine the frequency of a socially learned behavior in the population. Social learning opportunities are filtered through shared function to produce the incidence of a behavior in the population, with effects of differences in age and sex superimposed on the entire system. The striped and gray arrows represent pathways that favor horizontal and vertical transmission, respectively.

For animals that engage in relatively little group foraging, the degree to which a socially learned foraging behavior is shared should be driven by habitat similarity and access to similar prey items. Identifying the functions of socially learned behaviors in their ecological and social contexts should increase our understanding of (1) patterns of transmission (vertical, oblique, and horizontal patterns of social learning; see Cavalli-Sforza et al. 1982); (2) the frequency of the behavior within a population; (3) the extent of individual specialization; and (4) whether the behavior serves a social function, i.e., is cultural (Figure 7.1).

Several factors influence whether social learning occurs and the patterning of information transmission within a population. First, individuals must have opportunities to acquire information from others. The degree of social tolerance within a population can significantly affect the probability of social learning (Coussi-Korbel and Fragaszy 1995; Boesch and Tomasello 1998; van Schaik et al. 1999; van Schaik 2003) by allowing individuals opportunities to observe the behavior of others in the group or the products of their behavior. Although social learning can occur when animals do not directly interact, as in local and stimulus enhancement, stable, egalitarian social groups should provide more social learning opportunities than unstable or hierarchical groups by allowing behavioral coordination in time and space (Coussi-Korbel and Fragaszy 1995). Highly despotic groups have fewer opportunities for social learning because individuals are prevented from closely attending to each others' actions (Coussi-Korbel and Fragaszy 1995; van Schaik et al. 1999).

Once opportunities for social learning are available, the spread of socially learned behaviors will depend on the adaptive value of the behavior and environmental variation (Figure 7.1). Although maladaptive information can be transmitted (e.g., Laland and Williams 1998; Giraldeau et al. 2002), long-lasting persistence of a socially learned behavior both in individuals and in populations is likely to depend on its usefulness to individuals that have acquired it from others (Galef 1995). For example, pinecone stripping in black rats (Aisner and Terkel 1992) and teaching of prey handling in meerkats (Thornton and McAuliffe 2006) provide clear benefits to naïve foragers. Opportunities for social learning and behavioral function also have implications for the patterns of transmission that would be favored (Figure 7.1). Different types of information have different values depending on social structures and ecological heterogeneity. Socially learned behaviors that perform a social role are likely to result in population- or group-wide cultures as a result of oblique or horizontal transmission (e.g., Whiten et al. 2005), even if individuals are largely solitary. For example, group-specific vocal communications in sperm whales (*Physeter macrocephalus*) are shared and enable maintenance of the social group (Rendell and Whitehead 2003), and motor behaviors such as hand-clasp grooming in chimpanzees (*Pan troglodytes*) may help strengthen bonds among members of a community even though some individuals spend significant amounts of time alone (Nakamura and Uehara 2004). Capuchins (*Cebus capucinus*) show substantial variation in social behaviors at different sites, from hand sniffing to social games (Perry et al. 2003a). These would qualify as cultures by our definition. However, socially learned behaviors that do not function primarily to maintain social cohesion (such as foraging) may still be widely shared, provided that individuals experience the same habitat either because resources are spatially uniform or the social group travels together (Figure 7.1).

When a population is subdivided such that smaller subgroups (e.g., matrilines) preferentially associate and have similar patterns of habitat use in a patchy environment, vertically transmitted "traditions" (*sensu* Fragaszy and Perry 2003b; Perry et al. 2003a) may be common, a situation that may characterize foraging traditions in bottlenose dolphins. Further along the continuum of sociality, primarily solitary individuals in a patchy environment may learn socially, but large groups of individuals are unlikely to share socially learned traits. Thus, spatial environmental heterogeneity has clear implications for modes of transmission (horizontal, vertical, oblique) and for how widespread a behavior should be based on the probability that individuals will share the same environment (Figure 7.1). Similar effects of temporal environmental variability on transmission mechanisms have been addressed previously (Laland et al. 1996; Laland and Kendal 2003).

Factors promoting individual differences in the development of socially learned or cultural behaviors are not limited to ecological heterogeneity. Life-history strategies may promote mechanisms that favor age and sex differences in socially learned behaviors. For example, biases to learn from a model similar to oneself would result in more copying of mothers by daughters than by sons, even if mothers and offspring of both sexes spend considerable time together, as we suggest for dolphins (Mann and Sargeant 2003) and as has been suggested for chimpanzees (Lonsdorf et al. 2004). Behavioral patterns specific to juveniles may also develop because of morphological traits (e.g., size) or social structure (e.g., dominance hierarchies).

Cultural explanations are often invoked by cetologists and primatologists when individual variation is partially accounted for by association (social preferences). In such cases, researchers rarely consider how habitat heterogeneity and niche partitioning may also explain individual and subgroup variation. Vertically transmitted traditions, for example, may help explain matrilineal patterns of behaviors within single populations, but habitat biases may be equally important because association and habitat use are invariably correlated. In particular, individual specialization, when individuals exploit narrower niches than the population (Bolnick et al. 2003) and consequently differ from one another in niche use, becomes important to understanding cultural and social learning differences within populations (Figure 7.1). Individual specialization has a long history of research in the fields of ecology and evolutionary biology (reviewed by Bolnick et al. 2003), but has thus far received relatively little attention in the context of culture (but see Estes et al. 2003; Tinker 2004). We argue at the end of this chapter that individual specialization contributes critically to defining and understanding patterns of social learning in a given population.

## Identifying and Measuring Social Learning

Most definitions of culture depend, at the very least, on demonstrations of social learning. Thus, methods used to study culture in populations of wild animals typically use social learning as the key criterion. At present, the method most commonly used to identify social learning is the "group contrast method" (Fragaszy and Perry 2003b; also called the "method of elimination" [van Schaik 2003] and the "ethnographic method" [Laland and Janik 2006]), which has been applied to chimpanzees (Boesch et al. 1994; Whiten et al. 1999), orangutans (*Pongo pygmaeus*) (van Schaik et al. 2003a), bottlenose dolphins (Krützen et al. 2005), and other cetaceans (Rendell and Whitehead 2001). Originally, the group contrast method was used to identify behaviors that varied between sites when such variation could not be explained by ecological or genetic differences (e.g., Boesch et al. 1994; McGrew et al. 1997; Whiten et al. 1999). More recently, the group contrast method has been employed to identify cultural patterns within populations when individuals differ in their behavioral repertoires (e.g., van Schaik et al. 2003b; Krützen et al. 2005).

Although the group contrast approach has succeeded in extending the cultural debate to nonhuman animals, the approach is undermined by at least two major weaknesses. First, the group contrast method requires exclusion of ecological and genetic explanations of behavioral variability, a task equivalent to proving the null hypothesis (Fragaszy and Perry 2003b). Relatively few studies collect sufficiently detailed ecological or genetic data before dismissing these factors as important. Second, attempting to find a single explanation for complex behavioral patterns and divorcing behaviors from their ecological context conflicts with basic tenets of ethology (Tinbergen 1963). Many behavioral variants are likely to be correlated with ecological differences, making it both conceptually and practically impossible to eliminate the role of ecological factors in producing behavioral differences between populations. Many behaviors discussed as possible cultural variants involve foraging (e.g., Rendell and Whitehead 2001; Fragaszy and Perry 2003a), a task critically tied to ecology. Thus, the group contrast method may either underestimate or overestimate the actual frequency of social learning in wild animals. The method helps identify areas for investigation, but in the end, finding ecological contributions to a behavior does not rule out social learning, and ruling out ecological contributions may be both extraordinarily difficult and unnecessary for the study of social learning.

Several promising alternative approaches to identifying the occurrence of social learning in wild animals have been proposed, although some approaches are similar to the group contrast method in that they require behavioral variation to tease apart mechanisms and may therefore underestimate the frequency of social learning. These techniques include: (1) use of transmission chains and diffusion curves (Day et al. 2001; Laland and Kendal 2003), (2) correlations between association and behavioral similarity (Perry et al. 2003b), (3) field experiments (Humle and Matsuzawa 2002; Reader et al. 2003; Galef 2004), (4) comparison of field data to experimental work on captive populations (Terkel 1996) or theoretical models (Laland and Kendal 2003; Dewar 2003, 2004), and (5) use of multifactorial models that measure the contributions of multiple factors to behavioral development and avoid the concept of exclusion (Sargeant 2005). Finally, with the exception of a few studies (e.g., pinecone stripping in black rats (*Rattus rattus*) [Aisner and Terkel 1992], termite fishing in chimpanzees [Lonsdorf et al. 2004], food processing in orangutans [Russon 2003], and prey handling in meerkats [Thornton and McAuliffe 2006]), behavioral development has received scant attention in field studies, although understanding development is critical to establishing ecological and social contexts for socially learned behaviors. Researchers are also moving beyond the question of whether animals have culture to the functional questions about social learning and its transmission mechanisms (e.g., "social learning strategies"; see Laland 2004). We have used correlations between association and behavioral similarity, multifactorial methods, and developmental approaches to investigate the diverse foraging tactics of dolphins in Shark Bay, Western Australia (Mann and Sargeant 2003; Sargeant 2005; Sargeant et al. 2005; Sargeant et al. 2007).

## Evidence for Social Learning in Wild Bottlenose Dolphins: When Is It Culture?

On the basis of life-history characteristics, social patterns, and ecological environments, bottlenose dolphins have been considered likely candidates for socially learned and cultural behaviors. They are large-brained (Marino 1998) and capable of vocal (Janik and Slater 1997) and motor imitation (Herman 2002b). Although dolphin foraging is generally solitary, dolphin calves often accompany their foraging mothers, remaining within meters of her, and therefore have many opportunities for social learning before they are weaned at 3 to 6 years of age (Mann et al. 2000). Delayed weaning offers calves years of exposure to the behaviors of their mothers as well as time to practice them. Dolphin females, in general, are rarely aggressive (Scott et al. 2005) and tolerate close inspection of captured prey by others (Mann et al. 2007). Such high social tolerance allows the close proximity thought to favor social transmission (Coussi-Korbel and Fragaszy 1995; van Schaik et al. 1999). Furthermore, calves begin foraging independently at 3 months of age and thus experience several years of overlap between closely observing their mothers' foraging behaviors and developing their own foraging skills. Because of the calves' uniquely high degree of exposure to their mothers' actions, matrilineal transmission of foraging tactics and other behaviors seems probable and may be an adaptive strategy allowing quicker and more efficient behavioral development than individual learning alone (e.g., Laland 2004). Dolphins are top predators in a patchy marine environment, and consequently fitness may be determined largely by food acquisition, especially in females. Thus, social learning from foraging mothers may be favored by natural selection. Calves may also learn social behaviors from a variety of others while still in their mothers' care because mothers constantly change their social companions (Gibson and Mann 2008a,b).

Evidence for social learning and possibly for culture in bottlenose dolphins comes from three main sources: (1) strong experimental evidence of vocal learning and whistle convergence in wild animals; (2) inter- and intrapopulation behavioral variation seemingly without genetic or ecological explanation; and (3) correlations between degree of association and behavioral similarity. In some cases, there is evidence for culture when socially learned behaviors seem to aid in social cohesion. However, most candidates for social learning in bottlenose dolphins involve foraging, where social functions are less apparent. Such functions may be present, but little effort has been devoted to identifying them, and supporting data are lacking. For example, do sponge carriers (described below) preferentially associate or cooperate with each other (controlling for habitat use and kinship) rather than with non-sponge carriers?

## Vocal Learning

Dolphins are clearly capable of vocal imitation, although not all vocal imitation is cultural. For example, dolphins may mimic the whistles of conspecifics (Tyack 1997; Tyack and Sayigh 1997; Janik 2000b), a behavior that involves social learning but would not constitute culture because the whistles are copied only at that instant and the mimicry is ephemeral. The stability of socially learned behavior is often considered or implied as a necessary condition for culture. Whistles may be used to communicate specific contexts (e.g., food over here; Janik 2000a) and to carry identity information (Janik et al. 2006), but generally are not consistently shared by a subpopulation. However, convergence of whistle types among allied males (Smolker and Pepper 1999; Watwood et al. 2004) seems to qualify as culture because the whistle types are consistently used, shared by a subpopulation, and are likely to promote social cohesion and group identity within alliances.

The largest delphinid, the killer whale (*Orcinus orca*), exhibits clear examples of vocal culture, with pod-specific dialects (Ford 1991; Deecke et al. 2000; Yurk et al. 2002) and interclan (but not intercommunity) whistle similarity (Riesch et al. 2006) that is believed to be used to maintain group cohesion (Miller et al. 2004). While similar vocal cultures have not been identified in dolphins, alliance-specific whistles can be considered cultural and suggest some continuity between socially learned behaviors shared by a few individuals and those shared by large subpopulations or entire populations.

*Interpopulation Variation in Motor Behaviors*

Motor and even program-level imitation has been documented in captive bottlenose dolphins (e.g., Tayler and Saayman 1973; Bauer and Johnson 1994; Herman 2002a), providing a basis for inferring that dolphins are capable of socially learning motor behaviors in the wild. Although no comprehensive survey of interpopulation variation in bottlenose dolphins similar to Whiten and colleagues' (1999) analysis of chimpanzee behaviors has yet been published, preliminary evidence suggests that the group contrast method may help identify candidates for socially learned behaviors. Diverse foraging behaviors have been reported in the literature, some of which appear distinctive either to particular populations or to subgroups within a population (e.g., Shane 1990; Connor et al. 2000; Nowacek 2002; Mann and Sargeant 2003; Gazda et al. 2005; Sargeant et al. 2005). For example, use of sandy beaches and estuarine mudflats to isolate and capture fish has been documented in several populations (reviewed by Silber and Fertl 1995; Sargeant et al. 2005), but not in other populations living in areas where dolphins have access to habitats favoring these techniques. Sponge carrying, a foraging tactic in which marine sponges are worn over the beak and presumed to act as tools, has been documented only in bottlenose dolphins of Shark Bay (Smolker et al. 1997). An exclusively genetic cause for such complex motor behaviors is unlikely (Krützen et al. 2005), and beaches and sponges are common to many delphinid habitats. However, as suggested by our critique of the group contrast method, inferences as to the roles of genetic and environmental variability in producing behavioral variability must be made with caution.

Interpreting behavioral differences among populations of bottlenose dolphins presents additional concerns. First, without a systematic survey based on complete reports of observed behaviors and not solely on published findings, it is not clear how many behaviors are limited to particular populations. Second, even if such a survey were to be conducted, it is unclear whether all observers are studying the same species (Laland and Janik 2006). The phylogenetic status of *Tursiops*, *Stenella*, and other genera in the Delphinidae family remains unresolved (LeDuc et al. 1999), and therefore we refer to our own study population as *Tursiops* sp. (M. Krützen, unpublished data).

## Intrapopulation Variation in Motor Behaviors

Within-population variation is well established in bottlenose dolphins and is exemplified by the foraging tactics of dolphins in Shark Bay. Sponge carrying is exclusive not only to Shark Bay, but also to a specific subset of the Shark Bay population (Mann and Sargeant 2003), most of whose members belong to the same matriline (Krützen et al. 2005). Dolphins in other populations also exhibit individual or subgroup variation in behavior. For example, only one of two sympatric communities of bottlenose dolphins in Moreton Bay, Australia, feeds on scraps thrown from trawlers (Chilvers and Corkeron 2001); social groups of dolphins in Cedar Key, Florida, differ in their use of a group foraging technique (Gazda et al. 2005); and individual differences in a variety of techniques used to catch fish have been documented in Sarasota, Florida (Nowacek 2002).

The group contrast method has been applied to such intrapopulation variation, taking advantage of the fact that individuals within populations may be less likely to have substantial genetic or ecological differences than members of different populations (Krützen et al. 2005). However, critics of the group contrast method will note that there may be genetic or environmental differences among subpopulations, and studies of intrapopulation variation do not necessarily provide significantly stronger evidence of culture than studies of interpopulation variation. For example, sponge carriers are closely related to each other (Krützen et al. 2005), and sponge carrying is correlated with increased use of deep channels where sponges grow in high density (Sargeant et al. 2007).

## Correlations between Foraging Similarity and Association

The strongest support for socially learned behaviors in wild bottlenose dolphins is based on correlations between measures of association and the use of particular foraging tactics. With this approach, ecological differences need not be fully excluded, but can be added to statistical models, provided sufficient variation exists to tease apart contributions of ecology and association. Such disassociation of ecology and propinquity is most easily accomplished by examining both parent-offspring similarity and offspring behavioral development in species where offspring receive substantial parental care after birth (Aisner and Terkel 1992; Guinet and Bouvier 1995; Estes et al. 2003).

Several anecdotal studies of bottlenose dolphins describing unusual foraging behaviors suggest that vertical transmission from mothers to offspring may be important in the development of young. For example, Nowacek (2002) reported that two relatively rare foraging tactics of bottlenose dolphins in Sarasota (fish whacking [whacking fish with tail flukes] and kerplunking [smacking tail flukes down onto the water's surface]) were used solely by a few mothers and their offspring. However, the most comprehensive study of matrilineal transmission of bottlenose dolphin foraging tactics has been our work in Shark Bay (Mann and Sargeant 2003), which we now discuss in greater detail.

## Cultures and Traditions in Bottlenose Dolphins of Shark Bay
### Factors Influencing Foraging Development

A bottlenose dolphin born in Shark Bay has much to learn, if it is to survive and reproduce. Forty-four percent of calves do not survive to 3 years of age, and 29 percent of adult females did not successfully wean young over a 10-year period (Mann et al. 2000). Access to food (Mann et al. 2000; Mann and Watson-Capps 2005) and predation by large tiger sharks (*Galeocerdo cuvier*) (Mann and Barnett 1999) are important in determining both calf survival and female reproductive success. Predation pressure may also act indirectly, causing habitat shifts that affect foraging (Heithaus and Dill 2002, 2006).

Bottlenose dolphins are well known for their diverse diets, consuming dozens of species of fish, as well as various cephalopods (squid, octopus) and crustaceans (shrimp, crabs) (Mead and Potter 1990; Barros and Wells 1998; Gannon and Waples 2004). Given the myriad highly mobile prey species and antipredator tactics dolphins must overcome, it is likely that dolphins must learn which species and foraging tactics are appropriate.

Multiple factors are likely to influence the development of calf foraging tactics (Figure 7.2). At the most basic level, the calf's foraging behavior will be directly influenced by the prey available in its habitat, and foraging appears to involve a period of learning. Dolphins do not forage immediately after birth, but are 3 to 4 months of age when they consume their first prey (Mann and Sargeant 2003). In the first few months of life, calves engage in what appears to be "practice" foraging, chasing and biting blades of seagrass (Mann and Smuts 1999). Calves spend most of their foraging time engaging in belly-up fish chases near the water surface, a behavior that we call snacking (Mann and Sargeant 2003). Snacking is not restricted to specific habitats and allows a young calf to hone

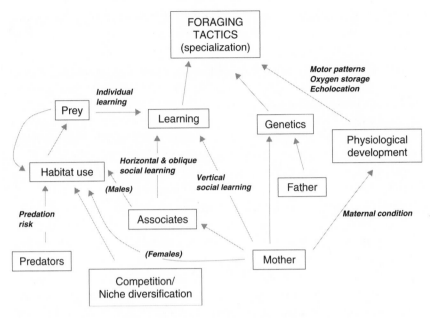

*Figure* 7.2   Factors likely to influence calf foraging development in bottlenose dolphins.

its hunting skills regardless of maternal habitat and prey distribution (Sargeant et al. 2007). However, snacking declines with age, and the calf must develop additional foraging tactics that are appropriate to its habitat, which is largely determined by its mother's foraging patterns (Mann and Sargeant 2003).

Newborn dolphins in Shark Bay spend, on average, more than 96 percent of their time within 10 meters of their mothers, although the amount of time they spend near their mother decreases as they mature (Mann and Smuts 1999). Calves eventually spend an average of 10 to 20 percent of their time further than 10 meters from their mothers (Mann and Watson-Capps 2005) but continue to associate closely with their mothers until weaning (Mann et al. 2000). Although out of visual range when separated by a few meters, calves and mothers can be in acoustic contact and receive foraging information during long-distance separations (Smolker et al. 1993). Post-weaning, females tend to stay in the same areas as their mothers, and males range slightly farther than females (Krützen et al. 2004b). However, into adulthood, the home ranges of both sexes continue to overlap with that of their mother. That is, none of the 56 male or female offspring we have monitored from birth to adulthood (with birth records since 1982) has completely left their mother's home range. Males form alliances to gain reproductive access to females (Connor et al. 1992), and thus the habitat use of juvenile and adult males may be more closely tied to that of their male associates than to that of their mothers. Habitat use may also be affected by intraspecific competition and prey availability, as dolphins forage predominantly in habitats that provide sufficient energetic benefits. Predation risk from tiger sharks is also known to influence habitat use of dolphins at multiple spatial scales (Heithaus and Dill 2002, 2006).

Foraging abilities and prey preferences could also be influenced by genetic traits inherited from either parent, by physical condition, and by physiological development. Oxygen stores in both blood and muscle are lower in calves than in adults, which could limit calves' diving abilities (Noren et al. 2001, 2002). Furthermore, maternal condition may be an important determinant of calf condition and development, and our data suggest that calves in better condition spend more time foraging than do calves in relatively poor condition (Mann and Watson-Capps 2005).

Opportunities for the social learning of foraging tactics may result from a calf's interaction with its mother or other associates (Mann et al. 2007) and may allow for any number of social learning mechanisms, including local enhancement via habitat selection. In sum, a calf's use of a foraging tactic is the result of a complex interaction of ecological, genetic, social, and developmental (age, sex) factors (Figure 7.2).

### Population-wide Cultures

Many of the diverse behaviors often thought to be socially learned by bottlenose dolphins are not common to all or even to most members of a population. Bottlenose dolphins have a fission-fusion social structure in which group compositions are very fluid and change as often as several times an hour (Smolker et al 1992; Connor et al. 2000). Within such a fission-fusion society, there are no fixed groups, but rather frequent temporary interactions with some individuals preferentially associating with one another. Individual variation in behavior is common, and many behaviors are linked to matrilines or subgroups within a population. However, some behaviors might qualify as population-wide cultures, including patterns of alliance formation (Connor et al. 1992).

In Shark Bay, several levels of alliance formation have been documented (Connor et al. 1992, 2000). Males form tight associations with one or two other males (first-order alliances) and work together to "herd" cycling females (Connor et al. 1992). Most males of prime reproductive age in Shark Bay form alliances (Connor et al. 2000), and males in alliances are more likely to father offspring than nonallied males (Krützen et al. 2004a). First-order alliance formation has been reported at three additional sites (Sarasota Bay, Florida, USA: Wells 1991, Owen et al. 2002; Port Stevens, Australia: Möller et al. 2001; Bahamas: Parsons et al. 2003), but not at others (Moray Firth, Scotland: Wilson 1995; New Zealand: Lusseau and Newman 2004).

Second-order alliances (alliances of alliances) have been reported only in Shark Bay. Still, a major difficulty in comparing across sites is that both observation conditions and intensity of observation differ markedly from one site to another. However, some very obvious behaviors, such as in-air head butts where two males leap toward each other and hit heads in air (much like head-to-head ramming in male ungulates), have been observed in New Zealand (Lusseau 2003), but not reported elsewhere. Patterns of alliance formation and specific agonistic behaviors (e.g., head butting) seem to qualify as cultural if they are socially learned and serve important social functions.

There is little evidence of population- or group-wide socially learned foraging behaviors in Shark Bay dolphins, possibly because dolphins forage in patchy habitats, usually independently of one another (Mann and Sargeant 2003), and individuals are likely to use different foraging tactics depending on the habitat they exploit. Occasionally, Shark Bay dolphins join large foraging groups and feed on aggregations of schooling fish. However, there is no clear evidence that this behavior is cultural or involves cooperation. Some cetaceans frequently feed in groups because they exploit large schools of fish or share sizeable prey (Connor 2000). For example, killer whales cooperatively hunt marine mammals (Baird and Dill 1996; Guinet et al. 2000). The number of individuals sharing a cultural foraging behavior is likely to be higher in killer whales than in bottlenose dolphins because killer whales exploit and share the same prey. Food-sharing has not been observed in Shark Bay bottlenose dolphins (Mann et al. 2007), but cooperative foraging on schooling fish has been observed elsewhere (Gazda et al. 2005). Feeding on clumped prey simultaneously, even cooperatively, also makes it possible for elements of foraging to take on social meaning as well. To cooperate effectively, individuals may engage in role specialization (e.g., Gazda et al. 2005) and communication, suggesting similarities with cultural behaviors that promote social cohesion.

### Traditions and Subcultures

The majority of socially learned behaviors identified in Shark Bay dolphins involve foraging tactics that vary among individuals and are shared by frequently associating individuals, particularly members of matrilines. A simple genetic cause for observed variation in foraging tactics is unlikely (see Krützen et al. 2005), although gene-environment interactions are possible.

Matrilineal transmission of sponge carrying has been indicated by both pedigree (Mann and Sargeant 2003) and genetic data (Krützen et al. 2005). However, ecological contributions are also evident, as sponge carriers make greater use of deep channels than non-sponge carriers (Sargeant et al. 2007). Overall, age, sex, maternal use of sponge carrying, and use of deep channels each appear to contribute to the development of sponge carrying in calves (Mann and Sargeant 2003; Sargeant 2005; Sargeant et al. 2007). Observational data do not allow determination of the social learning mechanisms involved in sponge carrying, and the coincidence of habitat use and sponge carrying could suggest that no social contribution is required for development of the tactic. However, because maternal habitat use largely controls that of calves during their period of dependency, individual learning by calves about prey in the environment would still be a result of local enhancement.

Maternal foraging patterns also predict calf foraging behaviors in multifactorial regression models, even after large-scale patterns of habitat use are controlled for (Sargeant 2005). For some behaviors, such as mill foraging (marked by irregular surfacings with frequent directional changes) and rooster-tail foraging (rapid swims near the water's surface), use by calves is predicted by the percentage of time that their respective mothers used each tactic in statistical models that also incorporate habitat use and other factors. Two additional tactics—provisioning by humans at Monkey Mia and begging from boats—show strict matrilineal transmission and could therefore be considered as possible traditions (Mann and Sargeant 2003). Such vertical social learning of foraging tactics appears to be relatively common, but there is little evidence of other modes of transmission. Because females tend to associate with their mothers and her social network after weaning, matrilineal transmission of foraging tactics is likely to facilitate rapid development of foraging tactics that are successful in female calves' preferred habitats.

Calves are attracted to and inspect fish catches by adults more than the converse, suggesting that calves have opportunities to learn about prey from adults other than their mothers. Calves are exposed to foraging patterns of adults other than their mothers during approximately 4 percent of their activity budgets (Mann et al. 2007). However, most foraging behaviors used by calves are also used by their mothers, and some obvious tactics attract the attention of nonoffspring but do not result in social transmission of the tactics. Thus, despite opportunities for social learning from nonmothers, no data currently show that calves incorporate tactics learned from nonmothers into their repertoires.

In some cases, foraging traditions may involve additional social functions that have not been well studied. For example, provisioned dolphins coordinate their visits with provisioning beaches and typically arrive in unison, because feeding is not initiated by rangers until several dolphins are present. Though currently only a matter of speculation, coordinated behaviors and increased association sometimes linked to foraging traditions suggest that unidentified social functions may exist and could be an area worthy of further study.

### Socially Learned Behavioral Variants

Bottlenose dolphins near the tip of Peron Peninsula in Shark Bay were first reported to chase fish along and onto sandy beaches in the 1980s. This conspicuous behavior subsequently attracted the attention of scientists, who began studying "beach hunting" in 1991. Despite observation of many dolphins in the general area, by 2004, only six adults and six calves (all born to three females engaging in the behavior) had been observed beach hunting (Sargeant et al. 2005). In fact, two beach-hunting adult females studied intensively over the course of 6 years were recorded in groups with a minimum of 25 individuals that did not beach hunt, including two females with whom the beach-hunting females had associated for at least 11 years (Sargeant et al. 2005). Thus, even though many animals are found both in appropriate habitats and associated with beach hunters, beach hunting is exceedingly rare in the population exhibiting it. However, beach-hunting females spend much more time near beaches where beach hunting is used and preferentially associate with each other (Sargeant et al. 2005).

The four beach-hunting adult females for whom genetic data are available do not have identical mitochondrial DNA haplotypes; this rules out the possibility of strict matrilineal transmission in all cases (Sargeant et al. 2005). Thus, like the matrilineal traditions discussed earlier, beach hunting has been transmitted vertically in some cases, but may be the result of horizontal transmission or of individual learning by additional females that frequent the beaches of Peron Peninsula.

## Social Learning and Individual Specialization

Unlike many other putative cultural behaviors that tend to be common to entire populations, the behavior of bottlenose dolphins typically exhibits marked intrapopulation variation (e.g., Nowacek 2002; Mann and Sargeant 2003; Gazda et al. 2005). For example, some foraging behaviors are restricted to two individuals in a population. The underpinnings of this behavioral variation are inextricably linked to the ecological concept of individual specialization (Estes et al. 2003; Tinker 2004).

Why individuals of a population exploit different resources is a long-standing question in ecology. Such individual specialization occurs when an individual exploits a narrower niche than that of the entire population, unexplained by age or sex differences (Bolnick et al. 2003). Individuals may benefit from having narrow niches (being specialists) if there are trade-offs or costs associated with exploiting many different resources (Roughgarden 1974; Price 1987; Holbrook and Schmitt 1992), while varying from others in niche use may function to reduce competition (McLaughlin et al. 1999; Bolnick 2001; Svanbäck and Persson 2004). At more proximate levels, differences in social learning, ecology, and phenotype determine which specific resources or combinations of resources an animal uses and are therefore more directly responsible for observed individual differences in behavior than competition or trade-offs.

Social learning and individual specialization are linked in multiple, and potentially reinforcing, ways. Most directly, social learning can be a proximate cause of individual specialization (Norton-Griffiths 1967; Partridge and Green 1985; Werner and Sherry 1987; Estes et al. 2003). Models of individual specialization have suggested that cultural transmission, particularly vertical transmission, can increase the number of alternative foraging tactics in a population (Tinker 2004). For example, when individuals socially learn different ecological information through vertical transmission, individuals will subsequently differ in their resource use.

In Shark Bay dolphins, vertically transmitted foraging tactics contribute to the generation of niche variation. Thus, some calves will use sponge carrying because their mothers use it, while other calves will not develop sponge carrying but will instead mirror their own mothers' use of beach hunting. Overall, vertical transmission has contributed to both individual variation in 13 foraging tactics identified to date, and to a high degree of niche variation in the population (Sargeant 2005). Such diversity in bottlenose dolphins may reflect the coevolution of problem-solving skills, social learning ability, and behavioral plasticity that allow for rapid responses to heterogeneous and fluctuating environments. Patterns of transmission have consequences for the level of individual specialization exhibited by a population (Figure 7.1); horizontal transmission is likely to result in predominantly shared behaviors, whereas matrilineal transmission would result in greater individual specialization in the population.

Although social learning has traditionally been viewed as a proximate mechanism generating patterns of individual specialization (e.g., a dolphin calf uses sponge carrying instead of beach hunting), individual specialization can also shape patterns of social learning and culture. For example, when both increased efficiency and decreased competition result from specialization, specialized individuals utilizing untapped niches should be favored. Such an advantage could promote the evolution of mechanisms reinforcing that niche variation, such as social learning within matrilines or subgroups of animals that exploit similar habitats. Alternatively, individuals should conform to cultures that aid in social cohesion, and would benefit more from learning mechanisms that promote conformity and rapid spread of information (e.g., horizontal transmission). Thus, individual specialization can favor matrilineal traditions.

The concept of individual specialization is likely to aid in our understanding of culture by providing adaptive explanations for patterns of transmission and the degree of behavioral homogeneity (Figure 7.1). In short, without individual specialization, matrilineal traditions would not exist, and the adaptive benefits of modes of transmission may be tied to the benefits of individual specialization.

## Summary

Although experimental data are required to identify social learning with certainty, a combination of captive and field studies strongly suggests the existence of socially learned behaviors in bottlenose dolphins, including both population-wide behaviors and behaviors that are shared by fewer individuals (behavioral variants, traditions). We have identified several areas that need clarification. Is social function helpful in discriminating between social learning and cultures? How widespread must a behavior be to be considered a "culture?" This distinction is highlighted by the prevalence of intrapopulation variation in bottlenose dolphin foraging behaviors. According to many definitions, because all socially learned behaviors necessarily involve at least two individuals (a subpopulation), all socially learned behaviors would be considered "cultural." We suggest (Figure 7.1) that a continuum of social learning can be refined by examining the spread of socially learned behaviors in a population and that behaviors that aid in social cohesion are qualitatively different and typically more widespread. The spread of such behaviors depends on the commonality of social and ecological contexts for the behavior.

The fission-fusion social structure of bottlenose dolphins, combined with prolonged dependency, high social tolerance, and variable prey habitats, result in correlations between maternal foraging, habitat use, and the emergent foraging behaviors of calves. When individuals share behavioral traits that are caused by several factors, the group contrast method is unlikely to identify cultural traits. Multifactorial models applied to developmental data that include developmental processes and ecological and social factors should elucidate factors driving the emergence of socially learned behavior better than the group contrast method.

We also propose that attending to the adaptive functions of individual specialization can be useful in understanding the prevalence of cultural variants within populations and that individual specialization may be useful in revealing the ecological context in which socially learned behaviors are likely to emerge. For example, habitat heterogeneity can promote individual specialization via proximate mechanisms such as sex-biased learning of maternal foraging tactics. Conditions that promote individual specialization should also favor matrilineal traditions or socially learned behavioral variants rather than population-wide patterns, just as social transmission provides the proximate mechanism for individual specialization. Thus, there exists a clear interdependence of social learning and individual specialization that deserves greater attention in studies of culture and may help reunite social learning with its ecological roots.

## ACKNOWLEDGMENTS

Sincere thanks to our colleagues from the Shark Bay Dolphin Research Project and the Shark Bay Ecosystem Research Project for collaborations and contributions to the long-term databases, and to Bennett Galef, Aaron Wirsing, and Jennifer Lewis for helpful comments on this chapter. We also thank our many field assistants, particularly Katherine Bracke, Kate Burmon, Lindsey Durbin, Allison Hough, Roger Lam, Nicole Marcell, Katie McAuliffe, Clint Poling, and Michelle Simeoni, who assisted in our examination of foraging behaviors. Funding was provided by Georgetown University, National Science Foundation grant #9753044, NSF-IOB grant #0316800 to Janet Mann, The Eppley Foundation for Research, and The Helen V. Brach Foundation, a National Science Foundation Graduate Research Fellowship, the Animal Behavior Society, Honda Motors, Magellan, and the Dolphins of Monkey Mia Research Foundation. We also thank the Monkey Mia Dolphin Resort, Monkey Mia Wildsights, and the Department of Environment and Consevation, W. Australia for field assistance.

# 8

# ANIMAL CULTURE:

## PROBLEMS AND SOLUTIONS

KEVIN N. LALAND, JEREMY R. KENDAL,

AND RACHEL L. KENDAL

## Animal Culture: Problems
### *Why Study Animal Culture?*

Many researchers who study animal cultures do so, at least in part, because they believe that their investigations will shed light on human cognitive evolution (Whiten et al. 1999; van Schaik, Ancrenaz et al. 2003; McGrew 2004). Animal social learning, it is argued, lies at the roots of human culture (see for instance, the title of Heyes and Galef's 1996 edited volume). If we could get to grips with termiting in chimpanzees or orangutan kiss-squeaks, we might gain some insight into homologous processes that led to the emergence of "full-blown" culture in humans, the conditions that favored the cognitive underpinnings of our own cultural capability, or the evolutionary trajectory of our cultural ancestors.

In contrast, we will argue that animal culture is inherently interesting, that is, that there are broader issues that validate investigating animal cultural processes over and above the light such study sheds on our own species. Although we share the belief that investigations of social learning and tradition in animals provide insights into the evolution of human culture, such insights are not restricted to identification of homologous characters, nor are they to be derived solely from the study of animals closely related to ourselves.

We will be referring to culture in a broad array of animals and suggest that unless we understand what culture may do for guppies, quail, and reef fishes, we cannot begin to comprehend *how* culture has transformed human existence. Culture is not just a property of humans; it is a fundamental cause of how humans got to be the way they are, a dynamic process that shapes psychological and material worlds (Boyd and Richerson 1985; Richerson and Boyd 2005). Human minds have evolved specifically to exploit the cultural realm (Richerson and Boyd 2005). But cultural transmission has changed how evolution operates in wrasses and grouse too (Warner 1988, 1990; Gibson and Bachman 1992).

Cultural transmission exhibits several properties that render it of interest to biologists. Perhaps the most obvious of these is that culture is a source of adaptive behavior; individuals can efficiently acquire solutions to problems such as what to eat and with whom to mate by copying others. But the fascination of culture also relates to a capability to propagate behavior in a manner that is to some degree independent of the ecological environment. For instance, it is a fundamental assumption of evolutionary biology that natural selection shapes organisms to reflect environmental conditions (Endler 1986). However, culture can violate this premise. A multitude of researchers, including evolutionary biologists, behavioral ecologists, and evolutionary psychologists, assume that a history of selection generates characters that are sufficiently close to optimal to render adaptationist reasoning and optimality models useful. However, without gainsaying the general validity of this assumption, cultural transmission can cause the characteristics of organisms to become partially disconnected from their environments. This is most obvious in humans, where, for instance, Guglielmino et al. (1995) found in a study of 277 African societies that most human behavioral and social traits correlated with cultural history (i.e., were handed down as traditions) rather than with a society's ecology. The same applies to animals. Bluehead wrasse *(Thalassoma bifasciatum)* mating sites, for instance, cannot be predicted from knowledge of environmental resource distributions (Warner 1988, 1990). Rather, removal and replacement experiments (described later) demonstrate that mating sites are maintained as traditions, with young fish and newcomers adopting the mating sites of residents.

Another interesting aspect of cultural transmission is that it can generate patterns of phenotypic variation in space. Evolutionary biologists and ecologists set out to understand the processes that underlie geographic variation in gene frequencies and phenotypic characters (Futuyma 1998). However, cultural processes, like gene-frequency clines, can generate geographic patterns in behavioral phenotypes. Among humans, Cavalli-Sforza and Wang (1986) studied differences in the languages of Micronesia. They found that the degree to which languages shared words declined according to a negative exponential of the distance between islands, in a manner equivalent to biological traits. Similar clines in behavioral characteristics have been reported for orangutan

175

behavior, birdsong, and whale vocalizations (Catchpole and Slater 1995; Janik and Slater 1997; van Schaik, Ancrenaz et al. 2003).

A third challenging feature of cultural transmission is that although it typically propagates adaptive behavior, both theory and empirical data suggest that under restricted circumstances arbitrary and even maladaptive information can spread. Once again, this is well documented in humans (Richerson and Boyd 2005). However, there are instances where arbitrary and maladaptive traits appear to spread among animals too. For example, when it is particularly costly for individuals to acquire information about resources, such as food or mates, copying others becomes an efficient compromise. Animal traditions may be maintained as Nash equilibria, in which it never pays any individual to abandon the tradition unilaterally; each is forced to do what others are doing, leaving populations locked into conventions that track changing environments inefficiently. For example, in animals that aggregate for protection, like shoaling fishes, taking the same route as others to a resource, such as a food site, offers fitness benefits even when the route is suboptimal, since going alone is dangerous (Laland and Williams 1998). Another case is informational cascades, where individuals base behavioral decisions on prior decisions of others (Giraldeau et al. 2002). For instance, among lekking sage grouse *(Centrocercus urophasianus),* the decisions of females that used social information to decide with whom to mate were less closely correlated with male traits indicating quality than were the decisions of females that made their own judgments about males (Gibson et al. 1991). Mate-choice copying will obscure the relationship between male quality and mating success and result in unpredictable "fads" in the characters that females find attractive and a lower intensity of sexual selection.

A fourth characteristic that renders cultural transmission of particular interest is that cultural traditions often affect the environment to modify the selection that acts on the population, an instance of niche construction (Odling-Smee et al. 2003). This is most obvious in humans, and a great deal of mathematical theory has investigated gene-culture coevolution (Boyd and Richerson 1985; Feldman and Laland 1996). By homogenizing behavior across a population and by allowing rapid changes in behavior, cultural niche construction typically increases rates of evolutionary change, although a reduction in rates of change is also possible (Feldman and Laland 1996; Laland et al. 2001). In other animals theoretical models of mate-choice copying reveal that learned preferences could plausibly coevolve with gene-based traits (Kirkpatrick and Dugatkin 1994; Laland 1994), models of bird-song suggest that song learning affects the selection of alleles that influence song acquisition and preference (Lachlan and Slater 1999), and other analyses have found that song learning could lead to the evolution of brood parasitism and facilitate speciation (Beltman et al. 2003, 2004).

Genetic data support these theoretical analyses by providing evidence of gene-culture coevolution. Recent statistical analyses of human genetic data reveal hundreds of human genes that show signals of very strong and recent selection, for example, in response to dairy farming, dispersal to cooler climates, and density-dependent infectious diseases (Wang et al. 2006; Voight et al. 2006). Such signals include high-frequency alleles in linkage disequilibrium and unusually long haplotypes of low diversity. In the last 100,000 years humans have been exposed to some tough challenges, virtually all of which have been precipitated or mediated by cultural transmission. For instance, humans have spread from East Africa around the globe, experienced an ice age, undergone a transition from hunter-gatherer to agricultural societies, and witnessed rapid increases in densities and new proximity of farmers to animal pathogens. To quote Wang et al. (2006, p. 140), authors of one recent statistical analysis of human genetic data: "*Homo sapiens* have undoubtedly undergone strong recent selection for many different phenotypes . . . it is tempting to speculate that *gene-culture interactions directly or indirectly shaped our genomic architecture*" (italics added). It is only a matter of time before similar genetic analyses are carried out on other cultural animals, and such analyses will surely show that cultural transmission has shaped the genome of other species too.

In summary, cultural processes in a broad range of animal species exhibit a number of properties that change the evolutionary dynamic, including detaching the behavior of animals from their ecological environments, generating geographic patterns in phenotypic characters, allowing arbitrary and even maladaptive characters to spread, influencing evolutionary rates and trajectories, and modifying selection to precipitate and direct evolutionary events. This different way of adapting and evolving is not unique to humans but is shared with many other species capable of social learning, including species only distantly related to ourselves. Animal culture is much more than a window onto human evolution.

*What Is the Most Useful Way to Conceptualize Culture?*

Human culture has proved to be a difficult concept to pin down (Kroeber and Kluckhohn 1952), and there exists little definitional consensus within the social sciences (Durham 1991). Accordingly, rather than ask, "What is culture?" we prefer to ask, "What is the most useful way to define culture?" Definitions must earn their keep. The best definition of culture will be the one that encourages the most useful work in understanding the underlying processes and consequences of culture. Hence on purely pragmatic grounds we favor a broad definition, since we are concerned that premature, overexacting distinctions will jeopardize the ability to see relationships between culturelike phenomena in diverse taxa. A narrow definition that restricts culture to our own species would act as a barrier to understanding the evolutionary roots of culture and the manner in which cultural processes have shaped humanity. Given the primitive state of knowledge of animal cultures and the current malaise in human culture studies (Bennett 1999; Bloch 2000), we believe that a broader definition is likely to be more stimulating (Laland and Hoppitt 2003). Down the line it may prove valuable to debate criteria by which cultural species are isolated or to make distinctions between different classes of culture, but such distinctions must be informed by empirical data. A broad definition encourages collection of such data; a narrow definition does not (see Galef this book for the polar opposite argument).

What, then, is the essence of culture? Virtually all definitions, however variable and whether restricted to humans or incorporating other animals, emphasize two key characteristics of culture. First, culture is built on socially transmitted information (Durham 1991; de Waal 2001). The term "culture" does not apply to inherited genetic information or the knowledge and skills that individuals acquire on their own. Second, culture is a source of both within-group similarity and between-group differences (Durham 1991; de Waal 2001). The spread of socially transmitted information can underlie group-typical behavior patterns, and these may vary from one population to the next because of differences in the learned information propagated. Accordingly, by "cultural traits" we mean those behavior patterns shared by members of a population that are to some degree reliant on socially learned and transmitted information. By "cultural variation" we mean those phenotypic differences between individuals, observed within or between populations, that are to some degree attributable to differences in what they learned socially. We treat "tradition" and "culture" as synonyms. We are conscious of the fact that many researchers (e.g. social anthropologists) would not be happy with such broad definitions, but we argue that they can nonetheless earn their keep within the biological domain.

### Why the Controversy over Animal Culture?

The main premise of this chapter is that the controversy over animal culture is to a large degree methodological in character. Disagreements abound largely because researchers lack appropriate means of validating claims of culture according to mutually acceptable criteria. As a consequence, judgments about which species exhibit culture vary with differing assessments of the plausibility of circumstantial evidence (Laland and Hoppitt 2003). Were we to devise perfect tools for identifying culture in nature, a major part of the controversy would disappear.

Methodological inadequacy is certainly not the only factor in the animal cultures debate. There are differences of opinion over definitions of animal culture (Whiten 2005). Some protagonists may draw attention to the social complexity and diversity of their animals for conservation-related reasons (e.g., McGrew 2004). Some skepticism may follow from the persistence of Cartesian thinking, which perpetuates the belief (in our view erroneous) in a chasmic divide between the mental abilities of humans and other animals.

Also central to the debate are differences of opinion concerning what are thought by some to be important properties or mechanisms of culture. There are those who assert that animal culture lacks some property (e.g., teaching, imitation, language) that they consider a key feature of human culture (Boyd and Richerson 1985; Galef 1992; Heyes 1993; Tomasello 1994). Human and animal cultures are therefore perceived to be qualitatively different, analogs but not homologs (Galef 1992; Tomasello 1994). But even here at least some of the criticisms are only tenable because protagonists lack the methodological tools to generate the kind of data that would satisfy the skeptics or demonstrate their criticisms to be right or wrong. Compelled, as they are, to rely on observational data, circumstantial evidence, and plausibility arguments, proponents of animal culture are easy targets for criticism. Without evidence to evaluate the skeptics' counterarguments, a broad array of criticisms are tenable. It should be no surprise, then, if the animal cultures debate occasionally suffers from ad hominem assertions on all sides.

179

We are forced to acknowledge that we find unconvincing both much of the evidence put forward in support of animal cultures and much of the criticism provided by animal culture skeptics. Let us deal with the skeptics first.

There are three principal reasons why we find most of the criticism of animal cultures uncompelling. First, many of the arguments are plausible but currently unsubstantiated hypotheses. For instance, Galef (1992, this book), Heyes (1993), Tomasello (1994, this book), and others claim that human culture is based on teaching and imitation, while animal traditions are not, and that because of this disparity in the underlying psychological mechanisms the similarities between animal and human "culture" are more superficial than real. Typically no data are presented to support this assertion; it is deemed self-apparent. Part of the problem is that there are currently no available methods for identifying "imitation" as a propagator of novel behavior patterns in natural social contexts in any animal species, including humans.

The hypothesis that human and animal cultures are reliant on different psychological mechanisms is certainly credible, but we think that there are few hard data, only circumstantial evidence, with which to evaluate this particular claim. Moreover, the circumstantial evidence is equally consistent with the existence of continuity in social learning mechanisms between humans and other animals. Both teaching and imitation are tricky terms that have associated with them more general or specific meanings, often with different levels of specificity applied to humans and other animals (Laland and Hoppitt 2003). Nobody knows how much human social learning is actually dependent on the narrowly defined forms of imitation (reproducing through observation the specific motor pattern of another individual) or teaching (costly modification of behavior specifically designed to instruct a tutee). The fact that humans are capable of such narrow imitation and teaching is no reason to regard these as the sole or even principal processes of information transmission that underlie human culture. It is conceivable to us that children learn how to get around their neighborhood, where the grocery store is, and which locations are dangerous by following parents around, like reef fishes. Plausibly, humans, like other primates (Tomasello 1990), may acquire many skills through emulation; we suspect that program-level imitation (copying of the underlying organization of behavior; Byrne and Russon 1998) is more common in humans than the narrow imitation of motor patterns. At the same time, impressive evidence for narrow imitation is starting to accumulate in animals (Akins and Zentall 1996; Whiten and Custance 1996; Whiten et al. 1996; Whiten 1998). Perhaps careful analyses of teaching and imitation will eventually leave our species looking considerably less distinct.

For Heyes (1993) and Tomasello (1994), teaching and imitation are significant because they are deemed to generate higher fidelity of cultural transmission among humans than in animals. Again, this prima facie credible claim, however well reasoned, is little more than a hypothesis (although at least it is one that can be evaluated empirically by using current methods). A rigorous comparison of the relative fidelity of human and animal social learning has yet to be undertaken, but at this stage the evidence of greater fidelity in human than in animal culture is far from overwhelming. Cavalli-Sforza et al.'s (1982) survey of patterns of cultural transmission reports average parent-offspring correlations for habits, entertainments, and sports to be as low as 0.07, 0.16, and 0.13, respectively. Boyd and Richerson (1985) report parent-offspring correlations ranging from 0.34 for attitudes toward feminism to 0.8–0.94 for political party affiliation, with values of around 0.5–0.6 common. Naturally, one can conceive of human cultural traits, say, ballet or classical music, that have been transmitted for centuries with strikingly high fidelity, but it is apparent that these are not representative of human culture in general. Conversely, the drinking of cream from doorstep milk bottles by British birds was first observed in 1921 in southern England, spread throughout much of Europe, and is still going strong in many regions. However, in an article ironically titled "Cultural transmission without imitation," Sherry and Galef (1984) provide experimental evidence that this trait is propagated through a simple local enhancement mechanism. Darwin (1871) described chimpanzee nut cracking on the basis of an observation made over 150 years ago; seemingly it has been faithfully transmitted over centuries. Can we really be sure that humans have higher cultural fidelity than animals? We are certainly open to the idea that teaching and imitation may be more important to human culture than to animal social learning, but we would like to reserve judgment until humans and other animals are evaluated according to the same standards.

Second, even if we were to accept that there is a qualitative difference between humans and other animals in some aspect of their cognition, how do we know that this is really a *key* difference? Galef, Heyes, and Tomasello claim that imitation and teaching are fundamental to what makes us distinctively human, but how do we know that the culture-relevant differences between humans and other animals are genuinely due to these mechanisms? Note that the aforementioned "interesting" properties of cultural transmission with which we began (e.g., the ability to propagate information through a population, or to modify selection) are not obviously linked to, or dependent on, particular psychological processes. There currently exist no empirical data with which we can compare the transmission fidelity or relative efficacy of, say, local enhancement and imitation to propagate behavior through a population, and there is every reason to assume that what are generally thought to be simpler mechanisms are more important to the spread of socially learned behavior in animals than are imitation and teaching (Galef 1988).

Tomasello (1994) also argues that much human culture is characterized by the "ratchet effect," with an increase in the complexity or efficiency of technology over time, which he claimed is not observed in animal social learning. We accept that there is currently little direct evidence within animal populations of additive, incremental improvement in behavior or technological advance (although a case can be made for leaf clipping and nut cracking in chimpanzees and sweet-potato washing in Japanese macaques; see Boesch 1993a; McGrew 2004). However, even if we agree that ratcheting constitutes a real difference between humans and other animals, what empirical evidence is there that teaching and imitation are what cause ratcheting? As far as we can see, none.

The skeptics' hypotheses have a common structure: "the key to human culture is property X, only observed in humans." A third reason for our dissatisfaction with their criticisms is that we are of the view that hypotheses of this type discourage comparative research, at least relative to hypotheses that stress such comparisons. For instance, in spite of the considerable attention that Galef's and Tomasello's arguments have attracted, there are still only a handful of studies of teaching in animals, while the considerable interest in animal imitation is multifaceted and long precedes Galef's and Tomasello's concerns (Galef 1988). We do not believe that the absence of research on teaching in animals is simply the result of a general absence of the phenomena in animals, although, to be fair, we cannot substantiate this impression. Rather, it appears to us that there are numerous hints of teachinglike phenomena in countless animals: tandem running in ants, the food calls of chickens and callitrichid monkeys, provisioning of disabled prey in domestic cats, cheetahs, and meerkats, intentional stranding in killer whales, and provisioning of infant chimpanzees, to name a few (Caro and Hauser 1992; Nicol and Pope 1996; Rousch and Snowdon 2000, 2001; Rendell and Whitehead 2001; Boesch 2003; Franks and Richardson 2006; Thornton and McAuliffe 2006; see Hoppitt et al. 2008 for a review). Given the little attention animal teaching currently garners, it seems to us highly probable that these examples represent just the tip of the iceberg. Scientific discoveries are made at a rate that is a positive function of the number of scientists who are addressing the topic. Unfortunately, the impression that animals do not teach discourages researchers from looking for such phenomena.

Nor do we see grounds for concluding that the processes that underpin animal and human cultures are not homologous. Comparative analyses of animal abilities reveal that many human behavioral and psychological traits have a long evolutionary history in common with other animals. Some human behavioral adaptations, such as a capacity to learn, evolved in our invertebrate ancestors (Bitterman 2000). Simple forms of social learning, such as stimulus and local enhancement, are probably widespread in vertebrates (Zentall and Galef 1988; Heyes and Galef 1996). An understanding of causal relationships may be common to mammals and birds (Dickinson and Balleine 2000). Much social knowledge, such as an understanding of third-party social relationships, probably evolved in our prehominid primate ancestors (Seyfarth and Cheney 2003). Narrow imitation and semantic communication seemingly evolved in prehominid apes (Whiten et al. 1996; Zuberbühler 2005). These capabilities are prerequisites for human culture, and understanding their origin sheds light on the history of our species. Human and animal cultures are reliant on countless homologous characters.

Moreover, comparative analyses of nonhomologous characters can be as informative as analyses of homologous traits. If cetaceans and humans share a capacity for true imitation as a result of convergent evolution, then analysis of the shared features of their life histories and ecologies would tell us much about the selection pressures that favor this capability.

A focus on the similarities between human and animal cultures potentially engenders a deep understanding of the roots of human culture; this claim holds even where such similarities involve analogous rather than homologous characters. We fear that a focus on the differences between human and animal cultures may sometimes discourage investigation of human evolutionary history, particularly in cases where humans are uniquely held to exhibit a capability with no obvious precursors.

However, there is one criticism of the argument for animal culture that we believe does carry some weight, and this is the one respect in which we find ourselves aligned with the animal culture skeptics. Quite simply, it behooves claimants of animal culture to demonstrate that their animals really do acquire cultural traits through social learning, or that putative within- and between-population cultural variation really is the manifestation of differential socially learned information. However, in virtually all cases claims of culture are made on the basis of weak circumstantial evidence rather than clear empirical demonstration or even reliable statistical inference. The hard evidence that their animals' cultures are socially learned is not yet there. Again, this is not surprising; there are currently no widely available, suitable methods for generating such hard evidence.

Consider, for example, the differences in the behavioral repertoires of chimpanzee populations (see McGrew this book; Whiten this book), which, for many, constitute the prime exemplar of animal culture. Building on earlier work by McGrew (1992), Whiten et al. (1999, p. 682) describe "39 different behavior patterns, including tool usage, grooming and courtship behaviors, [that] are customary or habitual in some communities but are absent in others where ecological explanations have been discounted" and characterize these as cultural variants. Highly credible though it may be, the case for chimpanzee culture rests entirely on a plausibility argument (Laland and Hoppitt 2003). Summarized, the argument is as follows: There is experimental evidence that chimpanzees are capable of social learning (Whiten and Custance 1996), including the seemingly more sophisticated forms of social learning, such as imitation (Whiten et al. 1996; Whiten 1998). Many group-typical behavior patterns, such as termiting, appear quite complex, and researchers envisage that they would be difficult to learn independently and cannot be genetically determined (see de Waal this book; Whiten this book; McGrew this book). Infant chimpanzees spend long periods of time apparently watching adults forage (Lonsdorf et al. 2004), and there are strong correlations between the behavior patterns that infants observe and those they later exhibit as adults (McGrew 1992, 2004; Whiten et al. 1999). Finally, captive studies demonstrate that alternative tool-using variants can be socially transmitted through small groups of chimpanzees (Whiten et al. 2005).

This body of circumstantial evidence is sufficient to convince many researchers that the group-typical behavior patterns reported for chimpanzees are learned socially. However, to our knowledge, in not one instance is there irrefutable evidence that a natural chimpanzee behavior is socially learned, and no demonstration in captivity, even of impressive phenomena such as sequential imitation or panpipes learning, will change this. We do not wish to single out chimpanzee research. The same concern holds for most putative orangutan, monkey, and cetacean cultural traditions. It remains possible that some other explanation, such as genetic or ecological variation, could account for the behavioral differences across populations.

Naturally, advocates of animal cultures have argued that ecological and genetic factors, as well as individual learning, are unlikely explanations for behavioral variation (McGrew 1992, 2004; Whiten et al. 1999; de Waal 1999; van Schaik, Ancrenaz et al. 2003). However, other researchers disagree (Galef 1992; Tomasello 1994; Laland and Hoppitt 2003). Once again, judgments vary on the basis of personal assessments of the plausibility of alternative explanations without any quantitative estimates of the likelihood of these explanations. The absence of any satisfactory, objective, widely accepted means to establish whether chimpanzee termiting, orangutan kiss-squeaking, capuchin hand sucking, or dolphin sponging are socially transmitted traits creates an intellectual vacuum, and in this vacuum subjective opinion reigns.

Consider, for illustration, the current dominant approach to identifying animal culture, labeled "the ethnographic method" (Wrangham, de Waal et al. 1994) or "method of exclusion" (Krützen et al. 2007. Elsewhere Laland and Janik (2006) have argued that this method suffers from a number of serious problems. Here we summarize these concerns briefly.

First, the ethnographic method tries to demonstrate the influence of a factor (culture) by excluding all alternative explanations (genetics, ecology, individual learning). This is logically impossible because the absence of a cause cannot be demonstrated in any absolute sense. It is simply infeasible to rule out the possibility that some unknown ecological or genetic factor explains the variance currently attributed to culture. Attempts to exclude other explanations are inevitably incomplete because only a small proportion of genetic or ecological variables can realistically be considered.

Worse, we suspect that were the ethnographic approach to be rigorously applied, it would reject most genuine cases of culture. Correlations between behavioral and ecological variables are to be expected because culture is a source of adaptive behavior, which enables animals to learn about and exploit environmental resources. Similarly, cultural and genetic covariance is also anticipated because animal learning is influenced by evolved predispositions and aptitudes. At the extreme, the reliance of the ethnographic method on ruling out alternatives would leave little to explain.

Another problem is that the ethnographic method appears to have encouraged the treatment of behavior as exclusively genetic, ecological, or cultural (Laland and Janik 2006). This categorical thinking evokes memories of the nature-nurture debate, associated with polarized views long since rejected by developmentalists. Clearly, genes, ecology, and learning all influence animal behavior, and each explains some variation. Such influences might be subtle. To identify cultural variation, it is not sufficient to rule out the possibility that the variation in behavior constitutes unlearned responses to differential selection; it is also necessary to consider the possibility that genetic variation precipitates different patterns of learning.

De Waal (1999, p. 635) judged genetic explanations for chimpanzee behavioral variation implausible on the grounds that "genes determine general abilities, such as tool use, but it is hard to imagine that they instruct apes how exactly to fish for ants." We agree. But even if we accept a vital role for learning in tool use by apes, it remains possible that differences between ape populations might be caused by alternative genetic influences on learned behavior. In light of experimental studies of woodpecker finches *(Geospiza pallida)* and New Caledonian crows *(Corvus moneduloides),* which revealed impressive tool use in birds reared with no opportunity to learn to use tools socially (Tebbich et al. 2001; Kenward et al. 2005), ignoring genetics is ill advised.

Could some ape behavioral variation be genetic in origin? Whiten et al. (1999) sampled across two chimpanzee subspecies, *verus* at the western sites and *schweinfurthii* in the east, while van Schaik, Ancrenaz et al. (2003) sampled two orangutan species in Borneo *(Pongo pygmaeus)* and Sumatra *(P. abelii).* A third of the chimpanzee "cultural" variants are observed in one subspecies alone, and half of the orangutan variants are seen only in one of the two species. Some chimpanzee subpopulations have been genetically isolated for hundreds of thousands of years, and *verus* and *schweinfurthii* occupy distinct branches of a neighbor-joining tree based on genetic distances for mtDNA haplotypes (Gagneux et al. 2001). Thus it is plausible that some behavioral differences between chimpanzees have a genetic origin.

186

Lycett et al. (2007) provide evidence that behavior patterns from *P. t. verus* and *P. t. schweinfurthii* combined fit less well to a phylogenetic tree than from *P. t. schweinfurthii* alone; a pattern they interpret as inconsistent with a genetic explanation. Unfortunately, although their analysis is an important first step in addressing the question of the role of genetics, it is only able to give a crude overview of the likely transmission mechanism across a number of behavior patterns, and does not preclude the possibility that a small number of the putative cultural variants may be underpinned by genetic, and indeed ecological, differences (see below).

There are similar concerns over the orangutan data, where a reported correlation between distance between sites and "cultural difference" (van Schaik, Ancrenaz et al. 2003) may merely reflect the well-established correlations between genetic and geographic distances, and it is not clear that this association would hold if analyses were restricted to a single species.

Of course, if behavioral differences between sites covary with genetic differences, this does not rule out culture as a source of some variability. Conceivably, chimpanzees and orangutans might have evolved predispositions to learn some associations more readily than others, as is observed in rhesus macaques *(Macaca mulatta)*, which acquire a fear of snakes through observation of conspecifics but cannot readily be conditioned to fear other objects, such as flowers (Mineka and Cook 1988). Detection of genetic covariates remains a serious problem for an ethnographic method that treats as cultural that variation that remains after genetic and ecological factors have been excluded. If advocates are employing the ethnographic method rigorously, they should remove some "cultural" variants. For instance, chimpanzee leaf grooming, which is observed at all *Pan troglodytes schweinfurthii* and at no *P. t. verus* sites, should be struck from the list. The same holds for the eating of slow loris *(Nycticebus coucang)* by orangutans, observed in all populations of Sumatran *Pongo abelii* but in no populations of Borneo orangutans *(P. pygmaeus)*. The failure to exclude such examples gives the unfortunate impression that genetic sources of variance are not being taken seriously.

More compelling evidence for culture comes from variation within chimpanzee subspecies. Methods of phylogenetic reconstruction applied to both *schweinfurthii* and *verus* have revealed no statistically supported subdivision within subspecies (Goldberg and Ruvolo 1997; Gagneux et al. 2001). Most candidate chimpanzee cultural variants differ in frequency between populations of the same subspecies, and these genetic

187

analyses render a genetic explanation for these differences less plausible. Future applications of the ethnographic method to chimpanzees would be better restricted to analyses within subspecies.

In addition, ecological differences could explain some of the variation in putative animal cultures (Galef 1992, 2003b; Tomasello 1994, 1999a; Laland and Janik 2006). This "ecological explanation" refers to cases in which behavioral variation reflects differential manifestations of phenotypic plasticity, be it learned or unlearned, in response to ecological variation. Although Whiten et al. (1999) consider the possibility of false positive reports attributable to an ecological explanation, only 3 of the 65 observed behavior patterns were removed because such an explanation could be given, which left the skeptics questioning how seriously this alternative was considered. Moreover, in this and all subsequent applications of the ethnographic method, readers have merely been told that there are no ecological confounds, without the listing of ecological factors that had been considered and without any statistical analysis of ecological covariation with behavior.

This problem is highlighted by a recent analysis of chimpanzee ant dipping (Humle and Matsuzawa 2002), which found that different methods were used within a single chimpanzee population for different kinds of ant. Conceivably, chimpanzees at different sites could individually be shaped by biting insects to use the strategy that results in the fewest bites, with variation in chimpanzee behavior a consequence of differences in ant aggression. Although the study does not rule out a cultural interpretation, it nonetheless puts ecological explanations back in the picture.

We think that use of the ethnographic method in isolation is the wrong way to approach the problem of discovering culture (Laland and Janik 2006). Clearly, behavioral differences can simultaneously result from genetic, ecological, and cultural variation. Researchers who are studying animal culture would be better advised to think in terms of partitioning variance between alternative sources rather than allocating behavior to genetic, ecological, and cultural categories. Ethnographic-method practitioners are not bad scientists. The field has been reluctantly forced into this way of thinking by the absence of any means for identifying social learning where it occurs in animal groups. If there were a way to demonstrate that social influences contribute to the development of, for example, chimpanzee ant dipping (McGrew 1992, 2004) and orangutan *Neesia* processing (van Schaik 2003), this aspect of the controversy would disappear. It is in this respect that the debate over animal cultures is a methodological debate. Animal social learning researchers lack satisfactory tools for identifying and quantifying social learning in natural and captive animal populations, and in the absence of such tools, controversy reigns.

This methodological problem is also a major impediment to establishing the breadth and significance of animal culture. By our definition, many hundreds of species of animals may eventually be shown to exhibit cultures, but at present in only a handful of species (two species of reef fish, some birds, and a whale or two) do we have satisfactory experimental demonstrations of this capability (Laland and Hoppitt 2003). However, the fact that there is currently better experimental evidence for culture in reef fishes than in chimpanzees demonstrates little more than that fishes have proved more amenable than chimpanzees to the relevant experimental manipulations. It is harder to demonstrate culture in some species than others. We need tools that can provide widely acceptable evidence for culture in *any* species.

## Animal Culture: Solutions

A potential resolution to the debate is through development of statistical and mathematical methods for detecting and quantifying social learning in animal populations. In spite of the upbeat heading to this section, what we describe here are not so much established answers to the aforementioned problems as strong candidates for possible solutions. The methods proposed are at a formative stage and are in need of refinement and evaluation. Nonetheless, in our judgment, the mathematical tools described here offer some of the best prospects for resolving the animal culture debate.

Before describing these tools, we will first outline an experimental method that provides the most satisfactory data with which to determine whether social learning underpins behavioral traditions. The method is not new, but it is underused and has already provided the clearest experimental demonstration of cultural variation in animals to date.

### Experimental Methods

Traditionally, social and asocial learning processes have been distinguished in the laboratory in paired demonstrator-observer designs (Galef 1988), but such experimentation is usually impractical in natural contexts. Moreover, such tests are designed primarily to investigate whether animals are capable of social learning, and such a capacity is only a prerequisite for

culture. Laboratory and captive animal studies have successfully investigated diffusion of learned behavior through populations (Lefebvre and
Palameta 1988; Whiten et al. 2005) or along chains of animals in a
Chinese-whispers-like design (Curio et al. 1978; Laland and Plotkin 1990,
1993; Galef and Allen 1995; Laland and Williams 1998). However, although such studies shed light on social transmission processes and the capabilities of particular species, they provide at best circumstantial support
for the argument that natural behavior is cultural.

One promising avenue is the use of field translocation experiments,
such as translocations of individuals between populations or of populations of animals between sites, which could, in principle, distinguish
between alternative explanations. For instance, if introduced animals
newly adopt the behavior of established residents, this is inconsistent
with an explanation in terms of genetic differences between populations.
Similarly, if an entire population is replaced, and the introduced individuals come to exhibit a behavior different from those of the former inhabitants, this would suggest that the variation does not result from shaping
to divergent ecological conditions.

This approach has been used successfully to demonstrate culture in
fishes. Helfman and Shultz (1984) translocated French grunts *(Haemu-
lon flavolineatum)* between populations and found that while those fish
placed into established populations adopted the same schooling sites and
migration routes as the residents, control fish introduced into regions
from which the residents had been removed did not adopt the behavior
of former residents. Helfman and Shultz's study is one of the most elegant demonstrations of animal culture to date.

Similar removal and translocation experiments of populations of
bluehead wrasses reveal that their mating sites cannot easily be predicted
from knowledge of the local ecology (Warner 1988, 1990). Rather, they
are maintained as traditions, with youngsters and newcomers learning
the routes to sites from experienced residents. The locations of fish
schooling, resting, and mating sites and migration routes have been experimentally shown to be cultural traits, and differences between populations in these characteristics can demonstrably be attributed to cultural
variation.

Unfortunately, applying this direct experimentation method to many
species, particularly apes, is likely to be neither feasible nor ethical. It
might be possible to exploit natural movements of animals between
populations or exchange of captive animals between holding facilities to

isolate instances of culture. However, many researchers who are studying animal cultures must work with the constraint of having only observational data on which to base their assessments, and it is here that the statistical methods described in the next section may be particularly useful.

## Statistical Methods

Here we describe statistical methods that have the potential to distinguish between socially contingent and asocial learning in natural populations or captive groups of animals (see Whitehead this book for a complementary statistical approach). The methods are only tools to guide our assessment, not magic bullets for the animal cultures debate. Many do not, for instance, distinguish learned from unlearned behavior patterns, and most methods would have to be deployed in conjunction with other (e.g., genetic) analyses in order to provide a comprehensive understanding of the causes of behavioral variation. Nonetheless, there is every reason to believe that if these methods are employed correctly, they can increase the reliability of our judgments about animal culture. They potentially provide objective and independent criteria by which the probability can be estimated that individual cases of behavioral variation should be deemed cultural.

### BOOTSTRAPPING SAMPLING DISTRIBUTIONS FOR THE ASOCIAL LEARNING OF NOVEL TASKS

Socially contingent learning should generate a greater-than-expected homogeneity in the behavior of individuals in a population. For instance, if the two observed methods of chimpanzee ant dipping (pull-through and direct mouthing) were equally profitable, and if each chimpanzee independently adopted a single method, we would expect that on average, approximately 50 percent of the individuals in a population would use each method. If, on the other hand, the method adopted by chimpanzees were biased toward the behavior of other chimpanzees in their population, we would expect to see distortions (here termed "option biases") away from the chance expectation. Social learning would cause a disproportionate number of chimpanzees to use a locally favored variant.

Potentially, then, the level of homogeneity of behavior within a population provides a metric that can be used to detect a social influence on learning. But how much homogeneity would have to be observed before we could reliably claim that a given observation is too extreme to have been generated by chance or through independent learning? Suppose

that we found that 80 percent of chimpanzees in one population were using the pull-through method; could we conclude that it must be socially
learned? If only we had access to the distribution specifying the probability that a given level of homogeneity in option choices could arise
through chance or asocial learning, we could compute the probability
that our data arose through social learning.

There is a statistical method for generating such a distribution, a re-
sampling method known as bootstrapping. There are two obvious
means by which researchers could bootstrap probability distributions
for the asocial learning of a target behavior. They could either use data
from the measured performance of isolated individual animals or, more
realistically, for tasks with two or more alternative solutions, use the
computed probability of levels of homogeneity based on derived probability distributions or on simulations using a mathematical model of
learning. Such analyses compute the probability of an option bias of a
given magnitude, in populations of the same size as the study populations, for a fixed probability of each individual learning the task, considered across the full range of learning probabilities (0 to 1). Once
these sampling distributions have been constructed, social learning can
be inferred if the observed level of behavioral homogeneity is in the tail
of the distribution.

This method was developed by Rachel Kendal, née Day (Day 2003).
We applied the method to data collected from groups of callitrichid
monkeys provided with novel extractive foraging tasks, each of which exhibited two equivalent means of obtaining a food reward (using different-
colored doors to access food). If the monkeys influence each other in their
choice of solution, then within each population there should be a disproportionate tendency for them to use the same door. Through the use
of computer simulations, the sampling distribution was estimated that
represented the probability that populations of monkeys behaving independently would exhibit the observed level of homogeneity in their option choices. This allowed us to infer social learning in the monkeys'
foraging where the null hypothesis, that chance or asocial learning generated that level of behavioral homogeneity, could be rejected at the
$\alpha = 0.05$ significance level. The analysis successfully distinguished social
learning from unlearned processes and asocial learning in a social context and found that some, but not all, callitrichid foraging diffusions
were reliant on social learning. In fact, the analysis revealed that monkeys differentially employed social learning on difficult tasks, where

the acquisition of asocial information was costly in terms of time and effort.

One exciting aspect of this method is that it is potentially widely applicable to animal populations, including captive and natural groups, even where population sizes are small and data sets are incomplete. Naturally occurring phenomena that might fit this context include chimpanzee termite fishing where either one end or both ends of the tool (e.g., a nonwoody stem) are used and grooming traditions that involve the clasping of either a branch or a conspecific's hand (McGrew, personal communication). This method is not restricted to cases where researchers are able to introduce a two-option task to many populations, and it is also applicable to instances where tasks have fewer or more than two options, where there is an unequal likelihood in the use of each option, and in the study of an apparently cultural phenomenon within a single group.

### SIGNATURES OF SOCIAL LEARNING

Another method for isolating cultural variation is to identify statistical "signatures" of cultural transmission, namely, rates, pathways, and patterns of spread that are diagnostic of social learning. Researchers have speculated whether the shape of the diffusion curve may reveal something about the learning processes involved. Theoretical models predict that the diffusion of cultural traits will typically exhibit a sigmoidal pattern over time (Cavalli-Sforza and Feldman 1981; Boyd and Richerson 1985), while asocial learning has been expected to result in a linear, nonacceleratory, or at least nonsigmoidal increase in frequency (Roper 1986; Galef 1991; Lefebvre 1995a, 1995b). Unfortunately, this particular signature is probably unlikely to be reliable, because other theoretical analyses reveal that asocial learning can generate acceleratory curves and social learning can generate deceleratory curves (Laland and Kendal 2003; Reader 2004).

Although further insights can be gained by constructing more realistic mathematical models (Laland and Kendal 2003), we and our collaborators are currently exploring an alternative approach that bases the search for signatures of social learning on empirical data drawn from diffusion experiments that manipulate the opportunity for social and asocial learning. Animals are presented with a novel task in small groups, and both the latency to solve and the order in which they solve the task are recorded. In the "asocial learning" condition (A) the researcher minimizes the opportunity for social learning within groups through use of visual barriers

193

around the task and by employing tasks with which only one animal at a time can interact. Animals in the social learning condition (S) are exposed to the same tasks but have unimpeded views of other animals manipulating the task apparatus and can fully rely on social learning. Subsequent regression and curve-fitting methods are employed to ascertain which combination of variables predicts the diffusions, and which mathematical functions provide the best fit to the data sets. The objective is to isolate factors observed in the potentially socially mediated diffusions of condition S that are not observed in the primarily asocially mediated "diffusions" of condition A. At this stage it is not known whether such signatures of social learning will be found, or what they will look like: perhaps a definitive rate, pathway, or pattern of spread. The hope is that if such signatures can be identified and their performance assessed and validated in natural contexts, they can be applied to observational data to draw inferences about learning processes employed.

### FITTING MODELS TO DIFFUSION DATA

Researchers can also develop mathematical models of asocial and social learning and fit these to behavioral data. We have used this approach to analyze the spread of novel foraging behavior in callitrichid monkeys and have found evidence for some but not other social learning processes (Kendal et al. 2007). The analysis quantified the effects of two social learning processes, stimulus enhancement and observational learning, and two asocial processes, intrinsic movement to the task and asocial learning of the task, on adoption of a novel extractive foraging behavior. We simulated the effect of these processes by using a model for the spread of a novel behavior and selecting the model parameters that provided the best fit to the monkey data. Our analysis revealed evidence for the two asocial processes and stimulus enhancement, but not for observational learning.

### FITTING MODELS TO INHERITANCE DATA

Another approach that we are currently exploring is to develop regression models to determine paths of influence on the behavioral repertoires of animals. We assume a model in which the behavioral phenotype of an individual is specified as a combination of normalized genetic, environmental, and cultural deviations. Mathematical expressions represent various forms of cultural transmission, including vertical (from parents), oblique (from the previous generation), and conformist transmission.

The analysis makes use of data on the entire behavioral repertoire of individual animals in a population. We plan to use data from a laboratory population of budgerigars for which the degree of relatedness of any pair of individuals is known. The models will be used to compute expected correlations in behavioral repertoire between pairs of individuals. These expectations will then be compared to observed correlations in the behavioral repertoires of animals, and this comparison will be used to estimate the magnitude of the model's parameters. For instance, if observers resemble their demonstrators, the cultural parameters will be nonzero. Similarly, if related pairs are found to be no more similar in their behavior on average than unrelated pairs, the genetic parameters will be set to zero. Eventually it should be possible to generate a model that provides the best fit to the empirical system, which will allow us to quantify the relative influence of genes, nontransmitted environment, and various forms of social learning on performance. We hope that once developed and tested on captive animals, the same methods can be applied to data from natural populations. Again, it will be necessary to assess performance of the method, and confidence in it will be enhanced by verification across methods.

### PARTIAL REGRESSION PERMUTATION TEST

Whitehead (this book) suggests using partial regression to test how much interindividual variation in behavior is caused by "social similarity," over and above any effects of ecological and/or genetic similarity. Here, social similarity is a measure of the opportunity for social transmission of information, typically measured by some form of association index (Laland and Kendal 2003). The interindividual variation in behavior, scored as behavioral similarity between each pair of individuals, is nonindependent, so Whitehead addresses the significance of the partial regression coefficient by bootstrapping a distribution of partial regression coefficients, where the similarity scores of the independent variables are randomized across individuals. Social learning, or more precisely, the effect of the social similarity, is invoked if the original partial regression coefficient lies beyond the critical point in the upper tail of the bootstrapped distribution (e.g., $P < 0.05$).

### PHYLOGENETIC ANALYSES

Lycett et al. (2007) applied cladistic methods to the 39 chimpanzee behavior patterns described by Whiten et al. (1999) as "cultural." They reasoned that if behavioral diversity is primarily the product of genetic differences between subspecies, then population data should show less phylogenetic structure when data from a single subspecies (*P.t. schweinfurthii*) are compared with data from two subspecies (*P. t. verus and P.t. schweinfurthii*) analyzed together. The analysis found lower levels of phylogenetic structure (lower retention index, or RI) associated with the full data set than the East Africa data set. While this finding is inconsistent with the hypothesis that the putative cultural variation is determined *primarily* by genetic differences between populations, whether it provides strong support for the culture hypothesis is a moot point, for several reasons. First, the method gives only a crude insight into the underlying transmission mechanisms averaged across all behavior patterns, and does not preclude the possibility that a small number of the behavioral variants are underpinned by genetic differences. Second, the low RIs for both maximum parsimony trees indicate considerable homoplasy in the data, an observation consistent with parallel independent evolution (or learning) in response to local ecology (or alternatively diverse cultural lineages for different behavior patterns). Third, it is not clear whether the maximum parsimony trees are significantly shorter (i.e., provide a better fit) than the genetic tree (although they are shorter than random trees; Lycett and Collard, pers. comm.). The Lycett et al. analysis is exciting, not so much because it resolves the debate, but because it points the way to future resolution. For instance, researchers could compare the maximum parsimony trees to a genetic tree, giving a more precise overall perspective on whether the behavioral variation fits a genetic explanation. They could also conduct more fine-grained analyses in which they compare each individual cladogram associated with a behavior to the genetic tree, and ask if each is significantly different (note, however, this approach would be vulnerable to false negatives where offspring learn from parents). They could investigate which behavior patterns are most responsible for the observed homoplasy, and go on to explore its causes. Such phylogenetic analyses have a great deal of potential, and may prove an important tool with which to resolve this debate.

## Concluding Remarks

We have argued that the controversy over animal cultures is largely methodological in character. Researchers currently lack tools for determining which animal behavior patterns are cultural according to any widely accepted criteria. Consequently, whether one deems a particular species "cultural" depends on personal judgments regarding circumstantial evidence. We believe that it will soon be possible to resolve this debate by using mathematical and statistical tools. Rather than relying on subjective assessments, researchers will be able to deploy objective, impartial criteria that will specify precise probabilities, relative to known sampling distributions, that a particular data set provides evidence for cultural transmission. It remains to be seen whether our optimism is justified. Much hard work is still necessary to develop the methods further, validate them, assess their robustness, and refine them. Nonetheless, we hope that within the next decade the field of animal cultures will enter a new era enlightened by rigorous quantitative analysis.

### ACKNOWLEDGMENTS

We are indebted to Will Hoppitt, Vincent Janik, and Henry Plotkin, who have each shaped our evaluation of the animal cultures debate, to Neeltje Boogert, Gillian Brown, Jeff Galef, and Jamie Tehrani for helpful comments on earlier drafts, and the Biotechnology and Biological Sciences Research Council and the Royal Society for financial support.

# 9

# THE QUESTION OF CHIMPANZEE CULTURE,

## PLUS POSTSCRIPT

## (CHIMPANZEE CULTURE, 2009)

MICHAEL TOMASELLO

## Introduction

In the early days of chimpanzee fieldwork researchers talked of "precultural" or "protocultural" behaviors (see, for example, the essays in Menzel 1973a). More recently the prefixes have been dropped and talk has turned to chimpanzee "culture," "cultural traditions," and "cultural transmission" (e.g., Goodall 1986; Nishida 1987; McGrew 1992; Boesch 1993b. The terminology has changed at least partly in response to the growing list of ontogenetically acquired, population-specific behavioral traditions observed in different chimpanzee communities. It is also likely that the meanings of terms such as "culture" and "cultural transmission" have broadened over the past two decades, perhaps inspired by the writings of evolutionary biologists who see culture in all kinds of social activities in all kinds of animal species (e.g., Bonner 1980; Boyd and Richerson 1985). This shift in meaning has led to some unproductive debates in which different researchers use the same words to talk about different things.

In Tomasello (1990) I argued that the most productive way to approach the question of chimpanzee culture is to investigate social learning processes. Different animal species engage in various and different types of social learning, and regardless of which of these we choose to call cultural, it is important to distinguish the different types if we hope to understand the way that different animal species organize their social activities, and the way the young of the species come to enter into these activities. The tools for making such distinctions have only recently become available. There is now accumulating a very large body of research on the social learning of all kinds of animal species, and this has resulted both in a wider range of empirical observations and in more detailed theoretical analyses of the various ways that animals may learn from one another (see the papers in Zentall and Galef 1988). Of special interest in the current context are studies of the social learning of primates, especially monkeys of various types (Cheney and Seyfarth 1990; Visalberghi and Fragaszy 1990), gorillas (Byrne 1992), orangutans (Russon and Galdikas 1993), chimpanzees (Tomasello 1990), and human beings of various ages (Meltzoff 1988). The key theoretical advance made by this group of researchers is the analysis of social learning into a number of related but distinct learning processes, among them: "social facilitation," "response facilitation," "matched dependent learning," "response matching," "local enhancement," "stimulus enhancement," "mimicking," "emulation learning," "action-outcome contingency learning," "imitation," "program-level imitation," "movement imitation," and "imitative learning" (see Whiten and Ham 1992 for a review). Although there is still much work to be done, this analysis now makes available to researchers a variety of distinctions and hypotheses that must be tested before precise learning processes may be attributed to a particular animal species or group.

Within this emerging paradigm, the question of chimpanzee culture may be reformulated initially as a straightforward question about which of various social learning processes chimpanzees employ in their acquisition and maintenance of population-specific behavioral traditions. In addition, however, recent approaches to the study of social learning have found that in order to identify specific types of social learning it is useful, if not necessary, to make cross-species comparisons. When the issue is cultural transmission, a particularly important comparison is of course to humans—quite simply because the concept of culture was specifically formulated to describe group differences in human behavior, and thus human behavioral traditions provide the prototypical case of cultural transmission. In this comparative perspective, then, answering the question of chimpanzee culture in terms of underlying social learning processes is facilitated by, and enlightened by, an explicit comparison of these to the social learning processes underlying human cultural traditions. In any case, my formulation of the question of chimpanzee culture employs both a focus on social learning processes and an explicit comparison to humans.

Given this overall formulation, my strategy in this brief chapter is to address the question of chimpanzee culture in three ways. First, I review recent research on the types of social learning that chimpanzees and human children employ in two domains of particular importance to discussions of chimpanzee culture: tool use and communicatory gesturing. Second, I attempt to clarify some of the issues involved in the study of chimpanzee social learning and cultural traditions through a brief look at some research on chimpanzees raised in humanlike cultural environments. Third, I identify some key characteristics of human cultural traditions and ask if chimpanzee behavioral traditions display these same characteristics.

## Tool Use

There are a number of population-specific tool-use traditions that have been documented for either one or a few different chimpanzee communities, for example, termite fishing, ant fishing, ant dipping, nut cracking, and leaf sponging (see McGrew 1992 for a review). Researchers such as Goodall (1986), Nishida (1987), Boesch (1993a), and McGrew (1992) claim that specific tool-use practices are "culturally transmitted" among the individuals of the various communities, and the first three of these use the process of "imitation" as a key part of their explanation. These pioneering field researchers have not as a matter of course employed all of the distinctions employed by the new wave of social learning researchers, however, and thus it is unclear from the perspective of this new and more differentiated point of view what specific social learning processes are at work.

Animals may learn in social situations in a number of ways. Especially important for current purposes is local or stimulus enhancement in which the behavior of one individual draws the attention of another to a locale or stimulus—which then increases the probability of successful individual discovery and learning. It might be, for example, that one chimpanzee's use of a stone tool to crack open nuts results in her leaving the stone hammer, an anvil, and some nuts and broken nut shells all in one place—which might facilitate individual learning of tool use by groupmates, especially in concert with local enhancement. In fact, based on detailed protocols of individual learning, this is precisely the hypothesis of Sumita et al. (1985) concerning the limited spread of nut cracking in their captive group. This finding is especially important given the many laboratory studies showing that chimpanzees *can* learn to use many types of tools individually, without any observation of others using tools (see Beck 1980 for a review). It is thus possible that in all of the reported cases of the rapid spread of a tool-use behavior (e.g., Menzel 1973b;

Hannah and McGrew 1987), what has happened is that one creative individual has made a discovery, which leaves propitious learning conditions that facilitate the individual discovery of others—sometimes made more salient through processes of local enhancement. This is as opposed to truly imitative learning in which the learner actually copies another's behavior or behavioral strategy. What is needed to tease apart the different possibilities is experimental manipulations.

Tomasello, Davis-Dasilva, Camak, and Bard (1987) trained an adult chimpanzee demonstrator to rake food items into her cage with a metal T-bar. When a food item was in the center of the serving platform, she learned simply to sweep or rake the item to within reach. When the food was located either along the side or against the back of the platform's raised edges, however, she had to employ more complex two-step procedures to be successful. Several young chimpanzees 4 to 6 years old were then exposed to this adult demonstrator as she employed all three of her strategies (the experimental group). Several other chimpanzees in this same age range were exposed to the demonstrator in an unoccupied state throughout (the control group). Results showed that experimental subjects learned to use the tool (after only a few trials in most cases) while control subjects mostly did not. It is important to note, however, that the experimental subjects employed a wide variety of raking-in procedures, and none of them learned either of the demonstrator's more complex, two-step procedures (even though they were trying and failing on these trials over 75 percent of the time). We interpreted these results as demonstrating that the chimpanzees in this study benefited from observation in learning to use the tool but that the form of this learning was not imitative learning, since the experimental subjects did not learn to copy the demonstrator's precise behaviors or behavioral strategies. Observation of the demonstrator, in our interpretation, served to draw the attention of the experimental subjects to the functional significance of the tool in obtaining the out-of-reach food—a kind of functionally based stimulus enhancement. They then used this knowledge of the tool's function in their subsequent individual attempts. We proposed that the social learning process in this case might best be called "emulation learning," since subjects attempted to reproduce the end result they observed (obtaining food), but without copying the behavioral methods of the demonstrator.

In a second experimental study, Nagell, Olguin, and Tomasello (1993) attempted to investigate the emulation hypothesis more directly. A human experimenter presented chimpanzees and 2-year-old human children with a rakelike tool and a desirable but out-of-reach object. The tool was such that it could be used in either of two ways leading to the same end result. For each species one group of subjects observed one method of tool use and another group of subjects observed the other method of tool use. The point of this design was that the stimulus enhancement of the tool was the same in both experimental conditions; only the precise methods of use were different. What we found was that whereas human children in general copied the method of the demonstrator in each of the two observation conditions, chimpanzees used the same method or methods no matter which demonstration they observed. This was despite generally equal levels of tool use and tool success for the two species. It should be noted that the children insisted on this reproduction of adult behavior even when it meant in one of the two conditions that they would be less successful than if they had simply used individual learning strategies, as the chimpanzees apparently did. We concluded from this pattern of results, once again, that chimpanzees in this task were paying attention to the general functional relations of tool and food and to the results obtained by the demonstrator, but they were not attending to the actual methods of tool use demonstrated; they were engaged in emulation learning. Human children focused on the demonstrator's actual methods of tool use (her behavior), and thus they were engaged in imitative learning.

What these studies suggest, I believe, is that chimpanzees and human children understand the instrumental behavior of conspecifics in different ways. For human children the goal or intention of the demonstrator is a central part of what they perceive, and thus her methods of tool use—the way she is attempting to accomplish that goal—become salient. For chimpanzees what is salient is the tool, the food, and their physical relation, with the intentional states of the demonstrator, and thus her methods, being either not perceived or irrelevant.

The major criticism of these experiments is that captive chimpanzees are not representative of wild chimpanzees—they may be "impoverished" in various ways relative to wild chimpanzees—and they thus tell us little about what is going on in the natural setting. One possibility is that captive chimpanzees have been impoverished with regard to tool use, and thus they do not have the capacity to understand through observation how certain tools work, since imitative learning in all species relies on a certain ability to understand the task at hand. However, in the Nagell et al. study the overall amount of tool use and tool success were the same for chimpanzees and children: their skill with the tool was the same; the only difference was in how they used the tool.

Another possibility is that captive chimpanzees are socially deficient relative to wild chimpanzees in ways that affect their social learning. However, in the Nagell et al. study mother-reared and nursery-reared chimpanzees performed in equivalent fashion. Although neither of these rearing conditions is the same as being raised in the wild, this equivalence shows that the lack of imitative learning was not due to some simple factor such as lack of adequate exposure to conspecifics. A related possibility is that chimpanzees do not attend to a human model—used in the Nagell et al. study—as they would to a conspecific. But the Tomasello et al. (1987) study used a chimpanzee demonstrator. While it is true that this demonstrator was not closely related to the subjects (which is also true of the human demonstrator for the children in the Nagell et al. study), it is important to note that there was a strong stimulus enhancement effect for chimpanzees in both of our studies: animals were attracted to the tool if a conspecific or human manipulated it in any way. This suggests that the subjects were indeed influenced in a positive way by demonstrators of both types. Despite all of this, it is still possible, of course, that the captive subjects in these two studies were socially deficient in precisely the dimension of interest, that is, because of their species-atypical social experiences during ontogeny they had not been exposed to the kinds of interactions necessary to develop skills of imitative learning. This is a possibility that cannot be ruled out at this time, but such atypicality was much less true of subjects in the gestural communication studies that I will now report.

## Gestural Communication

In the debate over chimpanzee culture and imitation the focus has been mainly on tool use and other instrumental behaviors, but potentially of just as great importance is gestural communication. Population differences in chimpanzee gestural signaling are well documented and provide evidence that at least some of their gestures are learned (Goodall 1986; Tomasello 1990). The best known of these are "leaf clipping" (Mahale K and Bossou communities) and "grooming handclasp" (Mahale K, Kibale, and Yerkes communities). Inspection of data for individuals in these communities, however, shows that in addition to differences among populations, there are marked individual differences within the communities in

203

how (toward what end) the signal is used. Although data relevant to the learning processes by which individuals might acquire these gestures have not been collected, McGrew and Tutin (1978) and Nishida (1987) have speculated that young chimpanzees might learn them by means of some form of "cultural transmission" such as imitation.

I am aware of no experimental studies of chimpanzee gestural communication in which natural communicatory signals are acquired under different learning conditions, but for the past 8 years we have been observing the gestural signals of a group of chimpanzees at the Yerkes Regional Primate Research Center Field Station in ways that allow for some inferences about learning processes. In a first study, we observed the infants and juveniles of the group (1 to 4 years old), with special emphasis on how they used their signals (Tomasello, George, Kruger, Farrar, and Evans 1985). Looking only for intentional signals accompanied by gaze alternation or response waiting (both indicating the expectation of a response), we found a number of striking developmental patterns. For example, we found that juveniles used many gestures not used by adults, that adults used some gestures not used by juveniles, and that some juvenile gestures for particular functions were replaced by more adultlike forms at later developmental periods. None of these patterns was consistent with the idea that gestural signals are "culturally transmitted" from adults to youngsters.

In a longitudinal follow-up to these observations, Tomasello, Gust, and Frost (1989) observed the same juvenile chimpanzees of the Yerkes group 4 years later (at 5 to 9 years old). In contrast to the hypothesis that imitative learning is the means by which young chimpanzees acquire their intentional gestures, we proposed an alternative hypothesis involving individual conventionalization (Smith 1977). In conventionalization a communicatory signal is created by two organisms shaping each other's behavior in repeated instances of a social interaction. For example, an infant may initiate nursing by going directly for the mother's nipple, perhaps grabbing and moving her arm in the process. In some future encounter the mother might anticipate the infant's desire at the first touch of her arm and so become receptive at that point—leading the infant to abbreviate its behavior to a touch on the arm with response waiting (so-called intention movements). In attempting to test this hypothesis, we noted two main pieces of evidence. First, in some cases there were a number of idiosyncratic signals that were used by only one individual. These signals could not have been learned by imitative processes and so must have been individually invented and conventionalized. Second, in other cases a number of signals used by the young chimpanzees in their interactions with adults involved signals that the youngsters had never had directed to themselves; for example, others never begged food or solicited tickling or nursing from youngsters. The youngsters' signals for these functions thus could not have been a product of their imitation of signals directed to them (so-called second-person imitation), and in many cases it was also extremely unlikely that they were imitated from other infants gesturing to conspecifics (so-called third-person imitation or "eavesdropping") because many were produced in close quarters between mother and child with little opportunity for others to observe.

We have recently completed observations representing a third longitudinal time point on this group (Tomasello, Call, Nagell, Olguin, and Carpenter 1994). In this study, however, a completely new generation of 1- to 4-year-old youngsters was the object of study. In order to investigate the question of potential learning processes, we made systematic comparisons of the concordance rates of all individuals with all other individuals across the three time points over the 8-year period. This analysis revealed quite clearly, by both qualitative and quantitative comparisons, that there was much individuality in the use of gestures, especially for nonplay gestures, with much individual variability both within and across generations. As before, there were a number of idiosyncratic gestures used by single individuals at each time point. Also notable is the fact that there was a fairly large gap between the two generations of this study (4 years), and many of the gestures learned by the youngsters of the younger generation were ones that youngsters of the older generation would have been using infrequently during the crucial learning period. It is also important that many of the gestures common to most or all individuals were gestures that are also learned by captive youngsters raised in peer groups with no opportunity to observe older conspecifics (e.g., Berdecio and Nash 1981). In these cases the commonality is presumably explained by commonalities in learning conditions—all chimpanzees desire to nurse and must do so in the same basic way, for example, and so the gestures they might conventionalize in this context are limited in number (much as human children have only a few basic ways for gesturing to be picked up).

It was concluded from this pattern of results that the young chimpanzees in the Yerkes group were not imitatively learning their communicatory gestures, but rather that they were individually conventionalizing them with one another. The explanation for this is analogous to the explanation for emulation learning in the case of tool use. Like emulation learning, conventionalization does not require individuals to understand the intentions of others in the same way as does imitative learning. Whereas to imitatively learn an "arm-raise" as a play solicit requires that an individual understand the intentions of a conspecific when it raises its arm, conventionalizing an "arm-raise" requires only an anticipation of the future behavior of a conspecific and the ability to translate this anticipation into an instrumental behavior. This contrasts sharply with the imitative learning skills that human children use in acquiring linguistic conventions, which in many cases require, in a very specific way, an understanding of the intentions of others (Tomasello 1992). My conclusion, therefore, is that in this domain as well chimpanzees are displaying much individual inventiveness and creative intelligence, but that they are not employing skills of imitative learning as human children typically do. They do not imitatively learn gestures from one another because they either do not perceive or do not attend to the intentions of others, again, as human children typically do.

The inferences about learning processes made on the basis of these studies of chimpanzee gestural communication are admittedly indirect. Experimental studies are planned. But the compensating factor is that the subjects of these studies have led lives that resemble the lives of wild chimpanzees to a much greater degree than the subjects of the tool-use experiments. The Yerkes Field Station group has been relatively stable over many years, and the physical setting is relatively diverse, so that the environment of developing youngsters in the group has been similar in many ways to the environment of developing wild chimpanzees. This is especially true with regard to peer interaction and play, which seem to occur very much as they do in the wild, and which were the contexts of use for the majority of the signals we observed and analyzed.

## Social Cognition and the Role of Human Interaction

It may be objected that there are a number of very convincing observations of chimpanzee imitation in the literature. It is interesting, however, that almost all of the clearest examples in the exhaustive review of Whiten and Ham (1992) concern chimpanzees that have had extensive amounts of human contact (the major exception being the "limp-walk" reported by de Waal 1982, in which one individual appeared to mimic the walk of an injured groupmate). In many cases this has taken the form of intentional instruction involving human encouragement of behavior and attention, and even direct reinforcement for imitation, that is, social interactions of the type human children are exposed to routinely (e.g., Hayes and Hayes 1952). This raises the possibility that imitative learning skills may be influenced, or even enabled, by certain kinds of social interaction during early ontogeny.

Confirmation for this point of view is provided by Tomasello, Savage-Rumbaugh, and Kruger (1993). In this study we compared the imitative learning abilities of mother-reared captive chimpanzees, enculturated chimpanzees (raised like human children and exposed to a languagelike system of communication), and 2-year-old human children. Each subject was shown 24 different and novel actions on objects, and each subject's behavior on each trial was scored as to whether it successfully reproduced (1) the end result of the demonstrated action and/or (2) the behavioral means used by the demonstrator. The major result was that the mother-reared chimpanzees reproduced both the end and means of the novel actions (i.e., imitatively learned them) hardly at all. In contrast, the enculturated chimpanzees and the human children imitatively learned the novel actions much more frequently, and they did not differ from one another in this learning. Interesting corroboration for this latter finding is the fact that earlier in their ontogeny these same enculturated chimpanzees seemed to learn many of their humanlike symbols by means of imitative learning (Savage-Rumbaugh 1990).

For the issue of chimpanzee culture, these results raise a very important question. Which group of captive chimpanzees is more representative of chimpanzees in their natural habitats: mother-reared or enculturated? Are enculturated chimpanzees simply displaying more species-typical imitative learning skills because their more enriched rearing conditions more closely resemble those of wild chimpanzees than do the impoverished rearing conditions of other captive chimpanzees? Or might it be the case that the humanlike socialization process experienced by enculturated chimpanzees differs significantly from the natural state and, in effect, helps create a set of species-*atypical* abilities more similar to those of humans? There can be no definitive answer to these questions at this time, but it is my hypothesis that a humanlike sociocultural environment is an essential component in the development of humanlike social-cognitive and imitative learning skills. This is true not only for chimpanzees but also for human beings—a human child raised in an environment lacking intentional interactions and other cultural characteristics also would not develop humanlike skills of imitative learning.

The reason this is true, in the current hypothesis, is that the understanding of the intentions of others, necessary for reproducing another's behavioral strategies, develops in, and only in, the context of certain kinds of interactions with others (Tomasello, Kruger, and Ratner 1993). More specifically, to come to understand others in terms of their intentions requires that the learner him- or herself be treated as an intentional agent in which another organism encourages attention to and specific behaviors toward some object of mutual interest—often reinforcing in some manner the learner's successful attempts in this direction. Such interactions are not sufficient, of course, as many animals are subjected to all kinds of human interaction, and even direct instruction, without developing humanlike skills of imitative learning (and the same is true of some human children, e.g., those with autism). The important point for current purposes is that, *in terms of these dimensions of social interaction,* captive chimpanzees raised by conspecifics are a better model for wild chimpanzees than are chimpanzees raised in humanlike cultural environments—since wild chimpanzees receive little in the way of direct instruction from conspecifics (cf. Boesch 1991 for some possible exceptions). It is thus my further hypothesis that the learning skills that chimpanzees develop in the wild in the absence of human interaction (i.e., skills involving individual learning supplemented by local enhancement, emulation learning, and conventionalization) are sufficient to create and maintain population-specific behavioral traditions, but they are not sufficient to create and maintain behavioral traditions displaying the key characteristics of human culture, to which I now turn.

## Human and Chimpanzee Behavioral Traditions

Documenting differences in the social learning and social-cognitive processes of chimpanzees and human beings does not by itself directly address the question of chimpanzee culture, of course. But I would contend that there are observable differences in the population-specific behavioral traditions of the two species that may plausibly be tied to these differences of social learning and social cognition. This connection makes all of the foregoing experimental work relevant to discussions of chimpanzee culture in the wild.

As a subtype in the general category of population-specific behavioral traditions, human cultural traditions may be said to have at least three additional characteristics (Tomasello, Kruger, and Ratner 1993). First, in all human societies there are some traditions that are practiced by virtually everyone in the society; any child who did not learn them simply would not be considered a normal member of the group. This is true of such things as language and religious rituals in many cultures, as well as more mundane subsistence behaviors having to do with food, dress, and the like. We may call this characteristic universality. Second, it is true of many human behavioral traditions that the methods employed by different persons—both within and across generations—show a high degree of similarity. This is true to some degree in concrete tasks such as using a hoe or weaving cloth, although there are often individual idiosyncrasies. In the case of social-conventional behaviors such as linguistic symbols or religious rituals, however, individual discovery and idiosyncratic use are not viable options; these behaviors simply would not be functional unless the methods of the mature users were reproduced somewhat faithfully. This second characteristic may thus be called uniformity. A third characteristic of human cultural traditions derives from the second and is perhaps most telling. Human cultural traditions often show an accumulation of modifications over generations (a.k.a. the ratchet effect). For example, hammerlike tools and the way they are used show a gradual increase in complexity over time in human history as they are modified again and again to meet novel exigencies (Basalla 1988). Thus at the same time that cultural traditions are passed on rather faithfully from one generation to the next, if a modification is made, that modified version is what is passed on to the next generation—so that many human traditions have what may be called a history.

There is some evidence for each of these characteristics in the behavioral traditions of chimpanzees, although the evidence is in many cases not completely convincing, and in all cases the candidate behaviors are very few in number and confined to only one or a few communities. With respect to universality, only a few communities have been observed for long-enough periods of time to come to reasonable conclusions. McGrew (personal communication) reports that virtually all of the noninfant members of the Kasakela community at Gombe fish for termites. It is also likely that virtually all of the noninfant individuals in the Mahale K community fish for ants (Nishida, personal communication). And virtually all of the physically capable members of the Taï Forest chimpanzee

209

group engage in nut cracking (Boesch 1993a. It is important to note, however, that the "grooming handclasp" and "leaf-clipping" behaviors of the Mahale K group are practiced by only some individuals (Nishida, personal communication). On the basis of their survey of the innovative behaviors of a number of primate species, including chimpanzees, Kummer and Goodall (1985) conclude that "only a few will be passed on to other individuals, and seldom will they spread through the whole troop" (p. 213). My overall conclusion is thus that the evidence for universality in some of the behavioral traditions of some chimpanzee communities is strong, but for any given community there are at most only one or a few population-specific behavioral traditions that all members engage in. This differs, at least quantitatively, from the many human traditions that are universal for particular cultures.

With regard to uniformity, a number of keen observers have remarked on the fact that individual chimpanzees often use their own creative techniques in all kinds of concrete tasks from termite fishing to nut cracking (e.g., Goodall 1986; Hannah and McGrew 1987; Sumita et al. 1985). This is as it should be, as slavishly copying the techniques of others is not always the best strategy in solving instrumental tasks (see Byrne 1992 on program-level imitation). More telling for current purposes, however, is chimpanzees' use of idiosyncratic techniques in social-conventional behaviors for which humans show such marked uniformity (e.g., in their choice and use of particular linguistic symbols within cultures). Much individuality has been reported, for example, in the communicatory gesturing observed in wild populations (Goodall 1986), and our own studies of the Yerkes group confirm the prevalence of marked individuality. These idiosyncrasies involve not only the way in which a particular signal is effected by individuals but also the use or nonuse of particular signals in particular contexts. Thus, despite the fact that there is no simple metric by which we might easily compare the degree of uniformity of behavior within groups and across species, there would once again seem to be a limited number of learned chimpanzee behaviors that show a widespread uniformity across all the individuals of a community—especially not strict uniformities of the type we see in human languages and other social-conventional behaviors.

Finally, in the case of practices with histories, Boesch (1993a) reports two examples that he believes show the ratchet effect in the chimpanzee group of the Taï Forest. To cite the clearest example: for some time adult males used leaf clipping only in the drumming context to indicate something about their movements, but then they began using it in the resting context as well. Some other group members then began using it in the resting context, some for the first time and only in this context. It is important to note in this observation, however, that what changes is the context of use of the behavior. There is no change—much less an accumulation of modifications—in the leaf-clipping behavior itself. This behavior thus seems, to me at least, a very weak candidate for the ratchet effect. It is possible, of course, that we simply have not observed chimpanzees in their natural habitats for long enough to know whether some of their practices do show the ratchet effect, since many human cultural traditions remain unchanged for long periods of time. Nevertheless, it may be said once again with some confidence that there are at best only a very few chimpanzee behaviors that have histories in the sense of an accumulation of modifications over time. It may be that there are none.

It is important that the differences observed in human and chimpanzee behavioral traditions, especially along the three dimensions just outlined, may very plausibly be attributed to differences in the social learning processes employed by the two species, especially those of the type documented by myself and others in experimental contexts—including also processes of instruction (Heyes 1993). It is my hypothesis that human cultural traditions have universality, uniformity, and history because human beings learn from one another in specific ways that lead to a high degree of fidelity of "transmission"—they learn to use their tools and symbols through imitative learning and instruction, for example (Tomasello, Kruger, and Ratner 1993). Chimpanzee societies have a "looser" structure because when they learn from one another, they employ to a much greater extent processes of socially enhanced, individual learning such as emulation learning and conventionalization (Tomasello 1990). My view is thus that the attribution of culture to chimpanzees has some basis in empirical fact: chimpanzees rely on social learning in some behavioral domains, and the behavioral traditions in those domains in some cases show some signs of universality and uniformity. There are nevertheless important differences in both social learning processes and the characteristics of their behavioral traditions (e.g., in terms of their histories), and these differences are precisely of the type that would lead to the unique cultural products of human societies, such as languages, complex tool use, and other cultural traditions and institutions.

## The Evolution of Culture

I am perfectly happy to say that chimpanzees have cultural traditions, as long as we may then go on to specify how these are similar to and different from the cultural traditions of other animal species, especially humans. It is my preference, however, to use other terms because human cultural traditions seem to have some unique qualities—they show universality, uniformity, and history in a manner and to a degree that makes them seem qualitatively different from the behavioral traditions of other species. It is very likely, in fact, that many species have behavioral or cultural traditions that show their own unique stamp. How the behavioral/cultural traditions of chimpanzees are similar to and different from those of other primate species is a question we have just begun to investigate systematically.

Following Galef (1992)—from whom I borrowed the title of this chapter—I also think there is a way to pose the question of chimpanzee culture that avoids much of this definitional indeterminacy. Quite simply, we may ask whether the behavioral traditions of various animal species are phylogenetically analogous or homologous. In the current context the question is whether some 5–10 million years ago the common ancestors of chimpanzees and humans possessed cultural traditions displaying universality, uniformity, and histories. I believe that they did not. And in fact much of the evidence from the study of human prehistory is that cultural traditions of the clearest kind—a full-blown language and other means of symbolic expression—may be of relatively recent origin, perhaps emerging only with *Homo sapiens sapiens* less than 100,000 years ago (e.g., see the papers in Mellars and Stringer 1989; Davidson and Noble 1989). Even if human cultural traditions as I have defined them emerged in earlier periods of human prehistory, as some researchers believe, that is still well after the divergence of the two species. In combination with the fact that current experimental and observational evidence suggests that chimpanzee and human social learning and social cognition are different as well, I believe that the evolutionary evidence supports the view that human and chimpanzee behavioral traditions are only analogous. The social-cognitive adaptations on which human culture and cultural learning depend came only after the differentiation of the two species.

212

It was true in 1994, and it is still true today, that most natural chimpanzee groups have a smallish number of population-specific behavioral traditions. It was true in 1994, and it is still true today, that these share some features with human cultural traditions. But science progresses, and we now know much more about how and why this is so—even though there is still much to be learned, of course. In this brief postscript I update my 1994 chapter by reporting what are, in my opinion, the most important new findings and assessing where we stand today on the question of chimpanzee culture.

## Chimpanzee Behavioral Traditions

In 1999 Whiten et al. reported the results of discussions among the major chimpanzee fieldworkers relevant to the question of chimpanzee culture. These fieldworkers reported observations of interesting chimpanzee behaviors and checked whether they occurred at other field sites. On the basis of these discussions and some systematic published data, several dozen population-specific behavioral traditions were identified as "cultural"— meaning that they were used by most members of a population, were not used by most other populations, and were most likely due to social learning (because they did not seem to be due to ecological factors).

Everyone agrees that this was an important first step in characterizing the nature of chimpanzee behavioral traditions, and work of this nature continues (by Whiten, Boesch, and others). Perhaps the most difficult issue is how to deal with behaviors that are widespread in some but not all populations, with these populations being so widely dispersed that there seem to be multiple origins for the behavior. This turns out to be characteristic of two of the best-known and best-studied chimpanzee traditions.

213

First, the so-called grooming handclasp (McGrew and Tutin 1978) has arisen in several populations independently, including at least one in captivity not even on the African continent (de Waal and Seres 1997). Second, nut cracking was always thought to occur only in West Africa on the west side of the Sassandra River, but it has recently been found 1,700 kilometers to the east, with many non-nut-cracking populations in between (Morgan and Abwe 2006). The most plausible explanation is that we are dealing here with behaviors that are fairly readily invented by individuals, and they spread within groups by some form of social learning—with the within-group spreading being facilitated in some way by the ease of individual invention.

But such ecological surveys are only a first step. To determine the nature of the traditions, more systematic studies are needed. The best example of this is the study by Humle and Matsuzawa (2002) on the ant dipping of the Bossou chimpanzees. Ant dipping was at one time used by many fieldworkers as the best example of chimpanzee "culture" because it involved different groups engaging in the same basic foraging activity: poking sticks into ant nests to capture and eat ants. Chimpanzees at Taï and chimpanzees at Gombe dip for the same species of ant using different techniques: at Taï they use shorter wands and bite the ants off the wand directly (Boesch and Boesch 1990), whereas at Gombe they use longer wands and typically (though not always) pull the ants off it with their other hand before eating them (McGrew 1974). Humle and Matsuzawa observed that the chimpanzees at Bossou sometimes used both techniques. The choice of technique was driven in the first instance by the length of the wand: biting from shorter ones and pulling ants off longer ones. In turn, the length of the wand was driven mainly by the aggressiveness of the ants—different species of ants were differentially aggressive, and all ants were more aggressive at the nest than when on the move— such that longer tools were used (to avoid being bitten) when the ants were more aggressive. Importantly for determining the source of these behaviors, when investigators compared the techniques used by three mother-infant pairs in different situations, there was no relationship.

Humle and Matsuzawa's work is perhaps the most important field study for the question of chimpanzee culture ever done because it is the first study to systematically compare different techniques for doing things to different ecologies on a population-wide basis. Field surveys are important for identifying candidate behaviors, but more such systematic, quantitative studies are needed if we are to characterize the source and nature of chimpanzee behavioral traditions more precisely. I am convinced by the existing data that some chimpanzee behavioral traditions will survive such scrutiny and be shown to be basically independent of local ecology. But this must be demonstrated systematically and quantitatively for each case independently. Another important method is exposing captive individuals to materials from the wild and seeing what they do with them. Thus Huffman and Hirata (2004) found that giving medicinal leaves—whose use was thought to be socially transmitted in the wild—to naïve individuals in captivity resulted in several using them in ways similar to those of wild chimpanzees, thus undermining the social transmission hypothesis.

It is also interesting and important that behavioral traditions of this same general type have now been reported by fieldworkers who are investigating many other animal species, both primate and nonprimate, for example, orangutans (van Schaik, Ancrenaz et al. 2003), capuchin monkeys (Perry, Baker et al. 2003), and whales and dolphins (Rendell and Whitehead 2001), among others. This raises the question how the naturally occurring behavioral traditions of chimpanzees compare with these and whether the chimpanzee versions are any closer to human cultural traditions than are those of other mammalian species.

## Chimpanzee Social Learning

Following Tomasello (1990), in the 1994 chapter I argued that it was not very productive to wrangle over the definition of culture. It would be more productive to focus on the social learning processes that underlie behavioral traditions and that give them their particular characteristics with regard to such things as how they are transmitted and maintained across generations. Humle and Matsuzawa's field study is important because it speaks directly to the learning processes involved.

Since 1994, we have learned much more about chimpanzee social learning and the cognitive skills that underlie it, mostly from laboratory experiments (see Whiten et al. 2004 for a review). Importantly, my particular hypothesis from 1994—that chimpanzees do not engage in true imitative learning because they do not analyze behavior in terms of its goals as differentiated from its behavioral means—has been proved false. A number of studies now have documented that chimpanzees do understand the actions of others in terms of goals; specifically, they differentiate intended from accidental actions that both produce the same external

215

result, and they differentiate trying (and failing) to transform an object from acting on it in similar nontrying ways (Call and Tomasello 1998; Call et al. 2004). And at least some individuals use this understanding in social learning situations; Tomasello and Carpenter (2005) found that young, enculturated chimpanzees reproduce only intended and not accidental actions, and they produce a demonstrator's desired outcome even when the demonstration was of a failed attempt (and they do not do this in other control conditions). But although this study does show that chimpanzees can focus on a demonstrator's internally represented goal in a social learning situation—her desired outcome rather than the actual outcome—it does not speak to whether they are concerned with the behaviors or behavioral techniques used as means to achieve those goals.

The most systematic program of research over the past dozen years or so is that of Whiten and colleagues. Whiten et al. (1996) showed that chimpanzees will choose the way of opening a box they observe rather than some other plausible way of opening it, and Whiten et al. (2005) even showed that other observing chimpanzees will follow the original learner in a "transmission chain" across individuals (see also Horner et al. 2006). These studies thus demonstrate a plausible mechanism for the social transmission of behavioral traditions in wild chimpanzee populations, at least for those that involve tools and/or other concrete objects—a not-inconsiderable achievement. But two facts about these experiments are important. The first is that they leave open the possibility that individuals are learning about how the box works—perhaps supplemented by an understanding of the demonstrator's goal—without attending much or at all to the behavioral techniques used. Indeed, in Whiten et al.'s (1996) study clear results emerged only when investigators looked at the demonstrator's and learner's behavior in terms of the result it produced on the box, not in terms of particular modeled actions. It is thus an open question whether the apes would have learned the same thing if they had simply observed the box opening itself in a particular way without any demonstrator (see Tennie et al. 2006 and Hopper et al. 2006, for use of this so-called ghost control condition).

The second important fact is that the Whiten et al. (1996) study also had a comparison group of 3-year-old human children, and they produced the demonstrated actions much more faithfully than did the chimpanzees. This result was corroborated by Call et al. (2005), who found that chimpanzees preferentially focused on the outcomes of problem-solving activities, whereas human children preferentially focused on the actions of a demonstrator. Most important, Horner and Whiten (2005) found that observer chimpanzees tended to ignore irrelevant actions on a box when their causal ineffectiveness was clear, but tended to produce them when their causal effectiveness was unclear—again suggesting that chimpanzees are focused to some degree on the desired outcome (the goal) of the demonstrator in assessing what they themselves should do to solve the problem. But in this study, as well as in the Nagell et al. (1993) study reported in my 1994 chapter, the human children paid much more attention to the actions of the demonstrator, even ignoring the apparent causal relations that governed the problem in order to imitate the adult—not an intelligent strategy, perhaps, but one more focused on demonstrator actions. In all the studies in which chimpanzees and human children have been compared, therefore, the clear result is that the human children are much more focused on the actual actions of the demonstrator, whereas the chimpanzees are much more focused on the outcome of her actions—either the actual outcome (the result) or the desired outcome (her goal). Indeed, at the moment the nonhuman species that have demonstrated the most attentiveness to actions are various bird species (e.g., Zentall 2004).

Further support for this interpretation of results comes from studies in the other behavioral domain highlighted in my 1994 chapter, namely, gestures. The most important new study in this domain is that of Tomasello et al. (1997), who systematically compared the gestures of two different groups of captive chimpanzees (with extensive longitudinal data on one group available as well). In brief, there was no evidence for the social transmission of gestures within groups, because there were just as many differences among individuals within each group as between the two groups. In addition, Tomasello et al. reported an experiment in which one chimpanzee was taught a novel gesture and put back in the group to demonstrate it (on two different occasions using two different gestures and demonstrators). The other members of the group did not pick up this gesture, which again suggests that chimpanzees do not socially transmit their gestures, but rather they are learned individually via ritualization. It is possible that individuals raised or trained by humans might imitate gestural actions, since Custance et al. (1995) were able to train individuals over a period of several months to reproduce some demonstrated actions in the so-called do-as-I-do paradigm, and Tomasello, Savage-Rumbaugh, and Kruger (1993) found that enculturated apes were much better at following demonstrations of actions on objects than were unenculturated apes.

217

A reasonable new hypothesis to explain all the recent data on chimpanzee social learning might be as follows (see also Call and Carpenter 2002). Chimpanzees are able to understand to some degree the goal of a demonstrator's action, and as observers they tend to focus on that goal or else the actual outcome and pay little attention to the actions designed to achieve that goal. Being raised and/or trained by humans can lead chimpanzees to focus more on actions. Human children—perhaps through some kind of adaptation for cultural learning in general, which includes many "arbitrary" actions such as linguistic conventions for which reproducing exact behaviors is crucial—focus much more on the actions involved. It is important to note, however, that children also focus quite a bit on outcomes in concrete problem-solving situations (Nagell et al. 1993; Tennie et al. 2006), and so one might actually say it this way. In observing instrumental actions apes in general, including humans, tend to focus on the outcome, either produced or intended, but in some cases they analyze the action backward to the behavioral technique used to see how that outcome was achieved; human children engage in such analysis more naturally, more frequently, and perhaps more skillfully than do chimpanzees.

## Comparison with Humans

As I argued in the original chapter, because the whole concept of culture was invented to describe human populations, the comparison of chimpanzees with humans is important for a full understanding of chimpanzee behavioral traditions and social learning. In that chapter I argued that differences between the species in social learning processes lead to different observable characteristics of the two species' behavioral traditions. In particular, human behavioral traditions have a cumulative history, and some of them show a kind of "ratchet effect" of accumulating complexity over time. There is still, to my mind, no convincing demonstration of the ratchet effect or any other form of cumulative cultural evolution for chimpanzees or any other nonhuman animal (contra Matsuzawa and Yamakoshi 1996).

I still believe that the explanation for this difference in social traditions crucially involves processes of social learning and interaction, but I would now modify my original account in the following way. Whereas previously I thought that there was a distinct qualitative difference in the imitative learning of humans as compared with the emulation learning of chimpanzees, I now see this as more a matter of degree. As outlined earlier, humans seem to be more focused on actions than are chimpanzees, which are mainly focused on outcomes and goals. In addition, however, I would now also emphasize more the role of teaching. Teaching has always been an important part of my account (Tomasello 1990; Tomasello, Kruger, and Ratner 1993; Tomasello 1999a), but I now think that it is even more important in the human case, especially for those kinds of cultural conventions that cannot be invented on one's own but can only be imitated. Gergely and Csibra (2006) have recently elaborated an account that explains why the existence of relatively "opaque" cultural conventions (there is no causal structure, or else it is difficult to see this structure) requires that human adults be specifically adapted for pedagogy toward children and human children be specifically adapted for recognizing when adults are being pedagogical and imitating them (what Tomasello, Kruger, and Ratner 1993 called instructed learning). There has been no systematic study of chimpanzees engaged in anything resembling teaching since the observations of Boesch (1991), which have multiple interpretations, as discussed in the original chapter.

Finally, I would like to add two other ingredients. The first derives from an original analysis of Uzgiris (1981). She noted that human children imitate not only in order to acquire more effective behavioral strategies in instrumental situations but also for purely social reasons—to be like others. The tendency of human beings to follow fads and fashions and to conform are well known and well documented, and the proposal here, following Carpenter (2006), is that this represents a different and important motivation for social learning that may produce qualitatively different behaviors. For example, human infants have a greater tendency than do chimpanzees to copy the unnecessary "style" of an instrumental action (Carpenter and Tomasello in prep., and in acquiring linguistic conventions children are driven not just by communicative efficacy but also by the desire to do it as others do it (Tomasello 2003). This analysis would also explain why children in the studies cited earlier sometimes tended to imitate poor demonstrators when it would have been to their advantage to ignore them, and, in general, why children copy the actual actions of others more readily than do other apes. This so-called social function of imitation—simply to be like others—is clearly an important part of human culture and cultural transmission.

The second added ingredient was suggested by Bruner (1993), who emphasized that human culture persists and has the character it does not just because human children do what others do, but also because adults expect and even demand that they behave in certain ways: children understand that this is not just the way in which something *is* done but rather the way in which it *should* be done. In a recent study Rakoczy et al. (2008) found that 3-year-old human children not only copied the way that others did things, but when they observed a third party doing them in some other way, they objected and told them that they were doing it "wrong." Kelemen (1999) has also shown that young children learn very quickly that a particular artifact is "for" a particular function, and other uses of it may be considered "wrong." This normative dimension to human cultural traditions serves to further guarantee their faithful transmission across generations in a way that supports further ratcheting up in complexity across historical time.

It may very well be, then, that it is these three additional processes— teaching, social imitation, and normativity—along with a tendency of learners to focus on actions, as well as goals and outcomes, that give human cultural traditions their extraordinary stability and cumulativity.

## Conclusion

Science progresses, and it is heartening to see such progress on the question of chimpanzee culture in the last dozen or more years. At least one hypothesis has fallen, much more is known about specific social learning processes, and, in general, there is much more consensus about basic issues than there was in 1994. This is not to say that there is total agreement, because I am sure that there are people who think that I am still being too hard on the chimpanzees. Indeed, I am among those who are regularly accused of "raising the bar" on chimpanzees; that is, as soon as we discover that something we thought was a human-chimpanzee difference turns out not to be (e.g., understanding goals), we then posit something else as different and uniquely human (e.g., social imitation, normativity). But this raising of the bar results from the simple fact that there are observable differences between chimpanzee and human societies in terms of such things as complex technologies, social institutions, and symbol systems, and those must be explained. If some hypothesis about these differences is wrong, then it is rightfully consigned to the trash heap. But then we must come up with some new hypothesis to explain the difference—and there is a difference.

ACKNOWLEDGMENTS

"The Question of Chimpanzee Culture" by Michael Tomasello was originally published in *Chimpanzee Cultures*, edited by Richard W. Wrangham, W. C. McGrew, and Paul G. Heltne, published by Harvard University Press, copyright © 1994 by the Chicago Academy of Sciences, and is used by permission. Thanks to Claudio Tennie for helpful comments on an earlier version of the postscript.

# 10

# CULTURE IN ANIMALS?

BENNETT G. GALEF

This chapter focuses on two questions: (1) What kinds of evidence suffice to establish that social learning contributes to maintenance of behavioral variation among allopatric populations of a species, and (2) given current evidence, should we accept the hypothesis that traditions observed in free-living, nonhuman animals (hereafter animals) are precursors of culture in *Homo sapiens*?

## Part 1: Do Animals Have Traditions?

Criticism . . . and the doubt out of which it arises are the prior conditions to progress of any sort.

(Wylie 1942)

All scientific hypotheses should be poked and prodded, tested and retested, and made to stand up to the available observations.

(White 2006, p. 472)

Titles of recent publications in prestigious journals boldly assert that differences in the behavioral repertoires of allopatric populations of a single species demonstrate culture in animals (e.g., Whiten et al. 1999, 2001; van Schaik, Ancrenaz et al. 2003). This use of the term "culture" to refer to variability in behavior among allopatric populations of animal species has proved contentious (Whiten 2005).

Some with an interest in animal behavior, often those trained in fields other than primatology, prefer to describe as cultural only behaviors in animals that result from the same behavioral processes presumed to support culture in humans. Tuition of the ignorant by the knowledgeable and

222

learning by imitation are frequently mentioned sources of culture in humans that appear to play, at best, a very limited role in the behavioral development of other animals (e.g., Galef 1992, 1996; Tomasello 1994, 2000, this book). Such emphasis in definitions of culture on behavioral variation resulting from imitation and tuition is based on logical arguments that indicate that only imitation and teaching can support the cultural "ratcheting" (the gradual accumulation of socially learned variants) that seems to many to be a critical characteristic of human cultures (e.g., Heyes 1993; Tomasello 1994; Richerson and Boyd 2005). This apparent inability in principle of the types of social learning that are commonly seen in animals to support cumulative behavioral change in a population provides the rationale for focusing attention on the social learning processes in discussions of animal "culture."

### Types of Social Learning

Like imitation, local enhancement (defined as "apparent imitation resulting from directing the animal's attention to a particular object or particular part of the environment"; Thorpe 1963, p. 134) can lead to uniformity in behavior among members of a group. However, although local enhancement directs the attention of individuals toward a particular portion of the environment local enhancement leaves each individual to then develop de novo its responses to that aspect of the environment that others have made salient. Consequently, local enhancement cannot produce cumulative behavioral change in a population (Heyes 1993).

Emulation, like local enhancement, does not involve learning about the particular behaviors in which others engage; rather, emulation involves learning about outcomes of those behaviors. For example, a focal individual may learn that a tool can be used to gain access to food by observing another use the tool to produce food, and the focal individual may then attempt to produce the same outcome by devising its own way of using the tool (Tomasello et al. 1987; Tomasello 1998, this book). A naïve individual's attention is directed toward a goal and, possibly, to the utility of a tool in achieving that goal, but as in the case of local enhancement, the individual must develop de novo its own means of achieving the goal, so again no ratcheting is possible.

In contrast, imitation, defined as "learning to do muscular acts from seeing [others] do them" (Thorndike 1898, p. 76), allows an individual to acquire the details of behavior developed by others (Heyes 1993). An imitator can directly incorporate a model's learned motor patterns into its own behavioral repertoire and then innovate, using the model's behavior as a platform or scaffolding (Wood et al. 1976) upon which to elaborate. Thus imitation (and a similar argument can be made concerning tuition) can support cumulative culture, whereas other forms of social learning cannot.

It has been proposed that "mechanisms are of secondary importance" (de Waal 1999, p. 636; de Waal and Bonnie, this book) in defining culture. I disagree. Distinguishing homology from analogy requires attention to details of mechanism, as well as to function, and distinguishing analogy from homology is critical to exploring hypotheses concerning the evolution of traits, whether those traits are morphological or behavioral.

The issues raised in the preceding paragraphs will become particularly important in the second part of this chapter, "Animal Tradition; Human Culture." I mention them here because analysis of the behavioral mechanisms that support social learning and traditions in animals has been the foundation of my own approach to the study of behavioral traditions and colors all that follows.

### Social Learning and "Culture"

Despite considerable controversy concerning use of the term "culture," all students of animal behavior seem to agree that involvement of social learning of some sort in the development of a behavior is a necessary condition for it to be defined as "cultural" (e.g., Galef 1988; Whiten and Ham 1992). Consequently, an obvious first step in discussion of "culture" in animals (which, for reasons explicated in part 2 of the present chapter, I shall refer to hereafter as "tradition") is to establish which purportedly traditional behaviors involve social learning in their development. If there is no social learning, there is no tradition. Unfortunately, although establishing that social learning has played a role in development of a behavioral variant is critical in determining whether that variant is traditional, showing that social learning is involved in development is often more difficult than it appears to be at first glance.

### Animal Traditions

In the following sections I discuss several purported instances of traditions in animals that illustrate the difficulty of determining whether social learning has actually played a role in development of patterns of behavior observed in some populations and not in others. I also indicate areas where real progress has been made in recent years in determining that some purported animal "traditions" are in fact traditional. As implied in the quotations at the start of this part, understanding comes from questioning the apparently obvious, not from uncritical acceptance of the richest interpretation of observations.

## Animals with Small Brains

Although free-living great apes and cetaceans provide many of the most intriguing examples of behavioral variation among allopatric populations that have been labeled "cultural," I shall begin by considering two examples less emotionally and politically charged than those that involve charismatic animals with unusually large ratios of brain to body weight. The first example explores in some detail the difficulty of determining whether social learning is important in generating population-specific behaviors in a free-living species even when its ecology and genetic structure are well described, if little is known about the development of behaviors of interest. The second example illustrates how, at least under special circumstances, a role for social learning in development of a behavior hypothesized to be traditional can be demonstrated conclusively.

### FINCHES OF THE GALÁPAGOS

*Taxonomy.* The Galápagos Archipelago is home to a unique group of 14 species of finch, the Geospizinae, that have been the subjects of innumerable studies of adaptive radiation. The most morphologically diverse of the Geospizinae is the sharp-beaked ground finch *(Geospiza difficilis).* On the basis originally of measurements of body parts of adult males and more recently of microsatellite DNA analyses, the *G. difficilis* resident on the two northernmost of the Galápagos Islands, Wolf Island and Darwin Island, have been classified as a distinct subspecies *(G. d. septentrionalis).* *Septentrionalis* males have longer wings and longer, more tapered beaks than males of the two other subspecies of *G. difficilis* (Lack 1969; Schluter and Grant 1984; Grant et al. 2000).

*Ecology.* *Septentrionalis* finches differ from other *G. difficilis* finches in more than genotype and morphology. Wolf and Darwin islands are relatively low lying and dry, and both lack forested areas (Grant et al. 2000). The islands also have no avian predators (owls and hawks) typically found elsewhere in the Galápagos (Bowman and Billeb 1965). Further, *G. difficilis* finches on Wolf and Darwin islands are the only members of their

225

species to inhabit islands where *Opuntia* cactus grow and ground finches that specialize in feeding on cactus (*G. scandens* and *G. conirostris*) are not present (Grant 1986).

*Behavior.* The behavioral repertoires of *G. difficilis* finches on Wolf and Darwin islands are quite different from those of others of their species. *Septentrionalis* finches are unusually tame (Bowman and Billeb 1965) and are the only *G. difficilis* finches to either eat *Opuntia* cactus regularly or probe *Opuntia* flowers for nectar and pollen (Lack 1969).

More spectacularly, only *G. difficilis* finches on Wolf and Darwin islands feed on the blood of living seabirds. The finches land on the backs of boobies (large white-bodied, dark-winged seabirds of the genus *Sula* that nest throughout the Galápagos), peck at the base of the boobies' wing feathers, and feed on blood that seeps from the wounds that they have created (Bowman and Billeb 1965; Figure 10.1). Also on Wolf and Darwin islands, but not elsewhere, *G. difficilis* finches pierce seabirds' eggs and eat the eggs' contents (Bowman and Billeb 1965; Koster and Koster 1983; Schluter and Grant 1984). Several *septentrionalis* finches were even once filmed "working together" to roll an egg 3 meters from a booby's nest, knocking the egg against a rock until it cracked and then feeding on it (Koster and Koster 1983, pp. 6–7).

If observation of patterns of behavior without obvious ecological correlates restricted to a single population of a species provides evidence of culture (Whiten et al. 1999), then *G. difficilis* finches are "cultural" birds indeed, only one of many avian species that have been reported to exhibit population-specific patterns of foraging behavior (e.g., Lefebvre and Bouchard 2003; Emery 2006). For example, the tools that New Caledonian crows *(Corvus moneduloides)* manufacture and use to forage for insects vary from one part of the island to another (Hunt 1996, 2000; Hunt and Gray 2002), and carrion crows *(Corvus corone)* in Japan have learned to use automobiles as nutcrackers (Nihei and Higuchi 2001).

*Blood feeding.* Only a handful of the several thousand extant avian species take blood from living animals. Three of that handful, *G. difficilis* and two species of mockingbird (*Nesomimus parvulus* and *N. macdonaldi*), are indigenous to the Galápagos. Consequently, it might be suspected that the mockingbirds learned blood feeding from the finches or vice versa by cross-species social learning, a mechanism for transmission of innovative behaviors

*Figure 10.1* A sharp-beaked ground finch *(G. difficilis)* on Wolf Island feeding on the blood of a seabird. Copyright David Parer and Elizabeth Parer-Cook/AUSCAPE. Modified and printed with permission.

that was widely accepted in the nineteenth century (e.g., Romanes 1882) but has seldom been considered since then (for exceptions, see Werner and Sherry 1987; Carlier and Lefebvre 1997; Lefebvre et al. 1997; for review, see Seppanen et al. 2007). However, although Galápagos mockingbirds and G. *difficilis* finches are coresident on six islands (Bowman and Billeb 1965; Grant et al. 2000), the finches feed on blood only on Wolf and Darwin islands (Bowman and Billeb 1965), and the mockingbirds only on Española and Santa Fé islands (Curry and Anderson 1987), where G. *difficilis* is currently absent.

*Discussion.* The relatively simple and well-described ecology of the Galápagos, together with detailed knowledge of the genetics, morphology, and behavior of G. *difficilis,* should make it relatively easy to determine which, if any, of the unusual behaviors of *septentrionalis* are in some way socially learned. However, without direct evidence that social learning plays a role in development of the unique behaviors of *septentrionalis,*

227

we are left to guess at the causes of observed behavioral diversity in *G. difficilis*.

Probably, *septentrionalis* feeding on *Opuntia* can be attributed to the absence of cactus-feeding competitors on Wolf and Darwin islands (Lack 1969). Removal of specialized competitors often results in "character release" in the less specialized of two sympatric species (e.g., Robinson et al. 1993). Of course, it remains possible that social learning has promoted diffusion of *Opuntia* feeding among *G. difficilis* on Wolf and Darwin islands.

Other ecological factors may play a role in emergence of blood feeding on Darwin and Wolf islands. Perhaps blood feeding is a response to demands of the low-lying, dry habitat where *septentrionalis* finches live. However, if so, explaining why only mockingbirds that reside on relatively elevated and damp Española and Santa Fé islands feed on blood is difficult, as is the observation that *G. difficilis* finches that live on Genovesa Island, which is relatively dry and low lying, do not feed on the blood of seabirds that nest there.

Perhaps the exceptional tameness of *septentrionalis* finches (possibly reflecting relaxed selection for wariness on predator-free Wolf and Darwin islands) allows *septentrionalis* finches to approach seabirds and feed on ectoparasitic hippoboscid flies that live among the seabirds' feathers. Accidental puncture of a seabird's skin while hunting flies could lead individuals to learn independently to feed on blood (Bowman and Billeb 1965).

Perhaps all *G. difficilis* finches have blood feeding in their behavioral repertoires, but seasonal loss of alternative sources of protein on Wolf and Darwin islands releases the behavior. Possibly some other ecological difference between the islands is critical in expression of blood feeding. Perhaps an interaction between the unique genotype of *septentrionalis* and the ecology of Darwin and Wolf islands results in blood feeding. Perhaps the unusually long bill of the finches on Wolf and Darwin islands is necessary to puncture the skin of seabirds or their eggs. Perhaps "the habit [of feeding on blood] is . . . learned and is transmitted by tradition from one generation to the next" (Bowman and Billeb 1965, p. 42). Perhaps many things. We just do not know. Simple observation of differences in the behavioral repertoires of allopatric populations of a species cannot, in itself, provide strong support for the hypothesis that social learning played a role in development of behavioral variants.

*Potential routes to understanding.* Cross-fostering of nestling *G. difficilis* finches from Wolf and Darwin islands and nestlings from one of the four islands where both *G. difficilis* and seabird colonies are found, but blood feeding is not, could prove informative. If some cross-fostered chicks behaved like their natural rather than their adoptive parents, then blood feeding could not be a result either of social learning or of interaction with particular environments. If cross-fostered chicks behaved like their adoptive parents, then genetic explanations of blood feeding would be excluded, but effects of social learning and environment would not be distinguished.

Nestling *G. difficilis* finches could also be transferred from Wolf and Darwin islands to low-lying, dry Genovesa Island, where no blood feeding occurs. If such transfers, but not those of nestlings from Wolf and Darwin islands to moist islands, resulted in blood feeding, evidence would be provided of a genotype-environment interaction responsible for development of the behavior.

Potentially most conclusively, adult *septentrionalis* finches could be transferred both to Genovesa and to one of the three other islands where *G. difficilis* finches are found, but blood feeding is not. If some members of populations of *G. difficilis* that received immigrants from Wolf and Darwin islands began to feed on the blood of seabirds, social learning would be proved to be an important contributor to current distribution of the behavior.

If, as is likely, moving *G. difficilis* finches from island to island proved impossible, options are more limited, but useful work could still be done. A small experiment carried out during the dry season (when arthropod prey are relatively infrequent on Wolf and Darwin islands) to examine the effect of providing high-protein liquids to some *septentrionalis* finches on frequency of occurrence of blood feeding might prove informative. If provisioned birds stopped harassing sea birds, while unprovisioned birds did not, an important ecological contribution to expression of blood feeding could be inferred.

Further, observations that cast light on the development of idiosyncratic feeding behaviors can be useful in identifying truly traditional behaviors. For example, individual Cocos finches *(Pinaroloxias inornata),* the only Geospizine found outside the Galápagos Archipelago, specialize in different foraging behaviors; some feed predominantly on insects gleaned from leaves, others on insects gleaned from branches, and so on. Foraging preferences are not correlated with either the time or place where foraging occurs or the age, sex, or morphology of the forager (Werner and Sherry 1987). Rather, Cocos finches appear to learn their idiosyncratic feeding behaviors socially. "Throughout the year, we repeatedly observed ($n = 20$) a juvenile finch follow an adult . . . and alternately watch the adult, then imitate its feeding, often in precisely the location vacated by the adult" (Werner and Sherry 1987, p. 5509). Providing a desirable food to Cocos finches in a container that required an unlikely behavior to open it and then observing spread through a marked population of container-opening behavior could provide evidence of a capacity for social learning of foraging specializations in the species but would not, of course, show that any naturally occurring foraging variant was in fact socially learned.

### WHAT CAN WE LEARN FROM ROOF RATS?

The power of controlled experiments to determine the causes of purportedly traditional behaviors is beautifully demonstrated in a series of studies by Terkel and his colleagues.

*Observation.* Aisner and Terkel (1992) discovered that black rats *(Rattus rattus)* living in the pine forests of Israel subsist on pine seeds that provide the sole source of nutriment in an otherwise sterile habitat. Extracting pine seeds by stripping scales from a pinecone and eating the seeds the scales protect permit rats in Israel to occupy an ecological niche occupied in other pine forests of the world by tree squirrels (Scuridae) that are absent from Israel.

*Experiments.* Rats captured in Israel's pine forests continue to strip pinecones in captivity, whereas rats captured elsewhere in Israel, and therefore unfamiliar with pinecones, fail to open pinecones even if they are offered insufficient alternative food. Laboratory studies have shown that to gain more energy from pine seeds than is expended in extracting them from cones, rats must take advantage of the structure of the cones. The rats must first strip scales from the base of a cone and then spiral up and around the cone's shaft to its apex, removing one overlapping scale after another (Terkel 1996; Figure 10.2).

Investigations of development of this energetically efficient method of feeding on pine seeds revealed that only 3 percent of rats reared in the laboratory learned by trial and error to strip pinecones efficiently even when they were provided for weeks with an insufficient amount of rat chow together with an excess of pinecones. The other 97 percent of

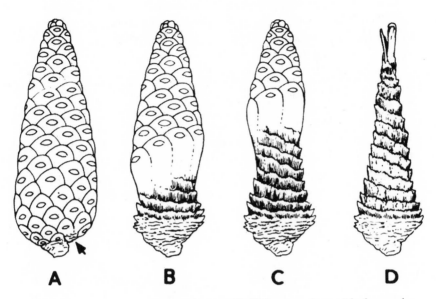

*Figure 10.2* Pinecones in different stages of efficient opening, with the number of rows stripped of scales increasing from left to right. From Terkel (1996); reprinted with permission of the author and Academic Press.

subjects either ignored cones or recovered seeds in a way that led to energy loss from eating pine seeds (Zohar and Terkel 1992).

Rats born in the laboratory to dams that efficiently stripped pinecones and foster reared by dams that did not failed to open cones efficiently, whereas more than 90 percent of pups reared by a foster mother that opened pinecones in their presence learned the efficient method of stripping cones (Aisner and Terkel 1992). Further experiments showed that 70 percent of young rats became efficient exploiters of pinecones following experience completing the stripping of pinecones started appropriately by either an adult rat or a human experimenter using pliers to copy the initial stages of an adult rat's pattern of scale removal (Terkel 1996; see Figure 10.2). When a rat mother stripped scales from a pinecone, young gathered around her and attempted to grab seeds as she uncovered them. As the young grew older, they snatched partially opened cones from a feeding adult and continued the process that the adult had started (Terkel 1996).

*Discussion.* Despite the overwhelming evidence that pinecone stripping is socially learned by young rats and is therefore traditional in forest rats,

231

occurrence of the behavior also reflects the absence from Israeli pine forests of tree squirrels, specialized competitors for pine seeds that exclude rats from pine forests elsewhere in the world, another possible example of character release similar to that discussed earlier in reference to *G. difficilis* finches on Wolf Island feeding on *Opuntia* cactus. Indeed, recent observations of black rats on the island of Cyprus, an area that, like Israel, is free of tree squirrels, indicate that they, like the rats of Israel, can survive in pine forests by feeding on pinecones (Landova et al. 2006). Thus the observed difference in the behavioral repertoires of Israel's rats and those living elsewhere has to be understood as an interaction between social learning and ecology, not as a result of social learning alone. As Laland, Kendal, and Kendal (this book) suggest, traditions are not purely socially determined. Both environmental variability and social learning contribute to observed variance in behavior. Second, although social learning clearly contributes to variability in the behavior of allopatric populations of rats, neither teaching nor imitation is involved in diffusion of this complex motor skill.

No other study of a traditional behavior exhibited by free-living members of any species, even humans, has been carried out with the rigor or elegance of Terkel's analysis of pinecone opening by rats (McGrew 1998). The combination of observation in the field and laboratory experiment provides all-but-incontrovertible evidence of social transmission of a complex motor skill from one generation to the next.

### NEW CALEDONIAN CROWS

The importance of studies of development in critical examination of purported instances of animal "culture" is particularly clearly revealed in Kenward et al.'s (2005) recent study of development of tool use in New Caledonian crows. These corvids manufacture tools from leaves and use these tools to retrieve food hidden in crevices. Differences in both the types of tools and number of types of tools found in different areas in New Caledonia suggest that there may be traditions of tool use in crow populations (Hunt 2000; Hunt and Gray 2002).

However, when Kenward et al. (2005) hand-reared four New Caledonian crows in captivity, giving the maturing birds no opportunity to interact with tool-using conspecifics, all four hand-reared birds developed the ability to use twig tools, and one of the four both cut a simple tool from a leaf and used it to obtain hidden food. At the time of Kenward et al's (2005) publication, none of the hand-reared crows had produced tools as sophisticated as some found in nature. A shortage in the United Kingdom, where the experiment was conducted, of appropriate leaves for tool manufacture by crows had prevented regular inspection of the ability of the still-maturing, hand-reared crows to manufacture sophisticated tools (Kacelnik, personal communication, 2006). Nevertheless, "The fact that an inherited predisposition can account for a complex behavior such as tool manufacture highlights the need for controlled investigation into behavioral ontogeny in other species that show culturally transmitted tool use" (Kenward et al. 2005, p. 121; see also Thouless et al. 1989; Tebbich et al. 2001).

## Animals with Large Brains

The same problems that bedevil purely observational studies of purported traditions in animals with small brains are common in studies of the better-known "traditions" of larger-brained species. The studies of "traditional behavior" of dolphins and chimpanzees, discussed later, surely provide evidence consistent with the hypothesis that population-specific behaviors are traditional in the populations that exhibit them. However, we still have some way to go before we know which purported traditions actually reflect social learning processes and which are not dependent on social learning for their development.

### TOOL USE IN BOTTLENOSE DOLPHINS

Some wild bottlenose dolphins (*Tursiops* sp.) resident in Shark Bay, Western Australia, carry marine sponges while foraging in deepwater channels (Mann and Sargeant 2003). The sponges are believed to be used to protect the rostrum of animals as they probe the sea floor to locate small, bottom-dwelling fishes, and sponge-using females have higher calving success than females that do not use sponges when foraging (Mann, personal communication, 2006).

Analyses of mitochondrial DNA of the Shark Bay dolphins show that sponge use occurs almost exclusively within a single matriline, in which most daughters (and a few sons) of sponge-carrying females adopt the habit (Krützen et al. 2005). Although a genetic explanation of such results seems plausible, Krützen et al. (2005, p. 8942) argue that examination of several possible modes of genetic inheritance makes it "extremely unlikely that a genetic propensity" is responsible for the observed distribution of the behavior. Further, the finding that only some of the many female dolphins that forage in the deepwater channels use sponges while foraging there makes it unlikely that exposure to deep channels in itself results in sponge use (Sargeant et al. 2007). The investigators conclude that social learning is the sole determinant of the observed variability in behavior in sponge use by dolphins in Shark Bay.

Although some researchers, including some experts in dolphin behavior, have proposed alternative explanations for the data from Shark Bay (see Laland and Janik 2006), the effort that investigators working in Shark Bay are making to examine effects of genotype (Krützen et al. 2005) and environment (Sargeant et al. 2007) in producing the observed distribution of sponge use in dolphins is worthy of emulation.

### TRADITIONS IN CHIMPANZEES?

In landmark articles Whiten et al. (1999, 2001) provided a list of 65 behaviors that vary in frequency of occurrence in seven geographically separate populations of chimpanzees *(Pan troglodytes)*, each studied for many years. As in *G. difficilis* and *R. rattus*, there are not only substantial ecological differences among sites where populations of interest reside, since chimpanzees live in habitats that range from dry, sparsely wooded savanna to moist forest, but also substantial genetic variation among populations.

Whiten et al. (1999, 2001) sort behaviors of potential interest into four categories: (a) patterns present at all sites, (b) patterns not achieving habitual frequency at any site, (c) patterns the absence of which can be explained by ecological factors, and (d) patterns customary or habitual at some sites but absent at others with no ecological explanation. Perhaps surprisingly, given the exceptionally broad range of habitats that chimpanzees inhabit, only three patterns of behavior are listed in category c (but see Whiten this book).

Differences across populations in frequency of occurrence of behaviors listed in category d are treated as "cultural" on the assumption that the observed differences in behavioral repertoires are a result neither of genetic differences between populations nor individual trial-and-error learning about differences in ecological circumstances. Presumably, at some time in the past a member of a population discovered a behavior in category d, and other members of the innovator's group subsequently acquired that behavior by interacting with the innovator. Members of groups that fail to exhibit a behavior in category d either never stumbled upon the relevant innovation or for some reason (other than a difference in ecology) did not copy an innovator.

However, there are reasons to question the assumption that some behaviors in category d are not a result of genetic or ecological differences between populations. First, 11 of the 39 behaviors in category d are common or habitual only in the genetically and geographically remote western subspecies, and 9 of the 39 behaviors in category d are common or habitual only among the three eastern subspecies.

Chimpanzees are currently classified in three subspecies: *P. t. verus* from West Africa, *P. t. troglodytes* from central Africa, and *P. t. schweinfurthii* from East Africa. A fourth subspecies *(P. t. vellerosus)* in Cameroon and northern Nigeria has been proposed. The western subspecies is more genetically distant from the central and eastern subspecies than the latter two are either from each other or from the potential fourth subspecies. Consequently, some consider the western subspecies to be a separate chimpanzee species (Morin et al. 1994; Gagneux et al. 1999). Of course, it is possible that such genetic differences among allopatric populations of chimpanzees are responsible for observed differences in their behavioral repertoires.

De Waal (1999, p. 635) has rejected genetic variation as a cause of variation in chimpanzee "cultural" behaviors, stating that "genes determine general abilities, such as tool use, but it is hard to imagine that they instruct apes how exactly to fish for ants." However, studies of the relationship between development of complex motor patterns in young chimpanzees and acquisition of behaviors once believed to result from "insight" (Koehler 1925) suggest that complex behaviors of chimpanzees, like those of New Caledonian crows, may develop in all individuals independent of social experience. Schiller (1952, 1957) found that behaviors like inserting one stick into another to form a longer stick and using that long stick to reach food too far outside a cage to be reached with either stick alone developed in all chimpanzees as they matured, even when no rewards could be acquired by joining sticks and using them to sweep outside a cage. Given such findings, it would not be surprising if chimpanzees with different genotypes were predisposed to develop different foraging behaviors.

Second, Whiten et al. (1999, 2001) do not discuss their interesting finding that 22 of the 39 behaviors assigned to category d are "common" or "habitual" in one or more populations but only "present" in others. For example, fishing for termites by using a leaf midrib is present at Bossou, common in the K group at Mahale, and absent at five sites (three with ecological explanation and two without). The challenge is to explain the difference in frequency of occurrence of such behaviors at the sites where it has been observed.

Possibly, by chance, long-term study of chimpanzees at Bossou and Mahale has taken place just when use of leaf midribs has saturated the population at Mahale and is beginning to spread among Bossou chimpanzees. Such chance occurrence is a reasonable explanation for any one of the 22 behaviors assigned to category d that are common or habitual in one group and only "present" in one or more others. However, chance occurrence is a very unlikely explanation for all 22 behaviors that share the characteristic of infrequent occurrence in some populations and frequent occurrence elsewhere. It seems more reasonable to suppose that the variable frequency of occurrence of these 22 behaviors in the various groups in which they are found results from differences in ecology (McBeath and McGrew 1982 provide evidence of effects of habitat on tool use by chimpanzees).

If ecological variables as yet undiscovered account for the differences in the frequencies with which a purportedly cultural behavior is expressed in populations that exhibit it, then ecological variables might also explain complete absence of the same behavior in other groups. Nonoccurrence is just the lower limit of uncommon occurrence.

*Ant dipping at Bossou.*   Much attention has been focused on different methods used by allopatric populations of chimpanzees to dip for driver ants, frequently described as one of the strongest examples of culture in chimpanzees (e.g., Boesch and Boesch 1990; McGrew 1992). At Gombe in Tanzania (East Africa) chimpanzees that are dipping for ants hold a long wand in one hand, introduce one end of the wand into an underground nest of driver ants, and then quickly withdraw the wand as the ants stream up it to attack. The chimpanzee then sweeps the length of the wand with its free hand, collecting the ants in a loose ball that it then pops into its mouth (McGrew 1992). In the Taï Forest in the Ivory Coast (West Africa), an ant-dipping chimpanzee uses a short stick to collect a small number of ants and then pulls the stick directly through its mouth. The Taï technique results in capture of far fewer ants per unit time than the Gombe technique (Boesch and Boesch 1990).

Boesch (1996a) examined ecological factors at Taï and Gombe that might favor different ant-dipping techniques, but could find none. However, Sugiyama (1995, p. 203) had previously proposed that the length of the wand, the dipping technique employed, and the working position of chimpanzees when ant dipping "must be determined by the characteristic features of the prey and, to some extent, tradition of the chimpanzee group."

It seems unlikely that the efficient foraging technique of ant dipping used at Gombe was never discovered at Taï. Chimpanzees at Bossou in Guinea (West Africa), like chimpanzees at Gombe, use long wands and use their hands to wipe ants from probes, which suggests that discovery of the Gombe technique is not a rare event. More intriguing, chimpanzees at Bossou not only use both the Taï and Gombe ant-dipping techniques but also use the Taï technique more frequently than the Gombe technique despite the alleged greater efficiency of the Gombe technique.

Humle and Matsuzawa (2002) investigated the role of nonsocial factors in determining use of short and long probes in ant dipping by chimpanzees at Bossou. They found that (1) chimpanzees at Bossou fed on several different species of ant, and (2) the aggressiveness of different ant species was correlated with the technique that chimpanzees employed in dipping for them. Use of long tools (and the corresponding use of the hand to remove ants from wands) was associated with dipping in risky contexts, for example, when feeding on black ants that delivered more painful bites (at least to humans) than did red ants, or when feeding on bivouacked ants that were more aggressive than were those at lower densities. A simple experiment in which humans used wands of different lengths to dip into nests of red and black ants revealed that black ants swarmed up the probe in greater numbers than did red ants. Not surprisingly, the chimpanzees at Bossou probe nests of black ants by using long wands and the pull-through-the-hand method and probe nests of red ants using the short tool and pull-through-the-mouth method.

Of course, the finding that the ant-dipping method is affected by prey behavior at Bossou does not show that the different ant-dipping techniques used by chimpanzees at Gombe and Taï reflect differences in the behavior of ants at the two sites. Indeed, results of experiments described in a very recent paper by Mobius et al. (2008) indicate that differences in the aggressiveness of preyed-upon ant species will not explain all of the differences in ant-dipping techniques used by chimpanzees in different areas . Nonetheless, the Humle and Matsuzawa (2002) finding does raise questions that need to be answered before cultural explanations of differences in ant-dipping techniques at Taï and Gombe are accepted.

That so obvious a potential explanation of differences in ant-dipping behavior as the nastiness of prey, previously mentioned in the literature

(Sugiyama 1995), escaped attention is particularly problematic because environmental correlates of variation in behavior can be far more subtle than those reflected in differences in aggressiveness of ants of different color. For example, seasonal variability in the rate at which chimpanzees at Gombe hunt for colobus monkeys has been attributed to seasonal variability in the frequency with which foraging chimpanzees encounter colobus troops. This encounter rate is, in turn, affected by seasonal patterns of fruit availability that determine both the distance traveled by foraging chimpanzee troops each day and the likelihood that both monkeys and apes simultaneously exploit the same resource (Stanford 1998).

*Development of termite fishing at Gombe.* Failure to detect ecological or genetic correlates of differences in the behavior of members of allopatric populations provides, at best, indirect evidence that those differences are cultural. Direct studies of development of behaviors of interest are clearly both needed and generally lacking. There are, however, exceptions.

Lonsdorf (2005; Lonsdorf et al. 2004) has reported results of a 4-year field study of development of termite fishing in wild chimpanzees at Gombe. Lonsdorf's analysis of her observations revealed, first, that male and female chimpanzees differ significantly both in the rate at which they acquire each behavioral component of termite fishing and in the age at which they become successful fishers for termites, with female chimpanzees achieving success, on average, more than 2 years before males. Infant (less than 4 years old) females spent significantly more time watching adults termite fishing than did infant males, who spent more time while at termite mounds playing than watching adults feed. Across both sexes time spent watching in the preceding year, but neither investigating mounds nor contacting mounds with tools, was correlated with age at first successful fishing. However, females, but not males, showed a distribution of the depths to which they dipped tools into termite mounds that was correlated with the distribution of dipping depths used by their respective mothers

Lonsdorf (2005, p. 681) concluded that "development of termite fishing includes social learning processes as well as individual trial-and-error learning . . . Male and female offspring learn from their mother that the termite mound is the object to which attention should be directed . . . Once the mother starts to termite-fish male and female learn that the goal of the behaviour is to capture termites . . . Male offspring then develop their own method of achieving this goal, while females learn something of the form of the behaviour . . . For both sexes, individual trial-and-error learning follows, as the offspring learns how to withdraw the tool without dislodging prey." There is thus evidence that social learning contributes to development of termite fishing in females but plays a lesser role (if any) in development of the same behavior in males. And if male chimpanzees learn independently to fish for termites, the role of social learning in development of termite fishing by females would be merely facultative rather than obligate, and termite fishing in chimpanzees would have little to do with "culture."

*Nut cracking in the Ivory Coast.* Observations of differences in the behavior of social groups separated by geographic barriers, particularly rivers, that prohibit social contact while perhaps permitting gene flow between groups provide another potential route to discriminating behaviors that are traditional from those that are not (Boesch et al. 1994; van Schaik 2004, 2006). For example, although wild orangutans *(Pongo pygmaeus)* on both sides of the impassable (for orangutans) Alas River in Sumatra feed on *Neesia* trees, only orangutans to the west of the river use twigs as tools to remove the fat-rich seeds from *Neesia* fruit, thus increasing their feeding efficiency (van Schaik 2004, 2006).

Similarly, chimpanzees to the west of the N'Zo-Sassandra River in the Ivory Coast use hammers and anvils to crack open nuts of five tree species, whereas those to the immediate east of the river do not (Boesch et al. 1994). This distribution of nut-cracking behavior could not be explained by differences in density of chimpanzees, density of nut-bearing trees, or frequency with which objects suitable for use as hammers or anvils are encountered. However, Morgan and Abwe (2006) have recently discovered troops of chimpanzees living some 100 miles to the east of the N'Zo-Sassandra River in Cameroon that use stone hammers and anvils to crack nuts. The finding raises potentially important questions concerning previous interpretations of the distribution of use of hammers and anvils in the Ivory Coast.

Inoue-Nakamura and Matsuzawa (1997, p. 172), in discussing results of their 4-year study of development of nut cracking by young chimpanzees at Bossou, concluded that imitation and teaching did not play a role in development of the behavior. Juveniles did not copy either the motor patterns of their mothers or other adult group members or the "way to relate nuts and stones." Instead, the authors conclude, young chimpanzees learned from observing the behavior of accomplished nutcrackers "the general functional relations of stones and nuts and . . . the goals obtained by the demonstrator." However, the only observation that Inoue-Nakamura and Matsuzawa (1997) provided in support of their proposal of even a limited role of social influence in acquisition of nut cracking by young chimpanzees was that infants frequently watched adults crack nuts and took pieces of nuts that others had cracked. Such observations may be consistent with the view that chimpanzees learn to crack nuts socially, but they prove little (Galef 1996).

Inoue-Nakamura and Matsuzawa (1997) also provided evidence inconsistent with the hypothesis that use of stones in nut cracking is socially learned. Adults frequently chased away juveniles that tried to take stones and nuts, and rearing by a mother who did not crack nuts did not slow acquisition of the behavior by her offspring. Indeed, Inoue-Nakamura and Matsuzawa (1997, p. 172) concluded their article by stating that "in summary, the members of the community provided only the infants with the opportunities to freely access stones and nuts. These opportunities could facilitate the individual experience of stone-nut manipulation and result in the apparent social transmission of the tool-use behavior among the wild chimpanzees." A role for social learning beyond local enhancement in development of nut cracking remains to be demonstrated.

*Social conventions in capuchin monkeys.* Perhaps the most convincing field evidence of traditional behaviors in nonhuman primates is provided by the work of Perry, Baker et al. (2003; Perry this book) on differences in the social behaviors of groups of capuchin monkeys *(Cebus capucinus)*. Because of the relatively brief lifespan of some capuchin social conventions, Perry, Baker et al. (2003) could document both the spread of idiosyncratic social behaviors from individual to individual within a group and the decline of those behaviors when key individuals either died or emigrated. For example, Gaupo, a subdominant, young adult male, introduced the "finger-in-mouth game" into group LB-AB. The game involved one monkey putting its finger into another's mouth. The recipient then clamped down on the inserted finger hard enough that the owner of the finger could not easily withdraw it and had to "go through various contortions" to pry open the recipient's mouth and free its finger. After a finger was freed, it was sometimes reintroduced, or the two players reversed roles, with the game continuing through several iterations.

Before 1993 all finger-in-mouth games involved Gaupo. In 1993 others began to play the game without Gaupo, and in time roughly half the dyads in the LB-AB group did so. Because adult male capuchins regularly move from one social group to another, a genetic basis for behavioral differences between groups seems unlikely, and such social conventions are sufficiently arbitrary that it seems all but impossible that members of a social group would have acquired them independently.

The tendency of adult male capuchins to migrate between groups has a further potentially useful consequence. With luck, it should be possible to determine whether a male that migrates into a new group brings idiosyncratic social conventions of his old group with him. If so, evidence of the traditional nature of social behaviors in capuchins would be irrefutable.

### Conclusion to Part 1

I expect that some, if not many, behavioral differences between populations of chimpanzees (Whiten et al. 1999), orangutans (van Schaik, Ancrenaz et al. 2003; van Schaik this book), monkeys (Perry this book), and various cetaceans (Mann and Sargeant 2003; Sargeant and Mann this book; Rendell and Whitehead 2001; Whitehead this book) currently discussed as traditions are, in fact, products of social learning of some kind. I would also wager that many purported traditional differences between allopatric populations of species are not, in fact, a consequence of social learning. It behooves us, as students of animal social learning, to do our best to go beyond a simple cataloguing of differences in the behavioral repertoires of allopatric populations to identify truly traditional differences in the behavior of free-living animals.

Identification of behavioral differences between populations is certainly an important and useful first step in the discovery of animal traditions. However, labeling a difference in the behavior of two populations as cultural is an assertion about how that behavior develops. Studies of development of population-specific behaviors are thus essential for determining which purported instances of culture truly reflect social influences on behavioral development. Developmental studies are few and far between (e.g., Whiten et al. 2005) and are still producing contradictory outcomes, sometimes providing evidence of the spread of techniques through a population and sometimes not (e.g., Hopper et al. 2007). Claims of "culture" in chimpanzees need to follow, not to precede, such investigations. Last, the question whether imitation rather than emulation is involved in laboratory demonstrations of social learning of foraging behaviors needs to be resolved (Tomasello 1999a, this book) so that, as discussed later, the relationship of animal traditions to human culture can be clarified.

## Part 2: Animal Tradition; Human Culture

No animal comes close to having humans' ability to build on previous discoveries and pass the improvements on. What determines those differences could help us understand how human culture evolved.

"What are the roots of human culture?" (2005, p. 99)

We must not overestimate the situation and say that "monkeys have culture" and then confuse it with human culture.

Hirata et al. (2001, p. 489).

Labeling ape behavior as "culture" simply means that you have to find a different word for what humans do.

Marks (2002, p. xvi)

### Similarities or Differences?

The first of the three preceding quotations is taken from a list of 100 "things we don't know that we need to know" proposed by the staff of *Science* magazine. The quotation suggests that at least with respect to the interests of the staff at *Science,* discussions of animal culture have been moving in quite the wrong direction. Attention has been focused almost entirely on what the similarities rather than the differences between human culture and animal traditions might tell us about the evolution of human culture.

Animal tradition and human culture serve similar functions. Both provide naïve individuals access to adaptive behaviors that others of their species have invented. Although interacting with others engaged in some behavior can facilitate acquisition of adaptive (and sometimes maladaptive) behaviors by animals from insects to apes (Heyes and Galef 1996), *Homo sapiens* has taken such social learning much further than has any other species.

Using different terms to refer to the products of social learning in nonhumans and humans simply reflects curiosity as to why, for example, our chimpanzee and bonobo cousins, after millions of years of experience, still sit naked, exposed to tropical downpours, while humans have gone on to build cathedrals and walk on the moon. We will never know why the ability of humans and animals to develop traditions has such different consequences unless we remain curious about differences, as well as similarities, between animal and human "culture." Whether "culture" in animals is referred to as animal culture, subculture (Kawamura 1959), preculture (Kawai 1965; Menzel 1973a), or tradition (Galef 1992) is a matter of indifference. However, until we know better, it might be wise to consider seriously the possibility that "culture" in our species and "culture" in other species are different phenomena. In politics, assuming differences where none exist can prove dangerous (de Waal 2001). In science, the reverse is as often the case.

Thinking about differences between traditions in animals and culture in humans need not reflect some sort of Western philosophical commitment to existence of an unbridgeable chasm between animals and humans that can be contrasted with an Eastern belief in the continuity of life (de Waal 2001). As suggested by the second quotation at the head of the present section, Japanese researchers introduced and still use the term "preculture" when discussing the traditions of monkeys at Koshima to indicate that in their view, monkey "culture" and human culture differ significantly (Kawai 1965; Watanabe 1994).

Demonstrating functional similarity is simply not sufficient. We need to know whether the "cultural" behaviors of animals and humans are products of similar underlying processes, and consequently whether animal tradition and human culture are analogs or homologs (e.g., Galef 1992; Tomasello 1994; Byrne et al. 2004). Using different terms to refer to the "culture" of animals and of humans focuses attention on the possibility (apparently a fact to the staff at *Science*) that there may be important mechanistic differences, as well as important functional similarities, between social learning in animals and in humans.

### Teaching

Caro and Hauser (1992) proposed that teaching be defined as occurring when (1) a teacher incurs some cost as a result of modifying its behavior when in the presence of a naïve individual, and (2) the modified behavior of the teacher causes the naïve individual to acquire some behavior more rapidly than it otherwise would. Caro and Hauser's definition thus treats teaching as an altruistic act and consequently brings teaching within the purview of neo-Darwinian approaches to the study of behavior (Galef et al. 2005).

A handful of possible instances of teaching by apes and cetaceans have been provided in the literature (e.g., Caro 1994; Rendell and White-head 2001). However, many find these examples unconvincing (e.g., Janik 2001; Maestripieri and Whitham 2001). Perhaps most informative is that with the exception of two instances of possible teaching of nut cracking by chimpanzees (Boesch 1991), there have been no reports of teaching in chimpanzees or bonobos despite tens of thousands of hours of observation (Matsuzawa 2001). Consequently, Franks and Richard-son's (2006) recent report that the ant *Temnothorax albipennis* exhibits behavior that exceeds the criteria for teaching proposed by Caro and Hauser (1992) came as something of a surprise.

In Franks and Richardson's (2006) experiment leader ants, but not their followers, knew where to find food. When a leader ran to food with a follower in attendance, the leader ran rapidly only after being tapped by the antennae of a follower. The consequent irregular movement of leaders running with followers resulted in a fourfold increase in the time leaders took to reach food (the cost to the teacher), and followers found food significantly sooner after engaging in running with a knowledgeable leader than did ants that searched for food on their own (the benefit to the pupil).

A second recent study, this one concerned with the role of social interaction in the development of predation in wild meerkats *(Suricata suricatta)*, provides similar evidence of behavior that meets the criteria of Caro and Hauser (1992). Meerkat helpers at the nest respond to changes in the begging calls of maturing meerkat pups by altering the frequency with which they provide the pups with disabled, potentially dangerous prey (scorpions). The experience of the pups with disabled scorpions accelerates the young meerkats' learning to handle intact scorpions without being stung or bitten (Thornton and McAuliffe 2006).

Both tandem runs by ants and provisioning of young meerkats with disabled prey meet the criteria Caro and Hauser (1992) proposed to define teaching. However, although the functional similarity of teaching in ants, meerkats, and humans is striking, teaching in ants and meerkats provides essentially no insight into evolutionary precursors of teaching in humans and would not even if ants, meerkats, and *Homo sapiens* were phylogenetically close. The behavioral mechanisms that support teaching in the three species appear so different from one another that it is all but impossible to conceive of one evolving into the other. Indeed, no one has yet proposed that teaching in ants and meerkats provides insight into the evolution of teaching in humans.

### The Evolution of Culture: Analogs and Homologs

The question whether chimpanzee tradition is the evolutionary precursor of human culture requires attention to the same issue as does the evolution of teaching. If the far more complex "culture" of humans than of chimpanzees reflects elaboration of behavioral processes that support traditions in chimpanzees, then chimpanzee tradition and human culture are homologous, and there is no reason to use different terms to refer to them. If, to the contrary, tradition in chimpanzees is the expression of fundamentally different behavioral processes than is human culture, then "culture" in the two species is analogous and tells us nothing about the evolutionary origins of human culture.

Much of human culture in the developed world clearly depends upon imitation, teaching (or pedagogy *sensu* Csibra and Gergely 2006), and language (the extent to which indigenous cultures depend on these processes for transmission of behavior from generation to generation remains to be determined). Present evidence suggests that chimpanzees essentially never teach and have no symbolic language unless they are taught one by humans.

Chimpanzees, like many other animals, are susceptible to effects of local enhancement and can emulate. How often and how precisely chimpanzees imitate (e.g., Whiten et al. 1996; Whiten this book) are both controversial (Horner and Whiten 2005; Call et al. 2005; Tomasello this book) and important in determining the relationship between human culture and chimpanzee tradition. As explicated more fully earlier in the present chapter, only a very few behavioral processes can support the cumulative culture that is characteristic of all human social groups. The total absence of such cumulative culture in even our closest relatives (Tomasello this book) suggests that there are fundamental differences between animal traditions and human culture.

### Are Animal Traditions the Evolutionary Precursors of Human Culture?

Perhaps, in time, we shall discover that differences between social learning processes in humans and animals are trivial rather than profound, and that small differences in capacity have resulted in huge differences in performance. Until then we need to ask both in what ways animal tradition and human culture are similar and in what ways they are different, not to sweep potential differences between culture in humans and traditions in animals under the rug in the name of Darwin or continuity. Using different terms to refer to the traditions of animals and the culture of humans should maintain a simultaneous focus on potential differences, as well as potential similarities. Both are of importance.

It is even possible that the cumulative culture that is characteristic of our own species requires either the linguistic capabilities unique to humans (Donald 1991) or the neural architecture that evolved to support human language. Such precursors of the social learning processes that support human culture may have emerged only in the ancestral hominid line that diverged from that of the great apes toward the end of the Miocene, some 7 million years ago. Szathmáry (2006, p. 307) has suggested that "it is perhaps no accident that cooperation in large non-kin groups, a developed theory of mind, tool use, teaching . . . and natural language go together in our species." Perhaps we should subtract tool use from, and add culture to, Szathmáry's list of possibly distinctively human characteristics. Despite the close phylogenetic relationship between *Homo sapiens* and the extant great apes, traditions of animals may have little to do with the evolution of human culture. We need to know.

# 11

## ARE NONHUMAN PRIMATES LIKELY TO
## EXHIBIT CULTURAL CAPACITIES LIKE
## THOSE OF HUMANS?

SUSAN PERRY

In the past 15 years extensive evidence for animal traditions has acc-umulated (Heyes and Galef 1996; Box and Gibson 1999; Fragaszy and Perry 2003a); nonetheless, no animal species is known to possess the richness of behavioral diversity and extreme reliance on social learning that is characteristic of human culture. Thus the question whether a particular species exhibits "culture" depends very much on details of the definition of culture. Most animal traditions researchers have chosen to focus on the role of social learning as being the key element of culture, ignoring those parts of cultural anthropologists' definitions that deal with exclusively human traits, such as language, complex moral systems, and religion, and refraining from making very specific requirements about temporal and geographic trait distribution within populations (Laland and Hoppitt 2003; McGrew 2003).

The definition of culture that is most often used by cultural primatologists is some approximation of the following: "geographic and temporal patterning of behavioral variation that owes its existence in part to social learning." Such a definition is useful both because of its simplicity and because of the ease with which it can be applied across species to gain a comparative perspective on the role of social learning across ecological and social contexts. This simple definition does not necessitate a very sophisticated knowledge of cognitive processes; it merely requires demonstration that social influences affect trait distribution.

I will argue later that there is more to human culture than social learning. Therefore, I tend to use the term "tradition," defined as "a behavior pattern shared by its practitioners due to some form of social learning," when I am conducting my own empirical research, so as to avoid other connotations of the word "culture" that may not apply to nonhuman animal traditions.

I have two goals in this chapter. First, I will review various methods used to detect culture or traditions in wild animals. Second, I will develop an argument about some important attributes of human culture that distinguish it from any homologous or analogous phenomena in nonhuman animals.

## What Sorts of Evidence Can Be Used to Detect Culture or Traditions in the Wild?

As with most topics in the social sciences, concrete claims about culture are hard to make. Even a simple assertion like "termiting is a socially learned trait in Gombe chimpanzees" is difficult to support conclusively. Instead, we are forced to be content with probabilistic statements based on complementary kinds of evidence. Some types of evidence are more convincing than others, and the more kinds of evidence that can be amassed, the more convincing the diagnosis. We can expect in this field that the same trait (e.g., termite fishing or nut cracking) will be variably diagnosed as socially or asocially learned during different phases of investigation, as the evidence accumulates.

As in any other academic debate, it is important to realize that a species may have (or is highly likely to have) "culture" (in the minimalist sense as defined earlier) even if the evidence is lacking, and to keep in mind that "culture" has not been rigorously demonstrated for most human populations, although intuitively we know that they must have it. The paucity of comparative data and the great expense and time commitment involved in documenting traditions make it improbable that there will ever be a definitive comparative database.

Currently the published literature is based on data collected with such variable methodologies that it is impossible to make meaningful statements about cultural repertoire size or the degree to which different species rely on social learning. All the following methods provide some useful insights for detecting culture in wild animal populations, but none alone, with the possible exception of the developmental approach (with a larger sample size than is generally used), is sufficient to make definitive claims regarding the presence of traditions in nature. The more sources of evidence that the researcher collects, the stronger the claim that can be made.

1. *Group contrasts approach.* The "group contrasts" or "process of elimination" approach is the method most often used for identifying

traditions in wild populations (chimpanzees: Whiten et al. 1999; orangutans: van Schaik, Ancrenaz et al. 2003; capuchins: Perry, Panger et al. 2003). In such an approach multiple sites are compared with regard to various behavioral traits, and behavioral differences between sites are inferred to be due to cultural processes if (a) there is no reason to believe that there are major genetic differences between sites, and (b) there are no notable ecological differences between sites. The advantage of this method is that it is easy to use, and it can help researchers focus future research efforts on traits that are likely to be due to social learning. However, assumptions about the plausibility of genetic and ecological influences can be erroneous, and because the group contrasts approach fails to provide any positive evidence for the presence of social learning, the method relies exclusively on negative evidence and plausibility arguments.

Last, the group contrasts approach would fail to detect true traditions when social learning in two separate sites produces convergent behavioral patterns, because the method infers traditions only from variation. For example, even if social learning were absolutely critical for the acquisition of some behavior pattern, such as processing a vital food resource or avoidance of a predator, cultural selection at multiple sites would be likely to produce similar outcomes. However, despite shortcomings, the group contrasts approach is a valuable starting point and is one useful source of evidence for detecting cultural variants in wild populations.

*2. Cross-sectional correspondence of social network (proximity) data with the distribution of behavioral traits.* Another source of evidence that bears on the hypothesis that social learning enhances the acquisition of particular traits is the mapping of behavioral variation onto social networks. Presumably, animals that spend more time together have more tolerant social relationships and should show a greater propensity to share behavioral characteristics. For example, in our capuchin work my collaborators and I were able to demonstrate that individuals who shared quirky food-processing techniques also tended to have high proximity scores; in general, they spent more time together than dyads that did not share techniques (Panger et al. 2002; see also O'Malley and Fedigan 2004). Similar analyses have also shown that spending more time together increases the probability that a pair of animals will share the same variant of a variable foraging trait (Perry and Ordoñez Jiménez 2006; Perry unpubl.).

Obviously, social networks and behavioral traits can only be correlated when there is within-group variation in the behavior. This method can tell us nothing about the role of social learning in the acquisition of traits that all group members share. However, discovery that increased proximity enhances the probability of sharing food-processing traits within a group enhances the plausibility of arguments that social learning influences traits that exhibit within-group homogeneity and between-group heterogeneity.

Of course, associations between proximity and the sharing of behavioral traits could be due to processes other than social learning. For example, if kin tend to spend more time together than do nonkin, then those animals that spend more time in proximity may share behavioral traits because of shared genes.

*3. Measurement of opportunities to learn socially.* Just because animals generally spend time together and exhibit the same behavior patterns does not mean that they are learning a particular skill from one another. Coordinating behavior in time and place by engaging in foraging behaviors in the same location may result in convergent behaviors because of exposure to similar environmental inputs for individual learning (Fragaszy, Visalberghi et al. 2004). Such basic coordination can result in shared traits even in the absence of visual attention toward others' behavior and qualifies as a simple form of social learning. However, precise copying of the motor details of behavior, such as would be expected for the learning of complex foraging techniques, cannot occur unless the animals attend to one another's behavior.

Measurement of opportunities to learn socially can enhance methods 1 and 2 by showing that individuals are in close proximity while practicing a particular behavior, and that they also visually attend to the behavior. Note that in this method there is no measurement of the final outcome (i.e., no reporting of the observer's behavior before and after observing the model), so there is no demonstration of actual learning, merely a demonstration that there was opportunity to learn.

My collaborators and I are measuring opportunities for social learning in capuchins. For each trait in question, we measure time spent in proximity to models of the behavior while the models are engaged in it, and we also record all instances of visual attention to the model while it is demonstrating the behavior. In a published study from this project (Perry and Ordoñez Jiménez 2006), we showed that juvenile monkeys attend more to models that are eating rare foods than to models that are eating common foods. Also, juvenile monkeys devoted more time to watching models foraging on foods that were difficult to process than models foraging on foods that require little or no processing before ingestion, even when handling time was controlled for. Similarly, Ottoni et al. (2005) have discovered that *C. apella* capuchins preferentially observe nut-cracking efforts of experienced models. These results suggest that capuchins are selective about whom they watch and may be actively focusing attention on individuals engaged in behaviors that the observer may not have learned (e.g., eating unfamiliar foods or foods that require complex processing).

*4. Developmental approach.* Probably the best method for assessing the role of social learning in the acquisition of behavioral patterns of wild animals is to take a developmental approach in which animals are studied from birth until the time when the behavior is firmly established. The obvious problem with this method, particularly when it is applied to species that have multiyear juvenile periods, is that it takes far more time, personnel, and money than are typically available to a primatologist. There is a trade-off between collecting sufficient data for each subject to have a meaningful behavioral profile of behavior in each developmental time period and including enough subjects to have a statistically meaningful sample size. To date, most studies have chosen to emphasize behavioral profiles at the expense of sample size.

Lonsdorf et al. (2004) conducted a study of sex differences in acquisition of termite-fishing skills in six juvenile wild chimpanzees and documented that when they were fishing for termites, the three daughters (but not the sons) demonstrated a preference for the same tool insertion length as their mothers. The female offspring also spent more time watching their mothers than did sons.

Russon (2003) detailed the changes in foraging tactics of a few immature rehabilitant orangutans over periods of 2 to 3 years. Russon produced detailed case studies that demonstrated that close associates tended to share foraging practices. A developmental approach is also being used at my own study site, where we are following 44 wild capuchin monkeys from birth to adolescence. Preliminary analysis of a single trait, *Luehea* foraging, indicates a significant role of social influence, as measured by differential exposure to alternative foraging techniques, on monkeys' decisions whether to pound or scrub their fruit to dislodge seeds (logistic regression, Nagelkerke $R^2=0.59$, $p=0.007$, $n=13$ for 2-year-old monkeys).

In theory, given sufficient time and resources, the developmental approach can address problems inherent in methods 1–3 described earlier. For example, one problem with method 2 (the cross-sectional approach) is that both social relationship/proximity data and data on expressed traits are taken at a time that may be far removed from when the behavior was learned. In the case of a dispersing sex, in particular, data on an individual's current social network may be completely meaningless because the behavior in question may have been learned before migration to this particular group. Also, there may be a critical period in development during which an animal is more receptive to learning a behavior, either independently or from others. This critical period is likely to be missed in a cross-sectional study. In a developmental study it should be possible to combine the best aspects of methods 2 and 3 such that data on social networks and opportunities for social learning are combined with data on the young animal's behavior as it changes in response to observations of others' behavior.

5. *Historical approaches: documentation of the temporal and geographic distributions of behaviors.* When social learning occurs, expansion in the number of performers of a behavior also occurs; however, it is not necessarily the case that an expansion of the number of performers implies that social learning is the mechanism of acquisition. For example, a shortage of some preferred food could lead many group members to begin including a novel food item in their diet, even if all individuals acquired the behavior individually. Therefore, historical approaches alone provide insufficient evidence of traditions. However, in combination with other types of data supporting the hypothesis that social cues are responsible for spread of a trait (e.g., Dewar's [2003] cue reliability approach or methods 2–4), historical approaches can provide valuable information about the transmission dynamics of behavioral traits.

If a population is observed intensively over a period of years, the spread of traits over time can be documented, even if corresponding data on social networks and detailed data on observation opportunities are not necessarily simultaneously collected in a systematic way. Such data are available for Japanese macaques, with varying degrees of preciseness about the order in which individuals acquired traits such as stone handling (Huffman 1996), fish eating (Watanabe 1989), bathing, potato washing, wheat washing, wheat eating, and candy eating (Kawamura 1959; Kawai 1965; Itani and Nishimura 1973).

If the behavior being documented is an individually performed behavior, as opposed to a collaborative social behavior, it is difficult to know from whom individuals learned the behavior or even to know with any certainty that the behavior was socially learned. If the behavior requires extremely close physical proximity between model and observer (e.g., Japanese macaque techniques for handling louse eggs; Tanaka 1998), it becomes more plausible that social learning has caused the spread of the behavior. A researcher can most easily invoke social learning as a mode of acquisition when a trait is a novel dyadic social ritual that requires focused and coordinated communication between the two performers for the trait to be performed (e.g., capuchin "games"; Perry, Baker et al. 2003). In such cases it is impossible to attribute trait acquisition strictly to nonsocial learning processes, and it is easy to tell who learned what from whom. For example, it was possible to construct fairly reliable social transmission chains for three "games" played by capuchin monkeys at Lomas Barbudal (Perry, Baker et al. 2003a). We documented up to three links in the transmission chains before the behaviors became extinguished from the group repertoire about 10 years after they were first observed. We could construct these transmission chains only because (a) the group was monitored almost daily over many years, particularly during the transmission phase of the behavior, and (b) dyads tended to play fairly regularly with one another for long periods before new members were added to the game-playing network, which made it unlikely that we would incorrectly attribute a transmission event to the wrong dyad.

6. *Experimental approaches.* Experiments can often yield types of information not obtainable by strictly observational studies. Not all species are equally amenable to manipulation, and because of their extreme xenophobia, individuals of most primate species cannot ethically be captured and moved into new social groupings. Consequently, it is no surprise that much of the most convincing research on animal traditions comes not from primates, but from far more tractable study species such as rodents (e.g., Galef 2003a).

One advantage of the lab over the field is that novel objects and situations can be presented, so that the researcher has a better chance of observing the moment of innovation and documenting all subsequent social transmission events. In contrast, fieldworkers have essentially no control over animals' exposure to learning opportunities and incomplete knowledge of their past exposure to events of interest. Documentation of transmission chains, for example, is far easier to accomplish in captive animals than in wild settings (e.g., guppies; Laland and Williams 1997).

The recent study by Whiten et al. (2005) exemplifies the advantages of captive experiments in demonstrating social learning capacities. In this study a novel apparatus (the panpipes, which could be operated in two different ways to obtain food) was presented to three groups of chimpanzees. In two groups there was a chimpanzee model trained specifically by humans in one of the two operating techniques. Chimpanzee groups with access to a model tended to conform to the technique practiced by the model and thus provided strong evidence for social learning. Such evidence of social learning in chimpanzees lends credence to the interpretation of social influence as a determinant of intersite behavioral variation in chimpanzees generally (Whiten et al. 1999), even though it cannot firmly establish that any particular behavior from the list of candidate cultural variants in wild chimpanzees is socially transmitted.

It is important to remember that although animals may show cognitive capacities to learn socially in captivity, the contexts in which they learn in nature will be different in that the motivation to learn, the degree of reliability of social versus asocial cues, and the degree of access to tolerant expert models will not only vary from species to species and from site to site but also according to the type of activity. For each behavior pattern seen in wild populations, evidence for the role of social learning must be collected separately.

Experiments on captive animals are an elegant and efficient way to provide insight into the cognitive capacities of study organisms. However, experiments can never substitute completely for observational data as a way of determining the influence of social learning in nature, particularly for animals that are as difficult to manipulate as nonhuman primates.

Observation of animals in their native habitats, developing under the sorts of social and ecological conditions in which their minds evolved, is critical to understand the evolution of cultural capacities. Only this kind of observation will lead to appreciation of the factors that promote or discourage reliance on social learning across different task types. Only by observing animals in nature can we obtain realistic measurements (via the historical approach) of cultural transmission dynamics in animals that live in a variety of circumstances. Consequently, even though observational data are time consuming both to collect and to analyze and are often dev-ilishly difficult to interpret, observational data collection will always remain a crucial tool for scientists interested in the evolution of cultural capacities.

## Is There More to Human Culture Than Just Social Learning?

Does a collection of social traditions in a particular group of animals qual-ify as a culture? McGrew (2003) and Whiten et al. (2003) discuss several ways in which human and nonhuman cultures have been proposed to vary, many of which concern (a) the issue of how many cultural variants are necessary, (b) the exact form of the social transmission mechanism (teaching and imitation being more central to human social learning), or (c) the "cultural content"—that is, the types of behaviors that are trans-mitted socially. In animal societies the ratio of foraging/technological tra-ditions to traditions that involve ways of conducting social interactions is quite high compared with human cultures. Because "cultural primatol-ogy" is still in its infancy, the verdict regarding the degree to which each of these proposed differences is a qualitative difference or a significant quan-titative difference is likely to shift as new data sets become available.

### Repertoire Size

Many scholars seem convinced that the size of the cultural repertoire is the most crucial factor that distinguishes human culture from animal "cultures." Even in the absence of a sophisticated measurement tech-nique for estimating repertoire size, it seems obvious that human cultural repertoires greatly exceed those of other animals. However, assessment of the relative cultural complexity of different animal species can also de-pend on the size of the cultural repertoire (e.g., Whiten et al. 1999, 2005. It is worth noting that there is currently no satisfactory method for counting traditions, and in my opinion, the current rankings of species' repertoire sizes are likely to be determined more by measurement errors than by the complexity of behavior of these species.

Cultural repertoire lists are most often created by asking researchers to rely on memories of their past studies rather than by designing specific protocols for measuring particular types of behavioral variation. Because each researcher has a different topical focus and different perceptual bi-ases, studies vary in the quality of relevant information collected. For ex-ample, a researcher who is studying vocal communication might have

**a**

Time interval

| Individual | 1 | 2 | 3 | 4 | 5 | 6 | 7 | 8 | 9 | 10 |
|---|---|---|---|---|---|---|---|---|---|---|
| A | P | P | P | P | P | P | P | P | P | P |
| B | P | P | P | P | P | P | P | P | P | P |
| C | P | P | P | P | P | P | P | P | P | P |
| D | P | P | P | P | P | P | P | P | P | P |
| E | T | P | P | T | T | P | P | P | T | T |
| F | S | S | S | S | S | S | S | S | S | S |
| G | P | P | P | P | P | P | P | P | P | T |
| H | P | P | P | P | P | P | P | P | P | S |
| I | P | P | P | P | P | P | P | P | P | P |
| J | P | P | P | P | P | P | P | S | P | P |
| K | P | P | P | P | P | P | P | P | P | P |
| L | P | T | P | P | P | P | P | P | P | P |
| M | P | P | P | P | P | P | P | P | P | P |
| N | P | P | P | P | P | P | P | P | T | P |
| O | P | P | P | P | P | P | P | P | P | P |
| P | P | P | P | P | P | S | P | P | P | P |
| Q | P | P | P | P | P | P | P | P | P | P |
| R | P | P | P | P | T | P | P | P | P | P |
| S | P | P | P | P | P | P | P | P | P | P |
| T | P | P | S | P | P | P | P | P | P | P |

*Figure 11.1* Distribution of food-processing techniques (P, S, and T) for 20 individuals (A=nT) in 10 time periods. Figure 11.1a shows the interpretation that results if 5 percent of all behaviors are sampled (those cells with bordered boxes are included in the sample). In this case all individuals are assumed to exhibit behavior P exclusively (i.e., pounding is customary). In Fig. 11.1b, 10 percent of all behaviors are sampled. In this scenario 92 percent of the population is characterized by behavior P, and 8 percent by behavior T (i.e., pounding is customary and T is present).

high-quality data on social signaling and a good ear for acoustic variation but poor recollection of foraging techniques and little or no knowledge of the foods her study species used. While noisy, conspicuous behavioral variants are likely to be noticed by anyone, more subtle behavioral variants are likely to be overlooked. Consequently, the cultural repertoire for a species is likely to depend on the number of researchers who are studying the species and the diversity of their interests.

**b**

| Individual | 1 | 2 | 3 | 4 | 5 | 6 | 7 | 8 | 9 | 10 |
|---|---|---|---|---|---|---|---|---|---|---|
| A | P | P | P | P | P | P | P | P | P | P |
| B | P | P | P | P | P | P | P | P | P | P |
| C | P | P | P | P | P | P | P | P | P | P |
| D | P | P | P | P | P | P | P | P | P | P |
| E | T | P | P | T | T | P | P | P | T | T |
| F | S | S | S | S | S | S | S | S | S | S |
| G | P | P | P | P | P | P | P | P | P | T |
| H | P | P | P | P | P | P | P | P | P | S |
| I | P | P | P | P | P | P | P | P | P | P |
| J | P | P | P | P | P | P | P | S | P | P |
| K | P | P | P | P | P | P | P | P | P | P |
| L | P | T | P | P | P | P | P | P | P | P |
| M | P | P | P | P | P | P | P | P | P | P |
| N | P | P | P | P | P | P | P | P | T | P |
| O | P | P | P | P | P | P | P | P | P | P |
| P | P | P | P | P | P | S | P | P | P | P |
| Q | P | P | P | P | P | P | P | P | P | P |
| R | P | P | P | P | T | P | P | P | P | P |
| S | P | P | P | P | P | P | P | P | P | P |
| T | P | P | S | P | P | P | P | P | P | P |

Time interval

*Figure 11.1*    *(continued)*

Also, each research project has a different behavioral sampling protocol, and in some cases only certain members of a study group will be subject to focal animal sampling. This means that many group members will be assumed not to engage in behavior types that they may in fact practice. Sampling errors can lead to errors in assessments of how many traditions are present.

If one uses Whiten's categorizations of "customary, habitual, present, or absent" to describe patterning of behavioral diversity, the determination of the number of traditions will depend quite heavily on the density of behavioral sampling both in terms of the number of observations of a task that are required before typing an individual and the number of individuals typed for a particular trait before a characterization is made for a population. The more a population is studied, the greater the chance a researcher has of discovering a larger range of diversity in both individual and group repertoires. If we assume a basic species tendency to create certain innovations rather than others, then the longer a population is studied, the more within-group diversity will be discovered, and the less often between-group differences will be seen, as some members of one population are discovered to occasionally exhibit traits that were previously thought to be practiced exclusively by another population.

It is hard to visualize the magnitude of the errors that can be produced by such sampling errors without doing a crude simulation. Assume that Figure 11.1 is an accurate portrayal of the frequency with which the animals in a population exhibit three variants of a food processing technique: pound (P), scrub (S), and tap (T). For simplicity's sake, assume that each animal performed the foraging task once in each time interval, and that the task was not performed at any other time. Out of 20 individuals, 10 perform the majority technique (pounding) exclusively, 1 performs the scrubbing technique exclusively, and 9 individuals exhibit a mixture of two of the three techniques. In Fig 11.1a the bordered boxes show what observations were included when 5 percent of the cells were sampled. In this case it appears that there is no variation in the sample whatsoever, and it would be concluded that there was no variation in food-processing techniques. Of course, there is an 11.5 percent chance of picking a non-pounding event on any draw of a particular cell in this table. However, in a real-world situation, if we assume that the most gregarious and tolerant individuals are most likely to adopt groupmates' behaviors (Coussi-Korbel and Fragaszy 1995; van Schaik 2003), then it is peripheral animals that are most likely to exhibit anomalous strategies, and precisely these animals are most likely to be overlooked in behavioral sampling, particularly if ad libitum sampling is the primary sampling technique.

In Figure 11.1b the sampling effort is doubled so that now 10 percent of all incidents become part of the database. Now two of the three behavioral variants are discovered. However, still only 1 member of the group (out of 12 sampled) is thought to have a different technique from the predominant pounding technique, and no individuals are discovered who have multiple techniques, despite the fact that 45 percent of all individuals in the population do so. Accurate estimates of the total number of traits represented in the sample do not occur until at least a quarter of all individuals and half of all time intervals are sampled. To arrive at even occasional accurate estimates of how many techniques were the *pre-*

*ferred* technique for at least one member of the population (rather than just a minority technique that was preferred by no one), at least half of all individuals and half of all time intervals need to be sampled.

The results of this simple simulation highlight the importance of extensive and rigorous behavioral sampling before statements are made about the nature of behavioral variation within and between groups. Errors of interpretation are particularly likely to arise when data from sites with little sampling effort are compared with data from sites with far greater sampling effort. In such a case behavioral variants that are present at both sites are likely to be overlooked at the site with less sampling, leading (when employing the "group contrasts" approach) to erroneous diagnosis of a "cultural variant."

### Emotional Salience of Cultural Traits and Linkage to Group Identity

I have always felt a slight uneasiness when using the word "culture" to describe collections of social traditions documented in capuchins, chimpanzees, and orangutans and have therefore tended to use the term "traditions" rather than "culture" in most of my writing. I firmly believe that social learning plays a role in producing much of the behavioral variation seen in nonhuman primates (even though convincing data sets that bear on this issue are still scarce). Consequently, if culture is defined as nothing more than "behavioral variation that owes its existence in part to social learning," then many animal species have culture. But I realize that when my cultural anthropologist colleagues hear talks about "culture" in nonhuman primates, they are fairly disdainful of these laundry lists of traits. Cultural anthropologists are so dismissive of the notion of animal "culture" that it is difficult to find one who thinks it worth his time to articulate his objections in print (but see Hill, this book). For example, an examination of the commentaries for studies of cetacean culture (Rendell and Whitehead 2001, in *Behavioral and Brain Sciences*), chimpanzee culture (Boesch and Tomasello 1998, in *Current Anthropology*), and traditions in capuchins (Perry, Baker, et al. 2003, in *Current Anthropology*) reveals only 4 responses by sociocultural anthropologists (2 by the same person) out of 50 total responses. Therefore, the "animal culture wars" (McGrew 2003) take place primarily in the corridors and seminar rooms of anthropology departments rather than in the published literature.

The dismissive attitude of cultural anthropologists does not stem primarily from concerns about human-nonhuman differences in social

259

learning mechanisms (e.g., true imitation versus social facilitation); rather, it stems from the perception that human culture, unlike any non-human facsimile, is both ideologically rich and embedded in complex social relationships governed by moral rules. Few cultural anthropologists today describe cultural traits and track their transmission (Ingold 2001); ironically, for this very reason it is often harder to attribute culture to humans than to many nonhumans if strict methodological criteria are required to demonstrate social learning. Rather than focus on transmission mechanisms and the spread of particular traits, modern-day cultural anthropologists are prone to focus on issues of group identity, expression of this identity by means of culture-specific symbols, and manipulation of cultural meanings to achieve social change.

Frustratingly, the definitions of culture that are most commonly cited in textbooks and other prominent sources (see reviews in McGrew 2003 and Laland and Hoppitt 2003) do not emphasize what appears to me to be the most significant qualitative differences that distinguish features of human culture from nonhuman animal traditions: the emotional salience that links particular symbols, artifacts, or behavioral traits to group identity, and the moral pressure to conform to a suite of traits. I do not argue that *all* socially learned traits in human cultures are tinged by this emotional and moral salience—some traits do appear to be neutral in that respect. However, many socially learned traits in human cultures are clustered into bundles that have associations with group identity.

Often the definitions of human culture found in academic texts do not express important aspects of human nature as neatly as works of literature do. Nowhere do I find this critical aspect of culture as simply explained as in a children's book, *The Butter Battle Book,* written by Theodore Geisel (Dr. Seuss; 1984). Here two ethnic groups, the Yooks and the Zooks, differ in their way of buttering bread. The Yooks eat their bread butter side up, and the Zooks eat theirs butter side down. The bread-buttering habit becomes the identifying symbol on their flags and the point of difference around which the two groups rally their within-group cooperation and fuel their hatred of the other group. Each group acquires a horror and loathing of the other's barbaric bread-buttering habits, such that bread-buttering techniques became symbolic of all of the group's other (presumably more significant) differences in behavior. Technological arms races ensue, walls are raised between the two groups, and further social stratification and specialization of labor develop as each tries to exterminate the other. In brief, a seemingly arbitrary social tradition with no apparent functional value assumes moral significance, and the emotions aroused by seeing and practicing this behavior are catalysts for enforcing within-group cooperation and between-group competition.

Could such "butter battles" occur in nonhuman primates as well? Would the sight of a nut-cracking chimp ever arouse moral indignation in a non-nut-cracking chimp because of the nut-cracking behavior (and not simply because of lack of familiarity with the animal)? Do capuchins that pound their *Luehea* fruits become morally outraged when they see capuchins scrubbing their *Luehea* fruits? Or are non-hand-sniffing capuchins appalled to see other capuchins insert their fingers in their comrades' nostrils?

Admittedly these are hard questions to answer, since they require knowledge of the internal emotional states of the animals. However, thus far, both my own observations and the published literature provide no indication that any of the traditions reported in the animal culture literature are associated with moral indignation that inspires punishment of deviants from traditional behaviors. Of course, lack of evidence is insufficient proof that a phenomenon does not exist, and some cultural primatologists hold out hope for the idea that "the ultimate function of culture for community-living apes is social identity" (McGrew 2003, p. 438).

When I first began studying traditions in capuchins, I was initially excited to find variability not only in the technological domain but also in the domain of social conventions. White-faced capuchins exhibit rich between-group behavioral variability with regard to their social behavior, including many bizarre and innovative rituals that spread within cliques of animals. Some of these include hand sniffing (in which two monkeys insert their fingers into each other's nostrils for lengthy periods), sucking of another monkey's body parts (e.g., tails, fingers, ears), inserting a partner's finger into one's own eye socket, and various games involving turn taking (see Perry, Baker et al. 2003 for further descriptions). One example of a game is the hair-biting game, in which a monkey bites a large clump of hair out of a partner's shoulder or face, and the partner must force the mouth open to retrieve his hair. The hair is passed from mouth to mouth in this manner until it is lost, and then the game begins again.

When I first began analyzing this data set and trying to infer the significance of these behaviors for the animals, I was intrigued by the thought that these behaviors might be ethnic markers of sorts—rituals performed by members of a group that would serve not only to strengthen their own bond but to advertise their solidarity and clique membership to nonclique members. But the more I thought about it, the more I realized that this "advertisement" hypothesis did not make much sense, given the context in which these behaviors occurred. Although the pairs that were practicing these bizarre rituals were intensely focused on one another's behavior, third parties showed absolutely no interest in the proceedings, which implied that they probably did not consider this to be useful information. Also, if the behaviors were designed as advertisements, they should have been designed to be conspicuous, but on the contrary, they were quiet, unobtrusive, and generally practiced outside the center of the group where the majority of socializing was taking place. These behaviors were never practiced when members of different social groups were present (i.e., during intergroup encounters or during the initial stages of immigration by a newcomer). Furthermore, what new and useful information would be gained by third parties who watched these dyadic rituals? Because observer monkeys could see daily grooming bouts, play bouts, and, most important, the conspicuous coalitionary displays that took place in polyadic aggression on a regular basis, they had visual and auditory access to far more telling information about who was in what clique. Capuchins are keen observers of their group-mates' behavior and use their knowledge of other monkeys' friendships and relative ranks to make intelligent decisions about whom to recruit to help them in fights (Perry et al. 2004). After rejecting the advertisement hypothesis, I concluded provisionally that these odd capuchin social conventions were dyadic bond-testing rituals that informed participants about their commitment to a particular relationship by providing a richer source of information about emotional commitment than could be provided by more stereotyped signals (see Perry, Baker et al. 2003 for further details).

*The function of ethnic markers in humans.* Understanding the functional significance of human ethnic markers can help us predict where else in nature such markers would be expected to occur. According to McElreath et al. (2003), behaviors as seemingly arbitrary as bread buttering may serve as indicators of shared membership in particular ethnic groups and hence of shared understandings regarding proper ways to behave. Superficial indicators, such as clothing styles, speech habits, or greeting gestures, indicate to newcomers to which ethnic group an individual belongs. McElreath et al.'s model assumes that cultural traits cluster in meaningful ways, such that arbitrary details of dress and manner are associated with (i.e., are acquired simultaneously with) more significant traits that are more crucial for maintaining high-functioning collaborative relationships, such as gender roles, subsistence techniques, and rules of conduct. Because more significant social norms and beliefs are not immediately obvious until particular critical situations arise, and individuals may not be aware that they have internalized these social norms, obvious and superficial ethnic markers are needed to help newcomers readily identify who is likely to be a useful and trustworthy ally and collaborator, and who is likely to have habits or allegiances that are at odds with one's own behavioral strategies.

McElreath et al.'s model assumes that social norms and ethnic markers will be tightly associated relative to other types of culturally inherited traits. Gil-White (2001) has argued that conformist transmission has caused local cultures to emerge that have stable equilibriums regarding clusters of social norms. Because it is costly to try to coordinate with people who do not have the same social norms (particularly when selecting marriage partners), people develop strong group identities that cause them to categorize members of other ethnic groups almost in the same way they would categorize members of different species.

*Defining social norms.* Social norms are widely discussed in the social sciences and are generally understood to be critical to understanding human social life. However, existing definitions vary widely in meaning. There is much variation among definitions regarding the issues of (a) how much conformity must be obtained for the pattern to qualify as a norm, (b) what psychological and social mechanisms are responsible for creating behavioral conformity (e.g., is third-party punishment necessarily the mechanism, or could other forms of behavior modification also qualify?) and (c) whether the behavior pattern must be cultural (i.e., variable across social groups and influenced by social learning) rather than a human universal. For purposes of this chapter, I will use the following definition of social norms: "standards of behavior that are based on widely shared beliefs [about] how individual group members ought to behave in a given situation" (Fehr and Fischbacher 2004a, p. 185).

Like humans, many species of nonhuman primates have complex social relationships and cooperate in long-term alliances. If ethnic markers enhance the ability to choose appropriate partners in coordination games, then why should not nonhuman primates also have ethnic markers?

There are two important aspects to the answer to this question. First, if ethnic markers are to be advantageous, there must be significant differences in social norms that affect collaborative potential. Second, there must be situations in which individuals are exposed to unknown individuals, such that ethnic markers serve as a useful shortcut to discovering the nature of these unfamiliar individuals' behavioral strategies. Despite numerous primatologists' tales of striking intergroup differences in the corridors of primatology conferences, there is very little published documentation of the degree to which social roles vary among conspecific groups of nonhuman primates. It remains both possible and plausible that there is important between-group variation in the degree to which certain types of social interactions (e.g., sexual coercion, food theft, refusal to submit to dominants, or interference in others' affiliative interactions) are tolerated, punished, or encouraged by group members.

If strict evidence of third-party punishment is taken to be a crucial part of the definition of a social norm, then there are, to my knowledge, no published accounts of social norms in the nonhuman primate literature, with the possible exception of Hauser's (1992) report of rhesus monkeys punishing noncallers to enforce food calling when food items are found. Even in that case it could be argued that the enforcers had a direct interest in enforcing the rule and were not truly "third parties." If less punitive forms of behavioral molding qualify (such as rewarding good behavior by conformists or withholding rewards from nonconformists), or if behavioral rules shaped by more interested parties qualify, then a few other examples of group-specific socially shaped behavior patterns can be termed social norms. Sapolsky and Share's (2004) account of a "culture of pacificity" in baboons and Kummer's (1995 pp. 116–124) account of patterns of variation in female consort behavior during translocation experiments in hybrid zones of anubis baboons and hamadryas baboons seem relevant. One primary difficulty in convincingly demonstrating social norms in nature is that it is virtually impossible to devise methods that would enable a clear demonstration of the social learning process (e.g., conformist transmission or teaching via punishment) that maintains the intergroup variation in behaviors related to social norms.

## Discussion

I have argued that culture, in its human sense, is more than a list of geographically variable traits: it includes both (a) socially learned ways of conducting social interactions and (b) a symbolic linkage between socially learned traits and a strong sense of group identity. Note that this description does not exclude any nonhuman animal from the "culture club" by definition. To resolve empirically whether a nonhuman species has culture in this sense, it will be necessary to produce (a) convincing documentation of between-group variation in social norms, (b) evidence of a strong sense of group identity, and (c) linkage between socially learned traits and group identity. What is the likelihood that any nonhuman species will clear these bars in another decade or two of research?

It seems likely to me that given enough research effort, there will be more reports of social norms in nonhuman animals. Among the Lomas Barbudal capuchins *(Cebus capucinus)*, for example, we see striking between-group differences in the degree to which sexual coercion is tolerated and the degree to which alpha males tolerate subordinate male-male affiliation.

White-faced capuchins are easily outraged and generally have pugnacious, meddlesome temperaments that cause them to intervene aggressively as third parties in interactions that do not directly concern them (e.g., food theft, same-sex sexual interactions, and play bouts). To date, no one has performed detailed analyses of such interventions to determine whether they support existing social norms and may be interpreted as a "sense of justice." "Policing" (i.e., impartial intervention in others' conflicts) is common in many primate species (see, for example, Flack et al.'s 2005 study of pig-tailed macaque policing), and it seems possible that species that exhibit policing would also exhibit social norms. In sum, the case for nonhuman primate social norms is far from closed, and it is possible that the lack of evidence for social norms in nonhuman primates is due primarily to the daunting methodological challenges to demonstrating existence of social norms rather than to a lack of such norms in nonhuman animal societies.

A strong sense of group identity is certainly not a feature of all animal societies, but it is a characteristic of many. To name just a few of many possible examples, white-faced capuchins (Perry 1996), vervets (Cheney and Seyfarth 1990), and chimpanzees (Wilson et al. 2001) react extremely strongly to the sight or sound of individuals who are not members of their own groups, sometimes launching unprovoked lethal attacks on foreign individuals (capuchins: Gros-Louis et al. 2003; chimpanzees: Goodall 1986). Spear-nosed bats (Boughman and Wilkinson 1998) even have group-specific calls used to preferentially coordinate foraging with group members.

Although it seems likely to me that some animal species will eventually be proved to exhibit some limited between-group variation in social norms and a strong sense of group identity, I am far more skeptical that there will ever be evidence for linkages between socially learned traits or symbols and group identity. Nonhuman primates, at least, lack the demographic characteristics necessary to favor evolution of ethnic markers.

Humans are probably unique among primates in the extent to which they engage in cooperative relationships with people they do not know on a personal basis. Human societies tend to have a fission-fusion structure in which kinship networks and political alliances with nonkin extend beyond the group of people who regularly interact to include people who may never directly encounter one another in the course of a lifetime (Rodseth et al. 1991). People who rarely, if ever, meet are involved in trade networks, exchange of marriage partners, and military alliances. Such far-reaching networks are probably impossible without language to convey messages between people who do not meet directly.

In contrast, nonhuman primates communicate and learn about one another via face-to-face interactions (although olfactory cues may be used to convey some simple messages in the absence of direct contact). With the exception of hamadryas baboons, nonhuman primate social structures tend to be fairly simple; they lack "supercoalitions" or multitiered structures in which smaller subgroupings of animals (e.g., nuclear families or other units consisting of animals that spend virtually all their time within sight or sound of one another) come together occasionally for specific collaborative purposes. Because most individuals in a nonhuman primate society already know virtually everything they need to know about one another via direct observation, there is no need for them to have ethnic markers, which are mere proxies for assessing cooperative potential. Those who are encountering unfamiliar individuals—migrating individuals, for example—would only benefit from seeing ethnic markers if those markers provided reliable cues about the individuals' propensity for being cooperative relative to individuals that lack those markers. Because there are no overarching cooperative structures that extend into neighboring groups of primates, there is no reason to expect such differences except, perhaps, along kinship lines.

It might behoove a migrating individual to be able to recognize previously unknown genetic relatives if they were closely related enough for kin selection to favor cooperation with these individuals over nonkin. However, presumably less falsifiable and malleable phenotypic cues, such as physical resemblance or olfactory cues, would be more reliable sources of information about kinship than would culturally created signals.

The best place to look for ethnic markers in nonhumans is overlap zones between communities of animals with the following characteristics: intense cooperative relationships, fission-fusion structures, symbolic capacities, group-specific social norms, and large individual home ranges. Under such circumstances individuals might interact frequently with conspecifics whose social propensities they do not know from direct observation. If cooperative relationships are important and social learning is involved in establishing group-specific ways of conducting social interactions, then ethnic markers might be useful. Therefore, I find it more plausible that ethnic markers exist in cetaceans, birds, or bats than in nonhuman primates.

Many birds, bats, and cetaceans shape vocal production via social learning (Janik and Slater 2003), and group-specific calls could plausibly be used as ethnic markers. Intense cooperative relationships have been documented in all these taxa (de Waal and Tyack 2003), although we do not know whether there is between-group variation in the social rules used to conduct cooperative relationships. Furthermore, birds, bats, and cetaceans lack the tight synchronization of group movement characteristic of many primates, and individuals have great freedom of movement and presumably some possibility of encountering unknown individuals on a regular basis. Regrettably, the social fluidity and ability to move long distances quickly (by swimming or flying) make it extraordinarily difficult to study details of social relationships. Consequently, I doubt that we will have a detailed picture of true cultural capacities in these species in the near future.

ACKNOWLEDGMENTS

I thank Jeff Galef and Kevin Laland for the invitation to write this chapter, Kim Hill and my colleagues in the UCLA Behavior, Evolution, and Culture Group for helpful discussions, and Joe Manson and Jeff Galef for commenting on the manuscript.

# 12

## ANIMAL "CULTURE"?

KIM HILL

A large fraction of all human behavioral variation is not simply the result of different individual adaptive responses to local ecological contingencies. Instead, the variation is partially determined by socially transmitted information that we call "culture." Although this is true for other species as well, the degree to which human behavior is influenced by socially transmitted information is exceptional. Most important, while "culture" in other animals consists of socially transmitted techniques and technological traditions needed to achieve specific adaptive ends, human culture consists of two additional types of socially transmitted information that may be absent in nonhuman species. First, learned social regulations of individual behavior in human societies shift the associated fitness costs and benefits in ways that affect all behaviors, from feeding to mating strategies. Second, humans universally develop systems of symbolic reinforcement of those regulations and show elaborate forms of display to signal adherence to a specific rule system.

Despite this rather obvious contrast, for most of my career as an anthropologist and behavioral ecologist, I would have argued that differences between humans and nonhumans in the importance of social learning were simply a matter of degree, not of kind. The diverse arguments of my cultural anthropology colleagues against this view did not dent my commitment to this position for many years. During this time I was mainly researching economic behavior and life-history patterns and felt that models borrowed from evolutionary ecology were quite productive for explaining human patterns in these areas regardless of "cultural" affinities (indeed, the optimality modeling approach that I employed was entirely culture free). Of course, I was aware that human behavior included components of social transmission, and these might even produce occasional maladaptive anomalies, but I felt that the same was likely true for other social animals. To my mind, humans showed no qualitatively "unique" patterns, and the theoretical and methodological contributions from the study of animal behavior were impressive.

Although I still believe that systematic biologically informed research into human behavior is highly productive, two things have changed my view on behavioral modeling so that I now believe that human culture should be considered in all explanations of intergroup behavioral variability in humans. First, I spent almost half of the past 30 years living with remote hunter-gatherers and also observing primates in the wild. Despite the obviously useful application of animal behavior models to understanding human patterns, it is difficult to deny that even "primitive" hunter-gatherer societies seem extraordinarily complex in comparison with social primates. Hunter-gatherers themselves, who are quite sophisticated in their knowledge of animal behavior (Blurton-Jones and Konner 1976), soundly reject the idea that animal and human behavior are determined by the same forces. For example, indigenous friends that I met in Amazonia were upset when they were told that scientists like me were using "animal behavior" models (their view of "sociobiology") to explain indigenous cultural patterns. The suggestion that there were strong continuities in human and animal behavior was obviously insulting to them, and I began to wonder why. Similar stories surfaced in other areas of the world. Mardu Australian Aborigines were reported to identify with animals but viewed them as beneath humans because they copulated incestuously and failed to adhere to laws of kinship. The Hiwi of Venezuela pointed out that it was offensive to suggest that humans were "just another animal" because "animals have no shame" (i.e., guilt associated with breaking social rules). In short, hunter-gatherers, despite their affinity with nature, seemed to universally reject the notion that humans and animals were part of some simple continuum. Instead (and contrary to popular romantic myths), hunter-gatherers that I have known for over thirty years clearly hold the view that humans are qualitatively different from animals.

Second, when I began researching material for a synthetic overview of modern hunter-gatherers, I became bothered by a recurrent problem. Although behavioral ecological models clearly had the potential to explain many interesting patterns in the hunter-gatherer ethnographic literature, it seemed to me that a statistical control procedure would be necessary to demonstrate that fact. Specifically, casual inspection of my growing comparative database suggested that the strongest predictor of almost any hunter-gatherer pattern, whether it be polygyny level, infanticide rates,

270

warfare, food taboos, postmarital residence patterns, child-rearing practices, puberty rituals, or body piercings, was "ethnolinguistic membership." Groups from the same language families were often remarkably
similar in some dimensions and in improbable ways even when they lived
in different ecologies and somewhat distant from each other (this is
equally obvious when we examine migrant groups dispersed throughout
the modern world today). Likewise, geographic proximity appeared to
be a strong predictor of specific behaviors irrespective of habitat type.
Entire world regions of hunter-gatherers (e.g., Australia, California)
showed enough commonalities in behavioral patterns despite habitat,
linguistic, or genetic differences that such geographic "clustering" would
have to be statistically controlled before testing explicit "ecological models" of hunter-gatherer behavior. The apparent need for statistical control of linguistic family and geographic region in order to test the effects
of ecology on behavior (cf. Borgerhoff Mulder 2001) led me to recon-
sider the role of culture in human behavior. I began to recognize that
most of my colleagues who studied human behavior from an adaptive
perspective (behavioral ecologists and evolutionary psychologists) had
mainly avoided the issue of "culture."

The question whether animals have "culture" is a semantic argument
at one level; the answer is easily yes or no depending on definitions. If
culture is nothing more than socially learned behavioral variation, then
some animals are surely cultural. But for evolutionary biologists, the real
question is, do we gain or lose in our ability to explain the special properties of *Homo sapiens* by employing the same term for the social traditions observed in animals and humans (Byrne et al. 2004)? My concern is
that the loose application of the term "culture" for all socially learned
behavior may obscure our ability to understand the evolution of what
appear to be very unique characteristics of *Homo sapiens*. If humans are
indeed as exceptional as they appear to be, we must describe more accurately their special characteristics rather than eagerly lump them with
other social vertebrates.

I believe that anthropologists who work on human culture and biologists (for convenience I refer to all scientists who work on animal behavior as "biologists" in this chapter) who work on nonhuman "culture"
have adopted very different understandings of the word "culture." This
may be primarily responsible for disagreements in which biologists insist
that animals do have culture, but anthropologists do not accept that
view. Of course, there is also room for a third view, which is most

271

intriguing. We may accept the more elaborate anthropological definition of culture but still argue that some animal species show characteristics that meet this definition.

I am aware that some very astute researchers believe that arguing over the definition of culture is a useless exercise. Although I sympathize with the need to avoid semantic haggling, I do think that a clear specification of unique human properties is the first step to explaining the origins and evolution of our uniqueness. This does not necessarily mean rigid adherence to a culture/no culture dichotomy, but it does mean attempting to identify the characteristics that produce human uniqueness. Therefore, I intend to outline specifically what traits I think must be observed in any species to claim that it has "culture" as it has been defined in humans for more than a century. Here I focus on behavioral patterns rather than cognitive mechanisms. For example, it may be the case that "true imitation" rather than "emulation," "social facilitation," or some other social learning mechanism is required to produce cumulative culture, but from an observational point of view the lack of cumulative cultural adaptation in nonhuman animals is one trait that clearly distinguishes them from humans. Likewise, it may be that "teaching" is required in order to allow the social transmission of divergent social norms, but the resultant lack of enforced social norms are what notably distinguishes animals from humans (see below).

## Culture versus Social Learning

Biologists often equate socially learned behavior and resultant "behavioral traditions" with "culture." When local groups of animals, especially primates, exhibit suites of different behaviors not obviously attributable to local ecology, the groups are commonly referred to as different "cultures" (e.g., Wrangham, McGrew et al. 1994; Boesch 1996a; McGrew 2004; Sapolsky 2006), and such labeling has even led to a new field, "cultural primatology." Biologists have been particularly enthralled with the social transmission of feeding techniques and technology, from the milk-bottle-opening titmice to sponge-using dolphins and tool-wielding great apes. Because these observations combine both tool use and social learning, they seem especially similar to popular notions of human culture. Despite the popularity of equating animal with human culture, however, there have long been dissenters to this view (e.g., Galef 1992; Tomasello 1994). I agree that we should be very cautious about assuming that animal traditions based on some type of social learning are equivalent to human culture. "Culture" as it has been defined and used in anthropology for more than a century consists of behaviors far more complex than socially learned tool use or the social transmission of foraging techniques or rates of aggression.

In the paragraphs that follow I will define "culture" from an anthropological perspective informed by evolutionary thinking. I suggest that animal "culture" probably differs from human culture in ways that have profound implications. First, socially learned information does not accumulate significantly in any nonhuman species (the so-called ratchet effect; Tomasello 1999a). Social learning allows human *populations* to successively modify and accumulate adaptive information over many generations, leading to the cumulative cultural evolution of adaptive techniques, technologies, and social institutions. Second, however, even if animals did accumulate socially learned technologies, they still would not be "cultural" in the human sense because critical components of the human culture complex are absent. Indeed, we may eventually find evidence of hominin species that show cumulative socially transmitted adaptations but are still not fully "cultural" in the same sense as modern humans. Specifically, the content of what is culturally transmitted may be unique in humans. The critical and unique components of "culture" are sufficient for humans to display unique behavioral patterns among living organisms, and they probably explain why both cultural anthropologists and Australian Aborigines set humans apart from all other species.

The earliest clear definition of "culture," provided by the founder of cultural anthropology, E. B. Tylor (1871, p. 1), is still the most widely cited: "That complex whole which includes knowledge, belief, art, law, morals, custom, and any other capabilities and habits acquired by man as a member of society." This definition does not simply equate social learning and behavioral traditions with "culture" but specifies types of socially learned information that compose the culture complex. Despite the fact that scholars had recognized since Aristotle (see Galef and Heyes 2004a, p. 1) that animals engaged in socially learned behavior, Tylor clearly felt that the label "culture" applied only to humans. Although I think that Tylor's insight was important, I also believe that modern evolutionary behavioral biologists can do even better. Tylor's definition has two weaknesses. First, it applies only to humans *by definition* and thus does not allow us to consider whether animals have "culture." Second, it includes behavior (i.e., "habits") as part of culture. As Cronk (1999) noted emphatically, anthropological definitions of culture cannot include behavior as a component of culture if culture is then invoked to explain behavior. It is obviously circular to propose that behavioral differences (a component of culture) can "explain" behavioral differences (behavior). Biologists are also sometimes careless when they write about culture and behavior. To (unfairly) pick on one of the editors of this volume, Laland and Janik (2006, p. 542) defined culture as "all group-typical *behavior patterns,* shared by members of animal communities, that are to some degree reliant on socially learned and transmitted information." They followed by suggesting that alternative explanations for behavioral differences in animals must include "consideration of the interplay between genes, ecology, and *culture.*" But, behavior cannot determine behavior. If culture is supposed to influence behavior along with genes and environment, we must then have a definition of culture that excludes behavior, and we must consistently use the term "culture" to refer only to nonbehavioral traits. I believe that most social scientists would probably be comfortable with a simple definition of culture such as: a historically derived set of shared ideas, values, norms, and beliefs that underlie behavior. Cultural "products" would then include behaviors and objects.

Despite some weaknesses in Tylor's early definition of culture, he pointed out (as have most anthropologists) that culture consists of more than one informational component—it is a complex. Human "culture" is more than just socially learned tool use or food-getting "habits." Tylor proposed that "culture" is a "complex" of traits that includes regulations on individual behavior ("law") and the development of symbolic reinforcement apparatuses for that regulation ("religion and morals").

Since Tylor's time dozens of definitions of "culture" have been proposed, and anthropologists have never settled on a single definition. For example, Cronk (1999, Table 1) cites 20 different definitions of culture drawn from currently employed introductory anthropology texts. A handful of these must be rejected because they proclaim that "culture" is human by definition (e.g., Ferraro et al. 1994, p. 18) or too easily admit nonhuman animals to the culture club: "The capacity to use tools and symbols" (Bohannan 1992, p. 320). But the definitions that I believe are most useful to evolutionary biologists stress the complete set of informational components mentioned by Tylor. For example, Crapo (1996, p. 17) defines culture as "a learned system of beliefs, feelings, and rules for living around which a group of people organizes their lives." Scupin (1992, p. 414) declares that culture is "a shared way of life that includes material products, values, beliefs, and norms, that are transmitted within a particular society." Haviland (1996, p. 32) defines culture as "a set of rules or standards shared by members of a society, which when acted upon by the members produce behavior that falls within a range of variation the members consider proper and acceptable." These definitions all emphasize behavioral regulation and its symbolic reinforcement as critical components of culture.

Some evolutionary anthropologists have provided definitions of culture that may be too general. For example, Boyd and Richerson (1985) and Cronk (1999) propose that culture is "socially transmitted information." By this definition animals clearly have culture, but we must ask if there are some classes of information that are socially transmitted only by humans and whether or not it matters. If animals have culture too, why do they not form imaginary social boundaries around arbitrary groups of individuals that fight thousand-year wars and recruit suicide bombers to kill "infidels" who advocate a different lifestyle to be imposed on all? Perhaps in order to understand unique human traits we need to recognize the specific types of socially transmitted information that are a universal part of human culture but absent from nonhumans.

## Human Culture

I believe that human culture can be distilled to three critical components. Since culture is information, these are three categories of information that are socially transmitted. These components are universally found in all human societies. Many vertebrates show the first component, but no species that I am aware of has been reported to socially transmit the second and third components. This, along with the apparently unique human ability to generate extensive cumulative cultural change distinguishes humans from all nonhuman species.

The three universal components of human culture are the following:

*1. Socially learned techniques, technology, and environmental information (traditions, beliefs).* This component of culture consists of socially learned information about techniques and technology, as well as facts of nature and causal understandings of phenomena in the world. Because the information learned may be correct or incorrect, most scholars refer to these bits of information as "beliefs." Ways of doing things (traditions) and understandings of the world (beliefs) are passed along through a variety of social transmission mechanisms but ultimately must be adopted by each individual that displays them. Although the transmission mechanisms evolve by natural selection and thus must increase fitness on average, these mechanisms will inevitably produce some maladaptive behaviors (Boyd and Richerson 1985).

2. *Regulations of individual behavior enforced by rewards and punishments (norms, conventions, institutions, laws).* These regulations set up the rules for the competitive game of life and constrain how resources and mates can be acquired. In evolutionary terms, norms change the fitness costs and benefits associated with behavioral alternatives. In human societies norms are institutionalized and apply to members of one abstract social category specifying the sanctioned rights and responsibilities toward members of other abstract social categories. This component of culture can frequently result in behaviors not predicted by acultural models. The different rules of any cultural group may arise spontaneously through response and counterresponse, or through some overt social bargaining process (this means that norms can be imposed to serve only the interests of a small group). Social rules of behavior are often explicitly developed to solve potential intragroup conflicts in the most efficient way and to facilitate group-beneficial cooperation in the face of public goods problems. Because many cultural rules encourage individual altruistic behavior that serves the common good and are backed by social punishment, this area of culture is closely tied to topics of "altruistic punishment," "prosociality," and "strong reciprocity" (Bowles and Gintis 1998, 2004; Gintis 2000; Fehr et al. 2002; Henrich and Henrich 2006). Strong reciprocators show "a propensity to reward those who have behaved cooperatively and correspondingly to punish those who have violated norms of acceptable behavior, even when reward and punishment cannot be justified in terms of self-regarding, outcome-oriented preferences" (Bowles and Gintis 1998, p. 1416). A defining feature of "strong reciprocity" is altruistic punishment: the willingness of individuals to punish noncooperators at a cost to themselves, even when the cooperative act cannot benefit them (or kin) directly (e.g., Fehr and Gachter 2002; Boyd et al. 2003; Fehr and Fischbacher 2004b).

Cultural rules are most likely to develop in order to regulate behaviors that can powerfully affect the fitness of a large number of group members. For example, among Ache hunter-gatherers there are strong cultural rules about cooperation during hunting and redistribution of the kill after a successful hunt (Kaplan and Hill 1985; Hill 2002). Those rules have tremendous potential impact on both overall group food production and the daily share that is received by each individual in the social group. Indeed, the most common cultural rules in hunter-gatherer societies are about dividing up "resources," such as potential mates (marriage rules), acquired food resources (sharing rules), and access to food resources (territoriality), and about regulating conflict (rules for ritual combat, warfare, and settling disputes).

*3. Symbolic means of reinforcing, and signaling adherence to, a specific rule system.* The third component of human culture is a system of signaling in the form of rituals and ethnic markers, which exists in conjunction with the preceding rules component. Cultural signaling sessions are public and emotionally charged, using nonverbal channels in a highly effective fashion to reinforce the regulatory status quo (ritual) and often implying that the norms are linked to supernatural rewards and punishments (religion) in addition to reinforcement by a large majority of peers. They are designed to produce an emotional investment in the continuation of the rules (morality). Enforced rules are internalized to form values when individual actors deduce that a specific rule system serves their interests. Developing juveniles and adults gradually become committed to their own rule system because they have a comparative advantage under that system. The elderly are expected to be most conservative because they become experts at playing the game of life according to a particular set of rules that are often complex and difficult to master and they would be at a disadvantage playing under a new set of rules. They learn the effective ways to bend the rules to their advantage, as well as to impose them on others in ways that favor their interests. Because "rule abiders" generally prefer to interact with others who play by the same set of rules, signaling adherence in the form of adornments, dialects, ritual participation, and other behaviors (ethnicity) emerges as a means to obtain social partners, allies, and mates (cf. Boyd and Richerson 1987; McElreath et al. 2003).

The general idea that human culture consists of these three components has been recognized for more than half a century. For example, A. R. Radcliffe-Browne's (1922) ethnography of the Andaman Islanders is divided into three sections corresponding to the three components of culture: (1) utilitarian customs; (2) moral customs; (3) ceremonial customs. To date there is no animal "ethnography" that describes a complex of these three components for any nonhuman species.

## Animal "Culture"

Animal behavioral variants produced by social learning appear to be limited to two types of traits:

1. Locally common socially learned techniques and technology, behavioral tendencies (e.g., preferred mate choice, preferred foraging patterns, predator-avoidance tactics, and typical levels of aggression), and information about the world (tradition). These correspond to the first component of human culture but there is no evidence of cultural evolution or complex cultural adaptation that cannot be invented in a single generation through individual learning.

2. Socially learned symbolic mating and territorial songs and dances and the like (display). These mark territorial "in-group" membership, species membership, or alliance membership, but they do not appear to signal adherence to any socially enforced system of behavioral regulations. Whether these displays are conceptually similar to human ritual and ethnicity is something that must be carefully examined. Some of these traits (e.g., birdsong) do appear to change systematically through time, leading to new variants that are then socially transmitted to subsequent generations. This can lead to increasing signaling complexity in a few cases such as the coordinated choral structure of local groups of plain-tailed wrens (Mann et al. 2006), but this ability has not translated into more complex cooperative behavior (seen in all human societies) that allows for the utilization of new habits, resources, elimination of competitor species, and so on.

Most commonly reported examples of socially learned traditions in animals concern feeding techniques and technology, social interaction such as grooming, and greeting patterns and territorial or mating displays. Any recent review of the social animal learning literature (e.g., this book) contains dozens of examples of these in a wide range of vertebrates. Less commonly reported but equally important are socially learned mate-choice preferences (e.g., Galef and White 2000; Witte and Noltemeir 2002), aggression patterns (e.g., Sapolsky 2006), and the social transmission of predator-avoidance behavior (e.g., Griffin 2004). There is no doubt that over time the list of specific socially learned behaviors observed in nature will grow quite large and include categories as diverse as niche exploitation, nest-building techniques, and parenting behavior. When all these are considered, however, only the observation of social traditions in greeting and mating or territorial displays fall outside the first component of human culture. But unlike the human cultural signals, these social displays do not signal adherence to a rule system in order to attract partners who adhere to the same set of behavioral regulations. Likewise, there are no animal rituals designed to reinforce a social rule system or elicit emotional commitment to a particular rule system (morality). Thus critical components of human culture appear to be lacking even in our closest phylogenetic relatives. For example, none of the 39 culturally variant traits reported in Whiten et al.'s (1999) article on chimpanzee "culture" is a clear example of the second or third components of human culture, but a similar study of regionally independent human populations would show hundreds of examples of cultural variation in these components. In short, animals show socially learned traditions, but there is no evidence in any animal for socially learned conventions, ethics, rituals, religion, or morality, which are critical and universal components of human culture.

## Examples of Hunter-Gatherer Cultural Regulations

I have suggested that animal social traditions do not constitute culture as defined by anthropologists because animals do not engage in socially learned regulation of behavior enforced by third-party punishment and reward. Although modern societies and large-scale state societies throughout history have developed an almost infinite number of formalized behavioral regulations codified in written laws, even the simplest hunter-gatherer societies illustrate this universal human pattern. Hunter-gatherers universally regulate access to valuable resources (e,g., food, mates) and regulate how competition for these resources may be legitimately expressed. And hunter-gatherers sometimes formally specify punishments or payments imposed on those who violate the norms. Here are some examples of the most common regulations reported in hunter-gatherer ethnographies, and many societies show nearly all the regulations mentioned on this list.

### Social Norms Regulating Behavior

1. Mate access
   a. Prohibitions and prescriptions (applied on the basis of age, kin, or ritual-group membership)
   b. Polygyny (degree allowed and who may practice it)

2. Food production
   a. Land use (territoriality)
   b. Specific resource rights (ownership of specific plants or animals)
   c. Niche specialization (informal trade unions)

3. Food redistribution
   a. Sharing (who receives, how much they receive, and, what body parts of some game species)
   b. Consumption taboos (applied on the basis of age, kin, or ritual-group membership).

4. Display rights (ritual participation)
   a. Mating (who may participate in organized displays that are important forums for mate choice)
   b. Other sociopolitical messages (who has the right to "broadcast" symbolic messages and in what context)

5. Access to kin and other allies
   a. Residence rules (who is allowed to reside with close kin)
   b. Activity and ritual regulations (who can be a member in some organized activities)

6. Political power
   a. Designated positions (reserved for specified age, sex, kin or ritual-group members).
   b. Transfer of power (rules of succession, turn taking, or context-specific leadership)

7. Regulation of violent conflict
   a. Within-group contests (ritual dueling, divining, and justice)
   b. Participation in social group defense (who may or must defend the group and in what contexts)

8. Regulation of life history
   a. Age at first reproduction (acceptable age for sexual relations and marriage)
   b. Investment in infants and juveniles (who must invest and in what contexts)
   c. Age- and sex-specific rights and restrictions (that change during the life course) in competition over resources

## Human-Specific Psychological Mechanisms of Culture

Because human and animal culture contain different components, we might expect that different psychological mechanisms underlie them. The first component of culture, socially learned techniques, technology, and information, is produced through at least half a dozen different learning and imitation mechanisms, only some of which may be distinct in animals and humans (e.g., Tomasello 1999a; Whiten et al. 2004; Byrne 2005). However, the second and third components of human culture, norms and their symbolic reinforcement, may emerge through additional processes including teaching, strong conformity bias, and cognitive mechanisms that add value judgments to the behavior of third parties. Cultural regulatory systems are backed by strong emotional responses to those who break them. Thus, when norms are transmitted, juveniles both accurately imitate the actions of adults, and also incorporate a moral sense that to do things otherwise is "wrong" (Tomasello, this volume). Humans experience feelings of anger, fairness, justice, indignation, guilt, and so on and categorize other humans as "jerks," "assholes," "self-centered," "egotists," "sleazeballs," "criminals," "villains," and so on when they violate social regulations. Individuals in different societies react to different behaviors (e.g., copulating with a clan member or an underaged partner, eating a taboo food, or speaking to a parent-in-law) as "disgusting," "revolting," "repulsive," "vile," "abhorrent," "deranged," and so on. There is little evidence that primates show similar emotional responses to deviants who fail to adhere to the local socially learned traditions.

Human social regulation of cooperation and competition has led to the codification of personal and property crimes and the recognition that some individuals are "victimized" by those who fail to adhere to social norms. Additionally, for humans, some "crimes" have no obvious victims (e.g., prostitution, drug use, consensual homosexuality), but still induce an emotional reaction and are punished by observers, most likely because people are invested in the rules themselves and in a belief that they can compete most successfully if certain behaviors are disallowed.

When a baboon is attacked by another larger baboon, it may feel fear and think about how to escape. Humans, in addition, categorize events of conflict (but not with computers or other animals) as "justified" or "unjustified" and have a strong emotional response to the latter. For example, experiments have shown that people experience a strong emotional reaction when human partners in the ultimatum game make "unfair" offers (highly unequal division of the original stake that favor the divider), but they do not show the same response to a computer that makes the same division (Sanfey et al. 2003). Instead, they simply accept the skewed offer if it will maximize their monetary gain. Chimpanzees seem to react the same way when playing each other in the ultimatum game (Jensen et al. 2007).

Humans, on the other hand, appear to have evolved an emotional mechanism to promote cultural behavior that we refer to as "conscience." Conscience is revealed in behavior as a concern for the welfare of others, or an "other-regarding preference" (Fehr and Fischbacher 2003). To date, it appears that most animals, including chimpanzees (Silk et al. 2005; Vonk et al. 2007), do not show prosocial tendencies but are instead "sociopaths/psychopaths." Hare (2006, p. 58) describes psychopaths as "intraspecies predators who use charm, manipulation, intimidation, and violence to control others and to satisfy their own selfish needs. Lacking in conscience and in feelings for others, they cold-bloodedly take what they want and do as they please, *violating social norms* and expectations without the slightest sense of guilt or regret." Such behavior, while perhaps typical in most animals, is considered pathological in humans and is estimated to be expressed by only about 1 to 4 percent of the population (Pitchford 2001; Stout 2005). Psychopaths show no remorse, empathy, anxiety, or guilt. These may be the emotional adaptations most associated with human culture and strong reciprocity, as well as the features of human cognition that are more important than imitation mechanisms for distinguishing human from animal culture.

Researchers have long recognized that there is a potential adaptive niche for sociopaths even in human societies that punish such behavior (Harpending and Sobus 1987; Mealey 1995). The continued prevalence of purely self-regarding rule breakers in all human societies probably indicates that alleles producing emotional indifference to norms exist over extensive time periods in a balanced ESS with more cooperative norm-abiding alleles (Bowles and Gintis 2004). Indeed, recent studies also suggest that a minority of humans show no tendency to engage in reciprocity/cooperation (Fischbacher et al. 2001; Kurzban and Houser 2005). But purely self-regarding individuals can be induced to behave as cooperators if they believe that the majority of other individuals in their society are strong reciprocators (Camerer and Fehr 2006).

The deep feelings of guilt and anger over injustice are emotional responses that are expressed in all human societies. The universality of these traits implies that their origin took place before the last common ancestor of modern humans. The existence of strong automatic emotional responses to breaking agreed-on social norms indicates that punishment of those who cheat on social norms has been a component of human natural history long enough for the emotional mechanisms to evolve. But since we know little about the rate of evolution for such mechanisms, we must rely on other evidence to ascertain when cooperative social norms first emerged.

## Even Neanderthals May Not Have Had "Culture"

Despite the fact that some biologists propose "culture" as a characteristic of numerous social vertebrates, I think it quite possible that complete "culture" as defined by anthropologists is very recent and unique to *Homo sapiens*. Some hominins may have evolved social learning mechanisms that produced cumulative cultural adaptations early on. Achuelian tools may be early signs of this (depending on whether hand axes were the goal or by-products of tool making), as is perhaps the ability to live at far northern latitudes by 1.7 million years ago (Vekua et al 2002; Zhu et al 2004). Throwing spears by 400,000 years ago (Thieme 1997) seem too complex to explain without progressive cultural adaptation, and the Mousterian tool kits used by Neanderthals indicate cumulative improvement on earlier technological traditions that culminates in improbable complexity—composite tools and hafting (d'Errico 2003). But none of these provides evidence for enforced social norms or the signaling that promotes them.

While direct evidence of social norms might be difficult to obtain, the ethnic signaling of adherence to a specific set of norms is more perceptible in the archeological record. Based on that evidence, complete "culture" including social norms and ritual and ethnic signaling, may have existed only for the past 160,000 years or so (Henshilwood and Marean 2002; Marean et al 2007; McBrearty and Stringer 2007) and only in *Homo sapiens*. However, the earliest date of ethnic marking is difficult to ascertain because the earliest evidence of symbolic display comes in the form of pigment use, something that may represent extrasomatic mating display of phenotypic quality (similar to bower birds) rather than ethnic marking.

283

Whether or not Neanderthals maintained social norms and ethnic marking is unclear. Many anthropologists have noted apparent cognitive differences between modern humans and Neanderthals. Wynn and Coolidge (2004, p. 468) note that Neanderthal technology is functionally as advanced as that of early humans, and hunting and gathering techniques were similar. "But . . . Neandertals never fully acquired the trappings of modern human culture . . . Such a pattern of coexistence for millennia without acculturation does not fit easily into anthropology's understanding of culture process." "It appears necessary to posit a cognitive/neurological difference between Neandertals and modern humans." Indeed, the lack of Neanderthal acculturation to human patterns (presuming likely exposure) is reminiscent of the chimpanzee ability to faithfully transmit food-getting techniques along a chain of social transmission within and between groups (Horner et al 2006; Whiten et al 2005) but the inability of chimpanzees to socially acquire a more efficient food-getting technique when exposed to a new model after they have already learned a less efficient one (Marshall-Pescini and Whiten in press). In any case, Neanderthals show only minor evidence of norms, ritual, and ethnicity, and the observed Neanderthal traditions might best be labeled "protocultural." After detailed examination of early modern human grave goods, Wynn and Coolidge (2004, p. 480) conclude that "such burials stand in stark contrast to Neandertal burials, where there are few, if any, convincing examples of grave goods." Evidence for Neanderthal art is ambiguous or lacking, although they may have used ochre or mineral pigments (d'Errico 2003), but again this is possibly extrasomatic mating display rather than ethnic signaling. If Neanderthals did develop simple social norms and practice incipient ethnic signaling, this does not seem to have led to extensive cooperation and between-group competitive advantage that produced spectacular cumulative cultural evolution as is typical in humans (Soltis, Boyd, and Richerson 1995; Boyd and Richerson 2002; Henrich 2004b). This lack might possibly explain the replacement of Neanderthals by *Homo sapiens* (Hill et al in press).

Whether Neanderthals possessed the evolved emotional underpinnings of cooperative morality is also difficult to assess. As Tattersall (2002, p. 129) points out, "We have no idea what the Neandertals were like temperamentally: whether they were aggressive or retiring; cooperative or individualistic; forthright or sneaky; trusting or suspicious; crude or lovable; or like our own species, all of the above." Although Neanderthals

were heavily dependent on game and probably shared their kill with social group members, so do many other species of social hunters, especially
the cooperative breeder species. But did Neanderthals develop extensive
cooperation between nonkin through enforced social regulations and did
they evolve the accompanying emotional mechanisms associated with
norm enforcement? If so, there is little archeological evidence that they
engaged in norm-promoting rituals and symbolic ethnic marking found in
all societies of *Homo sapiens*.

## Conclusion

Animal "traditions" appear to differ from human "culture" in two ways
(cf. Whiten, this volume). First, there is no significant accumulation of information through social learning in nonhuman animals, leading to cultural products of ever more increasingly adaptive complexity. Although
some animals have shown rudimentary cultural accumulation (e.g.,
crows: Hunt and Gray 2003; macaques, as described in McGrew 2004,
p. 23), the difference between animals and humans is likely to be rooted
in some deep cognitive differences. Humans alone appear to have a
strong innate taste for engaging in unrewarded imitation, especially in
childhood (but see McGregor et al. 2006) and an orientation toward
"shared intentionality," helping, conformity, and attaching moral judgments to others who are unwilling to conform (Werneken and Tomasello
2006; Tomasello, this volume). Finally, human cognitive mechanisms
that produce modifications (often much later in the lifespan) of previously copied behaviors may differ in ways that lead to a greatly expanded tendency to innovate the methods that can serve as models for
the next generation of social learning.

Second, there are important content differences in cultural traits. Animals do not engage in social transmission of certain types of information that result in the social regulation of behavior, or in symbolic
reinforcement of particular systems of rules and institutions that regulate
behavior. Animal species that exhibit local traditions but do not have
moral systems and do not reinforce social rules with symbolic display or
signal adherence to specific sets of norms do not have "culture" as it has
been defined by anthropologists. Local behavioral variants of nonhuman
species should thus be referred to as "traditions," not "culture."

In order to employ the term "culture" consistent with traditional usage and without the term carrying a separate meaning for humans and

for other animals, biologists must provide empirical evidence that is still lacking. For example, chimpanzees engage in meat sharing, and it is patterned, but for this to be "cultural," there must be observations of altruistic third-party punishment imposed on individuals who break meat-sharing conventions. Otherwise sharing patterns may simply represent a combination of self-interested foraging, reciprocity, unacceptable costs of resource defense, and costly signaling, all determined by the size and timing of prey acquisition by coresident group members (e.g., Gurven 2004). Finally, chimpanzee meat sharing would not be considered fully "cultural," in the anthropological sense unless chimpanzees engaged in symbolic display to signal which meat-sharing norm they adhere to, and engaged in symbolic activities designed to encourage group members to continue practicing the same norm in the future. If chimpanzees did regularly develop alternative socially transmitted sets of norms about meat sharing, described above, we might also expect evolved emotional responses of injustice, indignation, or anger upon observing another chimpanzee breaking those rules, (even when the observers were not directly affected). This, I believe would indeed make chimpanzees "cultural." (It would probably also qualify them as "human" for purposes of human-chimpanzee interaction—a goal that has long eluded animal rights activists.)

The view I present implies that "culture" has a moral component. I am aware of occasional reports of possible primate morality (e.g., de Waal 1996a), but retribution against defection in reciprocity is not equivalent to evidence of "morality." Selfishly amoral individuals are expected to reciprocate if it increases the likelihood of future fitness gains, and to object to defection against them. Thus, contingent reciprocity in some primates (e.g., Hauser et al. 2003) does not demonstrate commitment to a set of "moral" principles. Even observations such as the reported rejection of unequal rewards in capuchin monkeys (Brosnan and de Waal 2003) does not indicate the willingness to pay a cost to uphold a set of values (cf. Henrich 2004a). Only if primatologists report observations of nonkin altruistic punishment of norm violations will the proposal of primate morality be convincing. Morality is demonstrated by altruistic third-party punishment. To my knowledge the only potential examples of this come from specialized cooperative breeders and maybe eusocial insects where individuals who do not contribute to raising offspring are often punished, and possibly by individuals unrelated to the juveniles being raised (Clutton Brock 2006).

Culture is the apparent hallmark of human uniqueness, and whether any other species shares this trait, even in rudimentary form, is an extremely important evolutionary question. As scientists, accounting for exceptional human adaptability is one of our most exciting challenges. Culture is an obvious candidate for human uniqueness, but explanatory models that focus only on traditions and transmission mechanisms may not bring us the full understanding we seek. A focus on the unique qualities of transmitted information in human societies may also be critical. The study of complex adaptive behavioral rule systems, how they emerge, how they change, how they spread, how they are stabilized, and how they are enforced, will probably bring us closer to understanding the human-animal gap than the study of cultural techniques and technologies. A chimpanzee that can make a spear is still not human. Perhaps my hunter-gatherer friends are correct in their belief that humans are not just another animal. It would take some very rigorous science and a number of new observations to show that this commonsense view is incorrect.

# 13

# PEACEKEEPING IN THE CULTURE WARS

KIM STERELNY

They made a wasteland and called it peace.

Tacitus

*The Question of Animal Culture* is a splendid and challenging set of essays that as a whole indicate the importance of behavioral traditions in many species of animals. The establishment of these traditions makes local populations more homogeneous while it increases differentiation between these populations; traditions result in a distinctive metapopulation mosaic. That is, local populations have distinctive behavioral repertoires that contrast with the repertoires of other local populations of the same species. Moreover, there is a convincing, if not quite decisive, case that these local repertoires are socially sustained, although probably ecological (and perhaps genetic) factors are important too. The evidence is partly ethnographic, based on field studies of primates and cetaceans (although Laland et al. and Galef review data on other taxa too), but it is also experimental. These experimental data have led to a consensus that some nonhuman primates (and perhaps some cetaceans) are capable of observational learning. In sum, there is experimental evidence that these agents have the cognitive capacities necessary to stabilize traditions by agent-to-agent learning, and the ethnographic evidence suggests that in some species those traditions are important and numerous. No one is fully satisfied with this evidence, and so there is a good deal of discussion in this collection about how it can be strengthened. But as I see it, the take-home message of this collection is that socially stabilized local traditions are an established and arguably important feature of the lifeways of a fair range of animals. However, just how this phenomenological pattern should be incorporated into an evolutionary context, and what it means for our understanding of human evolution, are less clear.

The project of identifying the existence, limits, and evolutionary significance of animal cultures is not one of grand metaphysics. Skepticism about the idea that chimpanzees, orangutans, or cetaceans are encultured creatures is not based on a lingering Cartesian dichotomy between humans and the rest of nature or on an excessive faith in reductionist strategies in the natural sciences. Rather, as Michael Tomasello and Bennett Galef make clear, it is based on clear and observable differences between human lifeways and those of other animals. Human lifeways are based on an elaborate division of labor, extensive technology, normatively regulated interpersonal relationships, and extensive information about the local environment. No other animal seems to depend on this interconnected set of capacities. If that is right, this difference calls for an explanation, and one candidate explanation is the idea that we alone among living animals are encultured. Although this might not be the right explanation, it is not illicitly Cartesian or reductionist.

A number of essays in this collection mention "raising the bar": the complaint that as soon as an apparently distinctive human capacity is found not to be distinctive after all, it ceases to count as a criterion of culture. Of course, it is possible that the differences between human and nonhuman sociality are ones of degree (as Tomasello now suspects) or that they consist in a distinctive ensemble of capacities rather than in a unique, human-making adaptation. But if one project is to explain the difference between human and other primate lifeways, "raising the bar" need not be ad hoc. In the animal culture literature one core preoccupation has been that of explaining human distinctiveness, but this preoccupation has been developed in a new way in this collection. Earlier explorations of the similarities and differences between human and nonhuman social learning have been preoccupied with the problem of fidelity (Galef 1992; Tomasello 1999a, 1999b; Avital and Jablonka 2000; Alvard 2003). That is because a distinctive feature of human lifeways has been the evolution of complex adaptations to the specific challenges of local environments—adaptations that depend on cultural rather than genetic mediation of similarity across generations. Since the cumulative evolution of complex adaptation depends on high-fidelity inheritance, it was natural to focus on those forms of cultural transmission that support high-fidelity inheritance. Michael Tomasello, especially, saw debates about animal culture through this lens, arguing that specific forms of social learning were necessary for cumulative cultural evolution. In particular, in his picture true imitation plays a pivotal role. It is important because it enables complex adaptations to be built incrementally; imitation, by making cultural transmission accurate, makes hill-climbing solutions to adaptive challenges possible (Tomasello 1999a, 1999b). Tomasello's challenge has led to a productive decade in which researchers have probed the mechanisms of cultural transmission in nonhuman animals.

In this collection Galef, Perry, and Hill all emphasize another distinctive feature of human culture. Susan Perry and Kim Hill, in particular, are most concerned with the nature and explanation of variation. How are the differences between groups sustained and explained? What do those differences mean to their members? Human life is regulated by norms and by adherence to those norms. Moreover, those norms tend to differ from group to group: norms play an important role in making one human community different from others. We live in "symbolically marked" groups: we have rituals, modes of dress, and ways of speaking and acting whose function is to signal joint membership within a community and to distinguish one community from another. Human groups are not distinguished from one another just by different sets of behavioral regularities. Quite often these regularities are emotionally and normatively loaded. The way we do things is the way we should do things, not just the way that happens to work best around here. An account of the evolutionary origins of human culture has to give an account of the roots and the function of normativity and of group identity, and of the connection between them (see, for example, Joyce 2005).

As I noted earlier, in many of the essays in this collection, there is a focus on the differences between local populations. Such differences are seen as a potential signal of the operation of social learning, and social learning in turn is a core element of culture. The idea is that because different innovations will spread through different local groups, social learning will make local groups more homogeneous as members of the local population learn from one another, while it will increase metapopulation heterogeneity. No one thinks that this pattern of local homogeneity linked with metapopulation heterogeneity demonstrates the existence of culture in a species, but a major theme of this collection concerns what we need to add to this pattern to make that inference reasonable. However, for Perry, in particular, who is interested in the evolution of group identity and of normative regulation, this pattern is of interest not just as a signal of social learning but also as a potential prototype of culturally identified groups. Groups with distinctive repertoires of traditions might be prototypes of groups that are aware of themselves as groups and mark that awareness with learned signals of local identity. However, although Perry has documented many locally specific social signals in her capuchins, she herself doubts that these are part of a primitive form of symbolic group identification. For her, fine-scale capuchin ethnography seems to show that these signals are about building affiliations within groups; they are not signals to other groups.

I share her view that the roots of normativity will be much more elusive than the roots of fidelity. Fidelity is empirically tractable. As Whiten shows in this book, we can document the fidelity of transmission chains in chimpanzees. Normativity is much less scrutable because it has no overt ethnographic behavioral signature.[1] Perry thinks that a number of animal societies show a sense of group identity, but she doubts whether the sense of group identity is linked to socially learned signals of membership that distinguish one group from another and identify one group to another. I doubt that even group identity has been shown. It is one thing to distinguish familiar individuals from unfamiliar ones. It is quite another to think of the familiar individuals as forming a collective that also includes you. If experimental economics and the associated modeling literature are a guide, norms evolve with the elaboration of cooperation and the division of labor (Bowles and Gintis 2003; Fehr and Fischbacher 2003). I doubt that any of Perry's animal societies are cooperative enough to exemplify the earlier stages of the coevolution of norms, cooperation, and the division of labor. Kim Hill suspects that animal culture might model one critical element of human culture—the social amplification of skill—but not others. He may be right. Thus one skeptical reservation about the animal culture literature is that although it may throw some light on the evolution of socially mediated ecological competences, it is much less clear that it will shed light on the evolution of norm-guided action.

## Culture and Social Learning

Obviously, identifying a practice as cultural or traditional depends in part on what we mean by "culture" or "tradition." It is important to be explicit about definitions, about our criterion of a tradition or a cultural practice. But it does not follow from this that there is a clean distinction between a purely conventional task—choosing a criterion of culturally guided action—and the genuinely empirical project of assessing whether that criterion is met in the particular case of, say, chimpanzee ant dipping (for the contrary claim, see Whiten's essay). Definitions are not stipulative, because they embody a research community's assessment of explanatory targets and causal importance. For example, on one view, we should have a rich criterion of cultural transmission that includes the specific proximate mechanism of observational learning. That suggestion is motivated by the idea that human behavioral ecology must explain the cumulative growth of expertise, and that growth can only be explained by high-fidelity cultural transmission. Rich definitions of cultural transmission are tied to hypotheses about the causal basis of fidelity and to ideas about the role of fidelity in especially important forms of cultural change. Definition choosing is not stipulative; rather, definition choosing is hypothesis choosing. Philosophers sometimes write of "natural classification systems" (Ridley 1986; Ereshefsky 2001). A classification system is natural if the sorting trait coincides with other differences, including ones yet to be discovered. If a criterion of culture in animals or of tradition in animals is well chosen, those animals will share a distinctive causal syndrome. If chimpanzees and orangutans live in traditional groups, and gorillas do not, then the lifeways of chimpanzees and orangutans are importantly similar to one another and importantly different from those of gorillas in ways we have yet to discover.

I sketch this background in part because the status of definition is an explicit preoccupation of some of these essays. A common theme in these essays is that culture (or tradition) in animals consists in social learning plus some added component. There are various suggestions about the identity of that component, but something like a consensus has formed in this collection around the idea that a species is encultured if cultural transmission is both important and locally homogenizing. Social learning must be important to the development of a range of the animal's capacities. Most of the authors in this book do not consider a bird species encultured if it develops local song dialects, but no other trait, through social learning. But social learning must also be socially homogenizing, resulting in an ensemble of capacities that make local populations distinctive. Sargeant and Mann's essay is somewhat unusual in this context because they explore the possibility that cultural transmission increases local heterogeneity. If cultural transmission is strictly vertical rather than multisource (and thus a naïve individual's phenotype is influenced by many in the local group), it can promote family-level specialization.

Sargeant and Mann explore some foraging specializations among bottlenose dolphins that might exemplify this mechanism.

Somewhat surprisingly, the idea of social learning itself is treated as fairly unproblematic. It is true that in these essays the different species of social learning are recognized; they recognize the distinctions between social stimulus enhancement, emulation, program-level imitation, and observation learning. But these complications only touch on the equivocal nature of the concept of social learning. For one thing, despite Kevin Laland's role as a major theorist of niche construction, with the exception of a minor excursion by Laland and colleagues these essays do not explicitly discuss its role. Implicitly, then, this collection takes niche construction to be of peripheral importance; I think that it is central. Parents structure the environment in which the next generation learns, and in doing so, they can transform learning even if there has been no change in the internal cognitive machinery young animals bring to their world. In other words, the evolutionary transition from agents who learn mostly by individual exploration of their environment to agents that live in social worlds with stabilized traditions need not involve the transformation of the individual cognitive equipment of the agents. In a slogan: social learning is not an individual trait, it is an interaction. If this claim is right, it implies that learning regimes are evolutionarily labile; they might well vary even within a species. If this claim is true not just for social learning in general but for high-fidelity learning, then views of Tomasello and Galef are up for grabs too because they tie high-fidelity social learning to specific cognitive mechanisms. Let me say a bit to make these claims plausible.

Most simply, young animals often accompany their parents, and thus the environment that they explore in trial-and-error learning is structured by their parents' routines. This might well explain the transmission of technique from mother to calf among bottlenose dolphins. Sargeant and Mann remark that dolphins forage independently, except that juveniles accompany their mothers. They also note that differences in foraging techniques tend to be correlated with fine-structure differences in the adults' use of their environment. Those dolphins that use sponges to protect their rostrum while they forage in seafloor sediment tend to spend longer in deep channels than others; dolphins that exploit the shallow waters of beaches to catch their fish spend much more time in inshore beach settings. As they explore, the young of the sponge users will experience different environments from those of the calves of the beach exploiters. Likewise, there is some suggestion that female chimpanzees

293

both spend longer with their mothers and acquire the capacity to crack nuts faster than their brothers. If this is so, it is likely to be an instance of a quite general phenomenon.

In this collection there is a good deal of discussion of explicit teaching. On some views, it is very rare; on others, it is underreported. From a perspective that emphasizes the role of niche construction, though, teaching is just one way in which parents can structure the learning environment of their young. Another is supervised trial-and-error learning. If mothers monitor the exploratory behavior of their young, and if the young are responsive to signals of their mother's emotion state, mutual attention can reduce the cost of learning. The mother's signals of her emotion responses stand in for the world's error signals; signals of danger stand in for actual mishaps (Castro and Toro 2004). Similarly, if parents (or other adults) tolerate cadging as they process challenging resources, juveniles will be exposed to the intermediate products of skilled processing. If they can get their hands on a partly opened nut, a partly exposed fruit, or a partly shaped ant wand, their own processing problem will be an easier version of the one they will need to master as an adult. Among socially living animals, in which offspring accompany their parents for many years, and whose phenotypes are shaped by learning, it would be no surprise if mothers influenced the behavioral repertoires of their young profoundly but indirectly by organizing their learning experience. Although the interactions between learning and niche construction are not an explicit theme of these essays, this model fits many of the examples that are discussed explicitly. For example, de Waal and Bonnie suggest that information about social hierarchy in baboons, macaques, and rhesus monkeys is transmitted (and the hierarchy itself is stabilized) by maternal effects on the social exploration of a mother's young. Rhesus monkeys, for example, inherit from their mothers' association patterns: they play mostly with the young of their mothers' affiliatives, and those become their own affiliates. These primates learn their rank from their mothers' interventions (and noninterventions) in their own generational conflicts.

The essays in this book are mostly about large primates and larger cetaceans, though both Galef and Laland and colleagues review information about a much more diverse group of animals. The focal animals are large brained and long lived and act in complex and varied environments. We have good reason to believe that they are developmentally plastic. Learning plays a central role in the development of their behavior repertoire. It is very likely that much of this learning is hybrid learning:

they learn by trial-and-error exploration in an environment structured by parental behavior and perhaps with both their trials and their errors modified by their parents' actions. This evolutionary dynamic is plausible too: this route could begin as a pure side effect of the young accompanying the parents in their ordinary ecological activities. If the effects of this association on learning were adaptive for parent and offspring, it could then be fine-tuned by selection as the parents' activities become more salient to their young and vice versa. No expensive new cognitive technology is necessary to convert individual trial-and-error learning into parentally structured trial-and-error learning.

If this model fits the evolution of learning in primate or cetacean social environments, it is very likely that cognitive evolution in such environments also results in the modification of the internal cognitive machinery on which juvenile learning depends. For example, changes in perceptual salience and attentional focus are likely. De Waal and Bonnie's essay shows how important motivation is to what is learned. Thus changes in motivation can lead to changes in what is learned. Tomasello shows how crucial the focus of attention can be. In updating his classic (1994) essay on chimpanzee culture, he suggests that an important difference between chimpanzees and humans is in their attentional focus. For chimpanzees, the results of actions are highly salient and are likely to be the focus of their attention. Specific technique is less salient (although he now thinks that it is likely to be within the range of their competence). If this is right, a species might evolve the capacity for a rich, lifeway-transforming set of traditions through changes in motivation, social tolerance, attention, or perceptual salience. We do not have to suppose that there have been dramatic changes in internal machinery to explain dramatic transformations in what is learned. West-Eberhard (2003) develops the general evolutionary model into which this suggestion fits. She details a range of mechanisms through which environmental change interacts with mechanisms of adaptive plasticity to produce novel phenotypes, which are then genetically fine-tuned through further selection.

The uncritical appeal to social learning worries me because niche construction gets too little critical attention in these essays even though it undercuts a dichotomy between individual and social learning. I think that it is far from clear that there is a distinctive and identifiable form of learning, social learning, that contrasts with (say) individual trial-and-error learning in response to ecological circumstances. In turn, this dichotomized view of learning has been important in models of the evolution of learning. For example, it has been supposed that social learning is cheaper than trial-and-error learning but less reliable, both because of transmission error and because information has a limited shelf life (Boyd and Richerson 1996; Laland 2001). But I have a second and more serious worry about the distinction between social and individual learning.

## Social Learning, Learning about the Social, and the Method of Exclusion

In thinking about learning and its evolution, we need to distinguish between the content of the information (what the information is about) and the channel through which the learning agent has access to that information (social or asocial). Differing channels to the same content can differ in their reliability and cost. I can discover the presence of a leopard in high grass from the reports of others, from my visual inspection, or by a tactile encounter. This final channel is likely to be both very reliable and very expensive. This crucial distinction seems to be blurred in some of the essays in this collection. They shift between thinking of social learning as an information channel and thinking of it as an information content. A reliance on the method of exclusion has encouraged a focus on social skills rather than ecological skills. The idea is that if we focus on learned social signals, the pattern of local homogeneity and metapopulation heterogeneity cannot be explained as a reflection of ecological differences tracked by individual learning. But this changes the topic from how animals learn to what they learn, from channel to content. Think, for example, of whale song. Humpback whales acquire the appropriate seasonal and regional dialect for their particular population, and Whitehead remarks that ecological factors cannot possibly explain the difference between Pacific Ocean and Indian Ocean dialects. He says, correctly, that "there is no conceivable mechanism for such patterns other than animals listening to one another's songs and adjusting their own accordingly." But that is not because we know that the channel through which information flows is socially mediated. In this sense whale song would be learned socially if parents directed attention to songs and singing, for example, if they modeled simple versions of the songs, guided botched efforts before they annoyed large males, or reinforced more successful attempts. There is no evidence of anything like this. Rather, what is learned is social. Whales acquire information about the local dialect.

Whitehead's example shows that whales learn socially in the sense that they learn about their social environment. For this reason, in thinking about this example, none of the standard issues about an evolutionary transition from individual to social learning arise. There is no temptation to investigate the relative cost and reliability of individual versus social learning or to test whether the speed at which the local dialect changes is optimal for intergenerational or for individual learning. These evolutionary questions suppose that we are assessing the cost and reliability of alternative informational channels from a common content. But the channel/content distinction is not in play in the documentation of these phenomena. If there is an evolutionary question here, it is why songs are plastic, not why song learning depends on social input.

As far as I can see, the same shift in focus occurs in many of the primate examples. The information about local traditions in greeting, display, grooming signals, and the like shows (I agree) that these primates can learn about their social environment: they can learn social facts. This analysis is supported not just by documenting the existence of distinctive local traditions but with fine-scale evidence about the attentional focus of juveniles on adults and with evidence of association patterns. Animals that spend more time together are more similar in their patterns of behavior. In their essays Whiten, Perry, van Schaik, and Galef all underscore the importance of supplementing the method of exclusion with such fine-scale social data. But there is no evidence that these animals learn social facts socially.

In short, the existence of quite stable social traditions is compatible with each agent in these social worlds acquiring these skills by trial-and-error interaction with others, exploring its social world individually, and having its own patterns of action shaped by its successes and failures. Its exploratory behavior is shaped by the reaction of others as it pursues its own interests. This model can explain the existence of stable traditions. De Waal and Bonnie describe a couple of examples in which patterns of social interaction seem to be somewhat stabilized by individual accommodation to, and replication of, group practices. Rhesus monkeys brought up with dominant but more socially tolerant and reconciling stump-tailed macaques not only adopted their more peaceable ways while within mixed groups but also, when separated back into a single-species community, continued to show much higher reconciliation patterns. De Waal and Bonnie also cite a natural experiment reported by Sapolsky and Share (2004) in which a historical contingency removed the bullying males from a baboon community. Unsurprisingly, it became more peaceful. More surprisingly, the pattern of more peaceful interaction stabilized. Male juveniles that migrated in adjusted to its more peaceful ways. Individual agents learn about and adapt to their social environment (Sapolsky and Share 2004). The content of their learning is social.

I suspect that even Kevin Laland's favored example of tradition, mating sites in bluehead wrasses, is also an example of learning about the social. The mating sites of these wrasses cannot be predicted from local ecology: the sites change if the residents are removed and the location is repopulated. But mating sites, of course, are of direct fitness importance to newcomers. Contrast this case with that of a bar-tailed godwit that takes off from a mudflat in response to pied stilt alarm calls. These alarm calls are of no direct fitness interest to a godwit. Rather, the godwit is using the stilt as an auxiliary sense organ for detecting danger. Stilt behavior is merely an information channel for the godwit. The aggregation patterns of resident wrasses are not merely an information channel to the newcomers but an information target, for these newcomers need to interact with the residents. Their distribution is part of the fitness landscape, not just part of the informational environment.

In considering social information, we can distinguish between channel and content. We can learn social facts socially. I can discover that Alfred is a cheat through my own unfortunate experiences or by attending to his reputation. As we saw earlier in imaging scaffolded learning in a whale world, others' responses can simplify the social learning task of a naïve individual. But when we are considering nonhuman animals, drawing the content/channel distinction about social information will obviously be methodologically challenging, and that is why my reservations about the "method of exclusion" are rather different from those discussed in most of these essays. When we are considering ecological skills, I think that the method of exclusion is a decent, though fallible, indicator of the social mediation of learning. I think that it is much more problematic when we are considering learned social signals. When the information content is the agent's social environment, the method of exclusion cannot show that agents learn in a distinctive socially mediated or socially enhanced manner. The method of exclusion might show that agents learn about their social environment, but it cannot show how the agents learn about their social environment. So in contrast with the line of thought elsewhere in this book, I think that the method of exclusion is more powerful when the learning target is ecological than when it is social.

The method of exclusion is the idea that we can identify cultures or traditions in animal species by identifying distinct repertoires in distinct populations and by showing that neither ecological nor genetic differences explain those differences. There are complaints that this method does not deliver certainty, because we can never establish beyond doubt that different populations live in relevantly similar environments. Ecological differences can be subtle, and documenting the behavioral repertoires of different communities is a slow and difficult process. These difficulties in applying the method of exclusion are explored in these essays, and so there is a good deal on chimpanzee ant dipping. Chimpanzees eat ants by sticking twigs into their nests and eating the ants that attack the twigs, and chimpanzees are known to shape twigs for this task quite differently and to de-ant the twigs by using different techniques. These differences in technique and wand size were at one stage taken to be a clear case of a difference in ecological practice between different communities that had no ecological explanation. This is now more controversial (Humle and Matsuzawa 2002). It is true that ecological differences between sites can be unobtrusive, and that the method of exclusion cannot yield decisive evidence for the social causation of competence. But unlike the mathematical sciences, the natural sciences are not in the certainty trade. We cannot establish with certainty that moas are extinct in New Zealand; we cannot look everywhere simultaneously. Sadly, we still have excellent reason to believe that they have gone. So, too, very likely have the South Island kokako. But those are much smaller birds, and the North Island kokako, the sibling species, is shy. So our reasons for accepting that extinction hypothesis are not as strong. Likewise, if zoologists are explicit about the ecological triggering hypotheses they have considered, and if they are explicit about the evidential basis of the claim that two populations live in relevantly similar environments, we can estimate the strength of the eliminative inference.

A more serious objection to this method is that it undercounts traditions. As Laland and colleagues point out, we expect social learning to be crucial in establishing behavioral differences between populations that live in differing environments. That is why the capacity to learn socially has evolved. Since the adaptive value of the capacity for culture is to help agents adapt to their specific circumstances, we would expect social learning to interact with ecological differences to produce behavioral differences in just this way. Showing that environmental differences play a role in producing the differences in material culture between central Australian Aborigines and the Inuit does not preclude a role for culture. Environment and culture can interact to generate a competence; both might be necessary; neither might be sufficient. Carel van Schaik describes a plausible example of this interaction in discussing orangutan honey harvesting. An ecological difference between two sites (honey in tree holes was much more abundant in one site than in another) seems to have been socially magnified. In cases in which more than one adaptive response to the environment is possible, the social environment can determine the locally prevalent response by biasing the experimental and exploratory behavior of a naïve agent. Presumably, the method of exclusion detects cases like this, in which the contingencies of innovation history determine which of the adaptive responses to a common environment two particular populations find.

In short, the method of exclusion is conservative: it undercounts traditions. This conservatism is an advantage if the project is to identify a capacity, to show that traditions play some role in the life of a population. It is a problem if the project is to establish the extent to which the lifeways of a particular population depend on traditions, on social learning. The same, of course, is true of genetic causation: even if it is true that two populations differ genetically in ways that predispose one but not the other to acquire a specific behavior, social learning may well also be necessary for the development of that capacity. In many of the chimpanzee cases, differences in behavioral tradition covary with genetic differences (because different chimpanzee subspecies are involved). But that is no evidence that learning is not essential to establishing and maintaining those traditions.

Despite the complications and the need for caution, the method of exclusion can establish that ecological skills form a complex of traditions. In these ecological cases we have a clear and tractable difference between content and channel: what the information is about, and how the agent acquires the information. The ethnographic case for ecological skill is strengthened, of course, when it is supported by evidence that the species has the capacity to learn observationally, and where developmental and observational evidence shows that skill acquisition is not trivial. Thus Whiten's experimental program is very important. If we have reason to believe, via chimpanzee ethnography, that ant fishing by generation $N$ plays an essential role in generating the acquisition of ant-fishing competence of generation $N+1$, we know that generation $N$ was a channel through which information flowed. It somehow mediated the flow of information to the next generation, for the information the new generation acquires is about ants, not chimpanzees. I do not think that the method of exclusion shows that social signals are traditions, because that idea conflates content and channel. Once established, the stable reproduction of a local social repertoire does not look especially difficult to explain. In thinking about these cases and their significance, we need to focus on innovation and the establishment of the grooming handclasp and similar social signals. There may be a role for special cognitive adaptations in this process. How are such social signals established? An innovative foraging technique might be established by stimulus enhancement and niche construction. Animals that learn to tap into an unexploited resource will spend a good deal of time doing so, for it will be locally abundant, and there will be no local competitors. But that will change the fine-grained pattern of habitat use by their offspring and associates, who, no doubt, will be interested in what they have. But what are the early social dynamics of a new signal? How is it established?

## The Comparative Biology of Culture

Animals often act in environments that are informationally translucent. Relevant information exists in the environment, but the information is not freely available. The agent has to work for the information, acquire specific discrimination skills, take risks to acquire the information, or take the risk of acting on inadequate information. One possible response is social learning of the kind that sustains traditions. One project in behavioral ecology is to understand the costs, benefits, and preconditions of that adaptive response to the challenge of acquiring expensive but necessary information. This project is not at all tied to human distinctiveness and hence does not come with a culture concept that tags a key mechanism in explaining human distinctiveness. But as we have seen, a second reason for interest in animal culture is anthropocentric: to understand the distinctive trajectory of hominin evolution. It is clear that this explanatory project is important to many of these theorists. McGrew, de Waal, van Schaik, and Whiten (more equivocally) think that their claims about primate culture are important, in part, because if they are true, they tell us something about the evolution of culture in humans. But it is much less clear exactly how these animal

301

case studies are supposed to illuminate hominin evolution. Let me canvass two possibilities, neither, I think, satisfactory.

In developing and confirming an evolutionary hypothesis, comparative biology is almost always important. We can test evolutionary models of trait evolution by the patterns they predict in the relationship between trait and environment, For example, if intergenerational cultural learning is an adaptation to environments that change, but change slowly, we should find an appropriate pattern in the relationship between the capacity to learn from others and environmental lability. But to apply tests of this kind, we need to be able to reidentify the same trait in distinct taxa and to place our current pattern of trait/environment correspondence in a phylogenetic context. We need to be able to distinguish between cases in which a trait evolved in a particular type of environment from cases in which it merely failed to change once the species found itself in that environment. The dwarf forms of *Homo erectus* inherited most of their traits from their parental *erectus* species rather than evolving them on an island. Such inherited traits probably reflect the environment(s) of the widely distributed stem *erectus* species, not that of the hobbits. So, on the face of it, an evolutionary biology of culture requires us to map character states of culture onto the tree.

However, even if we had a well-developed evolutionary hypothesis that we wished to test comparatively, culture—or even social learning—is an awkward characteristic to place in phylogenetic context because culture itself is not a trait; rather, it emerges out of a mosaic of traits. Indeed, culture may not even be a trait of individual animals at all. In these essays it is quite often taken to be a characteristic of a local community rather than a trait of individual animals. That would not matter if culture as a characteristic of a collective were a symptom of a distinctive trait of the agents in the collective. It would not matter if (say) culture as a collective phenomenon reliably emerged from the ability and inclination to readily learn socially, for then we could track that trait through the tree. One of the points in calling social learning an interaction effect was to express my skepticism about the idea that there is a simple relationship between information flow at the group level and cognitive mechanisms at the individual level. Thus if the claims in the second and third sections of this essay are at all on the right track, traditions depend on the patterns of interactions within the group, not just on the internal cognitive machinery with which a naïve individual is endowed as he or she is confronted by an ecological challenge. Learning from the collective requires social tolerance: the naïve must be comfortable enough in the presence of others to attend to what they are doing rather than worrying about what their elders and betters are going to do to them. But it will often require much more. The relationship between mechanism and capacity is likely to be complex. The acquisition of the skill depends on how others act and respond, not just on the native equipment of the focal individual.

I have argued that hybrid learning, that is, socially structured or guided trial-and-error learning, is very important in human evolution. For example, learning a craft skill depends on a combination that includes the observation of skilled performance, practice, and the social correction of partially successful attempts (Sterelny 2006). My bet is that hybrid learning is important in the establishment of animal traditions too. Parents can supervise trial-and-error learning and intervene if juvenile play is becoming too dangerous. The role of the group, of course, is even more important when the skills that are being acquired are social rather than ecological, for these social responses will positively or negatively shape juveniles' exploration of their social world. I doubt that there is anything like a one-to-one correspondence between traditions in animal cultures and the mechanisms that support them. Yet it is these mechanisms, if anything, that can be mapped onto a tree. Of course, as Laland and his colleagues note, there are general, widely shared learning mechanisms that play an important role in human learning and are homologous with mechanisms in other animals. But those mechanisms are ancient and widely distributed, and for that reason their presence in human psychology does not tell us much that is distinctive about hominin evolution. Consider our closer relatives: the large-brained, tradition-shaped primates. Except perhaps for the capacity for observational learning, it is not clear that this group shares any distinctive cognitive mechanism whose evolution can be mapped onto a tree.

In brief, it is hard to apply comparative methods to emergent traits. And culture (and perhaps even social learning) is an emergent trait. A second idea is that our primate relatives tell us what early stages in hominin evolution were like, and that is important because it is hard to develop good models of those early phases of our evolution. The idea is that theories of human evolution face trajectory problems. It is not hard to assemble plausible models of the elaboration of language, pedagogy, cooperation, and the division of labor. For example, consider language. Once something like a pidgin system had evolved, it would have transformed hominin social life, and with it the selective landscape. The alingual would have been excluded from many benefits of coordination and information sharing. The adept would have gathered a disproportionate share of those same benefits. But how would such a system begin? Languagelike systems have setup costs, and they presuppose a cooperative social environment (since sharing information poses the same cooperation and defection problems as sharing any other resource). We might hope that animal culture can shed light on the evolution of human culture by providing models of early (and perhaps especially puzzling) stages in the hominin transition to ultrasociality. I wonder whether this hope will be fulfilled. The crucial question is whether these animal communities exemplify in a simple and more tractable form important aspects of the human adaptive complex: whether they are cooperative enough, are communicative enough, and have enough social complexity and division of labor to be reasonable models of early hominin evolution.

Overall, then, I think that understanding animal cultures is important in helping us develop empirically constrained models of the evolution of general and quite widely shared learning capacities—the project Laland, Kendal, and Kendal defend in their chapter. However, except by providing a baseline—a portrait of a generic great-ape mind—it is less clear that this project will contribute markedly to understanding our own strange dynamics.

## Note

1. Experimental economists use the willingness to spend resources on actions that make other players worse off as a behavioral signature of moralistic punishment. But this setup requires anonymous interactions and other controls to make sure that the sacrifice is indeed normatively driven punishment rather than investment in behavioral modification that will benefit the "punishing" agent at some future time. I suspect that it would be very experimentally challenging to develop primate experiments with similar controls. The literature on primate policing does not have the right control: dominants sometimes police subordinate squabbles without taking sides (Flack et al. 2005, 2006), but they themselves can benefit from a more peaceable and functional social world.

Addessi, E., and Visalberghi, E. 2001. Social facilitation of eating novel foods in tufted capuchin monkeys *(Cebus apella)*: Input provided, responses affected, and cognitive implications. *Animal Cognition,* 4, 297–303.

Aisner, R., and Terkel, J. 1992. Ontogeny of pine cone opening behaviour in the black rat, *Rattus rattus. Animal Behaviour,* 44, 327–336.

Akins, C. K., and Zentall, T. R. 1996. Imitative learning in male Japanese quail *(Coturnix japonica)* using the two action method. *Journal of Comparative Psychology,* 110, 316–320.

———. 1998. Imitation in Japanese quail: The role of reinforcement of demonstrator responding. *Psychonomic Bulletin and Review,* 5, 694–697.

Alp, R. 1997. "Stepping-sticks" and "seat-sticks": New types of tools used by wild chimpanzees *(Pan troglodytes)* in Sierra Leone. *American Journal of Primatology,* 51, 45–52.

Alvard, M. 2003. The adaptive nature of culture. *Evolutionary Anthropology,* 12, 136–149.

Anderson, J. R., Myowa-Yamakoshi, M., and Matsuzawa, T. 2004. Contagious yawning in chimpanzees. *Proceedings of the Royal Society of London, Series B,* 271(Suppl.), S468–S470.

Arcadi, A. C., Robert, D., and Mugurusi, F. 2004. A comparison of buttress drumming by male chimpanzees from two populations. *Primates,* 45, 135–139.

Asquith, P. J. 1996. Japanese science and Western hegemonies: Primatology and the limits set to questions. In: *Naked Science: Anthropological Inquiry into Boundaries, Power, and Knowledge* (Ed. by L. Nader), pp. 239–256. New York: Routledge.

Avital, E., and Jablonka, E. 2000. *Animal Traditions: Behavioural Inheritance in Evolution.* Cambridge: Cambridge University Press.

Backwell, L. R., and d'Errico, F. 2001. Evidence of termite foraging by Swartkrans early hominids. *Proceedings of the National Academy of Sciences of the United States of America,* 98, 1358–1363.

Baird, R. W., and Dill, L. M. 1996. Ecological and social determinants of group size in *transient* killer whales. *Behavioral Ecology,* 7, 408–416.

Baldwin, J. M. 1896. A new factor in evolution. *American Naturalist,* 30, 441–451, 536–553.

Bandura, A. 1977. *Social Learning Theory*. Englewood Cliffs, NJ: Prentice-Hall.

Barros, N. B., and Wells, R. S. 1998. Prey and feeding patterns of resident bottlenose dolphins *(Tursiops truncatus)* in Sarasota Bay, Florida. *Journal of Mammalogy*, 79, 1045–1059.

Basalla, G. 1988. *The Evolution of Technology*. Cambridge: Cambridge University Press.

Bauer, G. B., and Johnson, C. M. 1994. Trained motor imitation by bottle-nosed dolphins *(Tursiops truncatus)*. *Perceptual and Motor Skills*, 79, 1307–1315.

Beck, B. B. 1974. Baboons, chimpanzees and tools. *Journal of Human Evolution*, 3, 509–516.

———. 1980. *Animal Tool Behavior*. New York: Garland.

Bednarik, R. G. 2003. The earliest evidence for paleoart. *Rock Art Research*, 20, 3–28.

Beltman, J. B., Haccou, P., and ten Cate, C. 2003. The impact of learning foster species' song on the evolution of specialist avian brood parasitism. *Behavioral Ecology*, 14, 917–923.

———. 2004. Learning and colonization of new niches: A first step towards speciation. *Evolution*, 58, 35–46.

Benedict, R. 1935. *Patterns of Culture*. London: Routledge and Kegan Paul.

Bennett, J. W. 1999. Classic anthropology. *American Anthropologist*, 100, 951–956.

Berdecio, S., and Nash, A. 1981. Chimpanzee visual communication: Facial, gestural, and postural expressive movements in young, captive chimpanzees. Arizona State University Anthropological Research Paper No. 26, Tempe.

Biro, D., Inoue-Nakamura, N., Tonooka, R., Yamakoshi, G., Sousa, C., and Matsuzawa, T. 2003. Cultural innovation and transmission of tool use in wild chimpanzees: Evidence from field experiments. *Animal Cognition*, 6, 213–223.

Bitterman, M. E. 2000. Cognitive evolution: A psychological perspective. In: *The Evolution of Cognition* (Ed. by C. M. Heyes and L. Huber), pp. 61–79. Cambridge, MA: MIT Press.

Bjorklund, D. F., Bering, J. M., and Ragan, P. 2000. A two-year longitudinal study of deferred imitation of object manipulation in a juvenile chimpanzee *(Pan troglodytes)* and orangutan *(Pongo pygmaeus)*. *Developmental Psychobiology*, 37, 229–237.

Bloch, M. 2000. A well-disposed social anthropologist's problems with memes. In: *Darwinizing Culture* (Ed. by R. Aunger), pp. 189–204. Oxford: Oxford University Press.

Blurton-Jones, N., Konner, M. J. 1976. !Kung knowledge of animal behavior. In: *Studies of the !Kung San and their neighbors* (Ed. by R. Lee and I. DeVore). Cambridge, MA: Harvard University Press.

Boakes, R. 1984. *From Darwin to Behaviourism: Psychology and the Minds of Animals*. Cambridge: Cambridge University Press.

Boesch, C. 1991. Teaching among wild chimpanzees. *Animal Behaviour*, 41, 530–532.

———. 1993a. Toward a new image of culture in wild chimpanzees? *Behavioral and Brain Sciences*, 16, 514–515.

——. 1993b. Aspects of transmission of tool use in wild chimpanzees. In: *Tools, Language and Cognition in Human Evolution* (Ed. by K. Gibson and T. Imgold), pp. 171–183. Cambridge: Cambridge University Press.

——. 1994. Cooperative hunting in wild chimpanzees. *Animal Behaviour,* 48, 653–667.

——. 1996a. The emergence of cultures across wild chimpanzees. *Proceedings of the British Academy,* 88, 251–268.

——. 1996b. The emergence of cultures among wild chimpanzees. In *Evolution of Social Behaviour Patterns in Primates and Man* (Ed. by W. G. Runciman, J. Maynard-Smith, and R. I. M. Dunbar), pp. 251–268. Oxford: Oxford University Press.

——. 1996c. Three approaches for assessing chimpanzee culture. In: *Reaching into Thought: The Minds of the Great Apes* (Ed. by A. E. Russon, K. A. Bard, and S. T. Parker), pp. 404–429. Cambridge: Cambridge University Press.

——. 2001. Sacrileges are welcome in science! Opening a discussion about culture in animals. *Behavioral and Brain Sciences,* 24, 327–328.

——. 2003. Is culture a golden barrier between human and chimpanzee? *Evolutionary Anthropology,* 12, 82–91.

Boesch, C., and Boesch, H. 1990. Tool use and tool making in wild chimpanzees. *Folia Primatologica,* 54, 86–99.

Boesch, C., Marchesi, P., Marchesi, B., Fruth, B., and Joulian, F. 1994. Is nut cracking in wild chimpanzees a cultural behaviour? *Journal of Human Evolution,* 26, 325–338.

Boesch, C., and Tomasello, M. 1998. Chimpanzee and human cultures. *Current Anthropology,* 39, 591–614.

Bohannan, P. 1992. *We, the Alien.* Prospect Heights, IL: Waveland.

Bolnick, D. I. 2001. Intraspecific competition favours niche width expansion in *Drosophila melanogaster. Nature,* 410, 463–466.

Bolnick, D. I., Svanbäck, R., Fordyce, J. A., Yang, L. H., Davis, J. M., Hulsey, C. D., and Forister, M. L. 2003. The ecology of individuals: Incidence and implications of individual specialization. *American Naturalist,* 161, 1–28.

Bonner, J. T. 1980. *The Evolution of Culture in Animals.* Princeton, NJ: Princeton University Press.

Bonnie, K. E., and de Waal, F. B. M. 2006. Affiliation promotes the transmission of a social custom: Handclasp grooming among captive chimpanzees. *Primates,* 47, 27–34.

——. 2007. Copying without rewards: Socially influenced foraging decisions among brown capuchin monkeys. *Animal Cognition,* 10, 283–292.

Bonnie, K. E., Horner, V., Whiten, A., and de Waal, F. B. M. 2007. Spread of arbitrary conventions among chimpanzees: A controlled experiment. *Proceedings of the Royal Society of London, Series B,* 274, 367–372.

Boogert, N. J., Reader, S. M., Hoppitt, W., and Laland, K. N. 2008. The origin and spread of innovations in starlings. *Animal Behaviour,* 75, 1509–1518.

Borgerhoff Mulder, M. 2001. Using phylogenetically based comparative methods in anthropology: More questions than answers. *Evolutionary Anthropology,* 10, 99–111.

Boughman, J., and Wilkinson, G. S. 1998. Greater spear-nosed bats discriminate groupmates by vocalizations. *Animal Behaviour, 55,* 1717–1732.

Bowles, S., and Gintis, H. 1998. The evolution of strong reciprocity. Manuscript, University of Massachusetts, Amherst. http://ideas.repec.org/p/wop/safire/98-08-073e.html#download.

———. 2003. Origins of human cooperation. In: *Genetic and Cultural Evolution of Cooperation* (Ed. by P. Hammerstein), pp. 429–443. Cambridge, MA: MIT Press.

———. 2004. The evolution of strong reciprocity: Cooperation in a heterogeneous population. *Theoretical Population Biology, 65,* 17–28.

Bowman, R. I., and Billeb, S. L. 1965. Blood-eating in a Galápagos finch. *Living Bird, 4,* 29–44.

Box, H. O., and Gibson, K. R. 1999. *Mammalian Social Learning: Comparative and Ecological Perspectives.* Cambridge: Cambridge University Press.

Boyd, R., Borgerhoff-Mulder, M., Durham, W. H. and Richerson, P. J. 1997. Are cultural phylogenies possible? In: *Human by Nature: Between Biology and the Human Sciences* (Ed. by P. Weingart, P. J. Mitchell, P. J. Richerson, and S. Maasen), pp. 355–386. Mahwah, NJ: Lawrence Erlbaum Associates.

Boyd, R., Gintis, H., Bowles, S., and Richerson, P. J. 2003. The evolution of altruistic punishment. *Proceedings of the National Academy of Sciences of the United States of America, 100,* 3531–3535.

Boyd, R., and Richerson, P. J. 1985. *Culture and the Evolutionary Process.* Chicago: University of Chicago Press.

———. 1987. The evolution of ethnic markers. *Cultural Anthropology, 2,* 65–79.

———. 1996. Why culture is common but cultural evolution is rare. *Proceedings of the British Academy, 88,* 77–93.

———. 2002. Group beneficial norms can spread rapidly in a structured population. *Journal of Theoretical Biology, 215,* 287–296.

———. 2007. Culture, adaptation, and innateness. In: *The Innate Mind: Culture and Cognition* (Ed. by P. Carruthers, S. Laurence, and S. Stich), pp. 23–38. Oxford: Oxford University Press.

Boyd, R., and Silk, J. B. 2006. *How Humans Evolved.* 4th ed. New York: Norton.

Brosnan, S. F., and de Waal, F. B. M. 2003. Monkeys reject unequal pay. *Nature, 425,* 297–299.

———. 2004. Socially learned preferences for differentially rewarded tokens in the brown capuchin monkey *(Cebus apella). Journal of Comparative Psychology, 118,* 133–139.

———. 2005. Responses to a simple barter task in chimpanzees, *Pan troglodytes. Primates, 46,* 173–182.

Brown, D. E. 1991. *Human Universals.* New York: McGraw-Hill.

Bruner, J. 1993. Do we "acquire" culture or vice versa? *Behavioral and Brain Sciences, 16,* 515–516.

Byrne, R. W. 1992. The evolution of intelligence. In: *Behaviour and Evolution* (Ed. by P. Slater and T. Halliday), pp. 223–265. Cambridge: Cambridge University Press.

————. 2005. Detecting, understanding, and explaining animal imitation. In: *Perspectives on Imitation: From Mirror Neurons to Memes* (Ed. by S. Hurley and N. Chater), pp. 255–282. Cambridge, MA: MIT Press.

————. 2007. Culture in great apes: Using intricate complexity in feeding skills to trace the evolutionary origin of technical prowess. *Philosophical Transactions of the Royal Society of London, Series B,* 362, 577–585.

Byrne, R. W., Barnard, P. J., Davidson, I., Janik, V. M., McGrew, W. C., Miklosi, A., and Wiessner, P. 2004 Understanding culture across species. *Trends in Cognitive Sciences,* 8, 341–346.

Byrne, R. W., and Russon, A. E. 1998. Learning by imitation: A hierarchical approach. *Behavior and Brain Sciences,* 21, 667–721.

Byrne, R. W., and Whiten, A. 1988. *Machiavellian Intelligence: Social Expertise and the Evolution of Intellect in Monkeys, Apes and Humans.* Oxford: Clarendon Press.

Cairns, S. J., and Schwager, S. J. 1987. A comparison of association indices. *Animal Behaviour,* 35, 1454–1469.

Caldwell, M. C., and Caldwell, D. K. 1972. Vocal mimicry in the whistle mode by an Atlantic bottlenose dolphin. *Cetology,* 9, 1–8.

Call, J. 2001. Body imitation in an enculturated orangutan *(Pongo pygmaeus). Cybernetics and Systems,* 32, 97–119.

Call, J., and Carpenter, M. 2002. Three sources of information in social learning. In: *Imitation in Animals and Artifacts* (Ed. by K. Dautenhahn and C. Nehaniv), pp. 211–228. Cambridge, MA: MIT Press.

Call, J., Carpenter, M., and Tomasello, M. 2005. Copying results and copying actions in the process of social learning: Chimpanzees *(Pan troglodytes)* and human children *(Homo sapiens). Animal Cognition,* 8, 151–163.

Call, J., Hare, B., Carpenter, M., and Tomasello, M. 2004. Unwilling or unable: Chimpanzees' understanding of human intentional action. *Developmental Science,* 7, 488–498.

Call, J., and Tomasello, M. 1996. The effect of humans on the cognitive development of apes. In: *Reaching into Thought: The Minds of the Great Apes* (Ed. by A. E. Russon, K. A. Bard, and S. T. Parker), pp. 371–403. Cambridge: Cambridge University Press.

————. 1998. Distinguishing intentional from accidental actions in orangutans *(Pongo pygmaeus),* chimpanzees *(Pan troglodytes)* and human children *(Homo sapiens). Journal of Comparative Psychology,* 112, 192–206.

Camerer, C. F., and Fehr, E. 2006. When does "economic man" dominate social behavior? *Science,* 311, 47–52.

Carlier, P., and Lefebvre, L. 1997. Ecological Differences in Social Learning between Adjacent, Mixing, Populations of Zenaida Doves. *Ethology,* 103, 772–784.

Caro, T. M. 1994. *Cheetahs of the Serengeti Plains.* Chicago: University of Chicago Press.

Caro, T. M., and Hauser, M. D. 1992. Is there teaching in nonhuman animals? *Quarterly Review of Biology,* 67, 151–174.

Carpenter, M. 2006. Instrumental, social, and shared goals and intentions in imitation. In: *Imitation and the Development of the Social Mind: Lessons from Typical Development and Autism* (Ed. by S. J. Rogers and J. Williams), pp. 48–70. New York: Guilford.

Castro, L., and Toro, M. 2004. The evolution of culture: From primate social learning to human culture. *Proceedings of the National Academy of Sciences of the United States of America,* 101, 10235–10240.

Catchpole, C. K., and Slater, P. J. B. 1995. *Birdsong: Biological Themes and Variations.* Cambridge: Cambridge University Press.

Cavalli-Sforza, L. L., and Feldman, M. W. 1981. *Cultural Transmission and Evolution: A Quantitative Approach.* Princeton, NJ: Princeton University Press.

Cavalli-Sforza, L. L., Feldman, M. W., Chen, K. H., and Dornbusch, S. M. 1982. Theory and observation in cultural transmission. *Science,* 218, 19–27.

Cavalli-Sforza, L. L., and Wang, W. S.-Y. 1986. Spatial distance and lexical replacement. *Language,* 62, 38–55.

Chapais, B. 1988. Rank maintenance in female Japanese macaques: Experimental evidence for social dependency. *Behaviour,* 104, 41–59.

Cheney, D. L., and Seyfarth, R. M. 1990. *How Monkeys See the World.* Chicago: University of Chicago Press.

Chilvers, B. L., and Corkeron, P. J. 2001. Trawling and bottlenose dolphins' social structure. *Proceedings of the Royal Society of London, Series B: Biological Sciences,* 268, 1901–1905.

Clutton-Brock, T. H. 2006. Cooperative breeding in mammals. In: Cooperation in Primates and Humans (Ed. by P. M. Kappeler and C. van Schaik), pp. 173–190.

Connor, R. C. 2000. Group living in whales and dolphins. In: *Cetacean Societies: Field Studies of Dolphins and Whales* (Ed. by J. Mann, R. C. Connor, P. L. Tyack, and H. Whitehead), pp. 199–218. Chicago: University of Chicago Press.

Connor, R. C., Smolker, R. A., and Richards, A. F. 1992. Two levels of alliance formation among male bottlenose dolphins (*Tursiops* sp.). *Proceedings of the National Academy of Sciences of the United States of America,* 89, 987–990.

Connor, R. C., Wells, R. S., Mann, J., and Read, A. J. 2000. The bottlenose dolphin: Social relationships in a fission-fusion society. In: *Cetacean Societies: Field Studies of Dolphins and Whales* (Ed. by J. Mann, R. C. Connor, P. L. Tyack, and H. Whitehead), pp. 91–126. Chicago: University of Chicago Press.

Coolen, I., Day R. L., and Laland, K. N. 2003. Species difference in adaptive use of public information in sticklebacks. *Proceedings of the Royal Society, Series B,* 270, 2413–2419.

Cormier, L. A. 2003. *Kinship with Monkeys.* New York: Columbia University Press.

Coussi-Korbel, S., and Fragaszy, D. M. 1995. On the relation between social dynamics and social learning. *Animal Behaviour,* 50, 1441–1453.

Crapo, R. H. 1996. *Cultural Anthropology: Understanding Ourselves and Others.* Madison, WI: Brown and Benchmark.

Cronk, L. 1999. *That Complex Whole: Culture and the Evolution of Human Behavior.* Boulder, CO: Westview Press.

Csibra, G., and Gergely, G. 2006. Social learning and social cognition: The case for pedagogy. In: *Processes of Change in Brain and Cognitive Development* (Ed. by Y. Munakata and M. H. Johnson), pp. 249–274. Oxford: Oxford University Press.

Curio, E., Ernst, U., and Vieth, W. 1978. The adaptive significance of avian mobbing II. Cultural transmission of enemy recognition in blackbirds: Effectiveness and some constraints. *Zeitschrift für Tierpsychologie,* 48, 184–202.

Curry, R. L., and Anderson, D. J. 1987. Interisland variation in blood drinking by Galápagos mockingbirds. *Auk,* 104, 517–521.

Custance, D. M., Whiten, A., and Bard, K. A. 1995. Can young chimpanzees *(Pan troglodytes)* imitate arbitrary actions? Hayes and Hayes (1952) revisited. *Behaviour,* 132, 837–859.

Danchin, E., Giraldeau, L.-A., Vallone, T. J., and Wagner, R. H. 2004. Public information: From nosy neighbors to cultural evolution. *Science,* 305, 487–491.

Darwin, C. 1871. *The Descent of Man, and Selection in Relation to Sex.* London: Murray.

Davidson, I., and Noble, W. 1989. The archeology of perception: Traces of depiction and language. *Current Anthropology,* 30, 125–155.

Day, R. L. 2003. Innovation and social learning in monkeys and fish: Empirical findings and their application to reintroduction techniques. PhD dissertation, Cambridge University.

Day, R. L., Kendal, J. R., and Laland, K. N. 2001. Validating cultural transmission in cetaceans: Reply to Rendell and Whitehead. *Behavioral and Brain Sciences,* 24, 330–331.

Deecke, V. B., Ford, J. K. B., and Spong, P. 2000. Dialect change in resident killer whales: Implications for vocal learning and cultural transmission. *Animal Behaviour,* 60, 629–638.

Delgado, R., and van Schaik, C. P. 2000. The behavioral ecology and conservation of the orangutan *(Pongo pygmaeus):* A tale of two islands. *Evolutionary Anthropology,* 9, 201–218.

d'Errico, F. 2003. The invisible frontier: A multiple species model for the origin of behavioral modernity. *Evolutionary Anthropology,* 12, 188–202.

de Waal, F. B. M. 1982. *Chimpanzee Politics.* New York: Harper and Row.

———. 1988. The communicative repertoire of captive bonobos *(Pan paniscus),* compared to that of chimpanzees. *Behaviour,* 106, 183–251.

———. 1996a. *Good Natured: The Origins of Right and Wrong in Humans and Other Animals* Cambridge, MA: Harvard University Press.

———. 1996b. Macaque social culture: Development and perpetuation of affiliative networks. *Journal of Comparative Psychology,* 110, 147–154.

———. 1998. No imitation without identification. *Behavioral and Brain Sciences,* 21, 689.

———. 1999. Cultural primatology comes of age. *Nature,* 399, 635–636.

———. 2001. *The Ape and the Sushi Master: Cultural Reflections by a Primatologist.* New York: Basic Books.

————. 2003a. Silent invasion: Imanishi's primatology and cultural bias in science. *Animal Cognition*, 6, 293–299.

————. 2003b. Darwin's legacy and the study of primate visual communication. In: *Emotions Inside Out: 130 Years after Darwin's "The Expression of the Emotions in Man and Animals"* (Ed. by P. Ekman, J. J. Campos, R. J. Davidson, and F. B. M. de Waal), pp. 7–31. New York: New York Academy of Sciences.

————. 2007. The Russian doll model of empathy and imitation. In: *On Being Moved—From Mirror Neurons to Empathy* (Ed. by S. Bråten), pp. 49–69. Philadelphia: Benjamins.

de Waal, F. B. M., and Johanowicz, D. L. 1993. Modification of reconciliation behavior through social experience: An experiment with two macaque species. *Child Development*, 64, 897–908.

de Waal, F. B. M., and Seres, M. 1997. Propagation of handclasp grooming among captive chimpanzees. *American Journal of Primatology*, 43, 339–346.

de Waal, F. B. M., and Tyack, P. L. 2003. *Animal Social Complexity: Intelligence, Culture, and Individualized Societies*. Cambridge, MA: Harvard University Press.

Dewar, G. 2003. The cue reliability approach to social transmission: Designing tests for adaptive traditions. In: *The Biology of Traditions: Models and Evidence* (Ed. by D. M. Fragaszy and S. Perry), pp. 127–158. Cambridge: Cambridge University Press.

————. 2004. Social and asocial cues about new food: Cue reliability influences intake in rats. *Learning and Behavior*, 32, 82–89.

Dickinson, A., and Balleine, B. W. 2000. Causal cognition and goal-directed action. In *The Evolution of Cognition* (Ed. by C. M. Heyes and L. Huber), pp. 185–204. Cambridge, MA: MIT Press.

Dindo, M., and de Waal, F. B. M. 2007. Partner effects on food consumption in brown capuchin monkeys. *American Journal of Primatology*, 69, 448–456.

Donald, M. 1991. *Origins of the Human Mind: Three Stages in the Evolution of Culture and Cognition*. Cambridge, MA: Harvard University Press.

Dow, M. M., and de Waal, F. B. M. 1989. Assignment methods for the analysis of network subgroup interactions. *Social Networks*, 11, 237–255.

Dugatkin, L. A. 2000. *The Imitation Factor: Evolution beyond the Gene*. New York: Free Press.

Dunkel, L. P. 2006. Development of ecological competence in Bornean orangutans *(Pongo pygmaeus)*: With special reference to difficult-to-process food items. Diploma thesis, Anthropological Institute, University of Zürich.

Durban, J. W., and Parsons, K. M. Submitted. Quantifying clusters in social populations. *Animal Behaviour*.

Durham, W. H. 1991. *Coevolution: Genes, Culture, and Human Diversity*. Stanford, CA: Stanford University Press.

Efron, B., and Gong, G. 1983. A leisurely look at the bootstrap, the jackknife, and cross-validation. *American Statistician*, 37, 36–48.

Eibl-Eibesfeldt, I. 1989. *Human Ethology*. New York: Aldine.

Emery, N. J. 2006. Cognitive ornithology: The evolution of avian intelligence. *Philosophical Transactions of the Royal Society, Series B*, 361, 23–43.

Endler, J. A. 1986. *Natural Selection in the Wild*. Princeton, NJ: Princeton University Press.

Ereshefsky, M. 2001. *The Poverty of the Linnean Hierarchy*. Cambridge: Cambridge University Press.

Estes, J. A., Riedman, M. L., Staedler, M. M., Tinker, M. T., and Lyon, B. E. 2003. Individual variation in prey selection by sea otters: Patterns, causes, and implications. *Journal of Animal Ecology*, 72, 144–155.

Fadiga, L., and Craighero, L. 2007. Cues to the origin of language: From electrophysiological data on mirror neurons and motor representations. In: *On Being Moved—From Mirror Neurons to Empathy* (Ed. by S. Bråten), pp. 101–110. Philadelphia: Benjamins.

Fehr, E., and Fischbacher, U. 2003. The nature of human altruism. *Nature*, 425, 785–791.

———. 2004a. Social norms and human cooperation. *Trends in Cognitive Sciences*, 8, 185–190.

———. 2004b. Third-party punishment and social norms. *Evolution and Human Behavior*, 25, 63–87.

Fehr, E., Fischbacher, U., and Gächter, S. 2002. Strong reciprocity, human cooperation and the enforcement of social norms. *Human Nature*, 13, 1–25.

Fehr, E., and Gächter, S. 2002. Altruistic punishment in humans. *Nature*, 415, 137–140.

Feldman, M. W., and Laland, K. N. 1996. Gene-culture coevolutionary theory. *Trends in Ecology and Evolution*, 11, 453–457.

Ferrari, P. F., Maiolinia, C., Addessi, E., Fogassi, L., and Visalberghi, E. 2005. The observation and hearing of eating actions activates motor programs related to eating in macaque monkeys. *Behavioural Brain Research*, 161, 95–101.

Ferraro, G., Trevathan, W., and Levy, J. 1994. *Anthropology: An Applied Perspective*. Minneapolis: West.

Fischbacher, U., Gächter, S., and Fehr, E. 2001. Are people conditionally cooperative? Evidence from a public goods experiment. *Economics Letters*, 71, 397–404.

Fisher, J., and Hinde, R. 1949. The opening of milk bottles by birds. *British Birds*, 42, 347–357.

Flack, J. C., de Waal, F. B. M., and Krakauer, D. C. 2005. Social structure, robustness, and policing cost in a cognitively sophisticated species. *American Naturalist*, 165, 126–139.

Flack, J. C., Girvan, M., de Waal, F. B. M., and Krakauer, D. C. 2006. Policing stabilizes construction of social niches in primates. *Nature*, 439, 426–429.

Fogassi, L., Ferrari, P. F., Gesierich, B., Rozzi, S., Chersi, F., and Rizzolatti, G. 2005. Parietal lobe: From action organization to intention understanding. *Science*, 308, 662–667.

Ford, J. K. B. 1991. Vocal traditions among resident killer whales *(Orcinus orca)* in coastal waters of British Columbia. *Canadian Journal of Zoology*, 69, 1454–1483.

Ford, J. K. B., and Fisher, H. D. 1982. Killer whale *(Orcinus orca)* dialects as an indicator of stocks in British Columbia. *Reports of the International Whaling Commission*, 32, 671–679.

Fox, E. A., van Schaik, C. P., Sitompul, A., and Wright, D. N. 2004. Intra- and interpopulational differences in orangutan *(Pongo pygmaeus)* activity and diet: Implications for the invention of tool use. *American Journal of Physical Anthropology*, 125, 162–174.

Fox, M. A. 2001. Cetacean culture: Philosophical implications. *Behavioral and Brain Sciences*, 24, 333–334.

Fragaszy, D. M., Izar, P., Visalberghi, E., Ottoni, E. B., and De Oliveira, M. G. 2004. Wild capuchin monkeys *(Cebus libidinosus)* use anvils and stone pounding tools. *American Journal of Primatology*, 64, 359–366.

Fragaszy, D. M., and Perry, S., eds. 2003a. *The Biology of Traditions: Models and Evidence*. Cambridge: Cambridge University Press.

———. 2003b. Preface. In: *The Biology of Traditions: Models and Evidence* (Ed. by D. M. Fragaszy and S. Perry), pp. xiii–xvi. Cambridge: Cambridge University Press.

———. 2003c. Towards a biology of traditions. In: *The Biology of Traditions: Models and Evidence* (Ed. by D. M. Fragaszy and S. Perry), pp. 1–32. Cambridge: Cambridge University Press.

Fragaszy, D. M., Visalberghi, E., and Fedigan, L. M. 2004. *The Complete Capuchin: The Biology of the Genus* Cebus. Cambridge: Cambridge University Press.

Franks, N. R., and Richardson, T. 2006. Teaching in tandem-running ants. *Nature*, 439, 153.

Futuyma, D. J. 1998. *Evolutionary Biology*. Sunderland, MA: Sinauer.

Gagneux, P., Gonder, M. K., Goldberg, T. L., and Morin, P. A. 2001. Gene flow in wild chimpanzee populations: What genetic data tell us about chimpanzee movement over space and time. *Philosophical Transactions of the Royal Society, Series B*, 356, 889–897.

Gagneux, P., Wills, C., Gerloff, U., Tautz, D., Morin, P. A., Boesch, C., Fruth, B., Hohmann, G., Ryder, O. A., and Woodruff, D. S. 1999. Mitochondrial sequences show diverse evolutionary histories of African hominids. *Proceedings of the National Academy of Sciences of the United States of America*, 96, 5077–5082.

Galef, B. G., Jr. 1988. Imitation in animals: History, definition and interpretation of data from the psychological laboratory. In: *Social Learning: Psychological and Biological Perspectives* (Ed. by T. R. Zentall and B. G. Galef Jr.), pp. 3–28. Hillsdale, NJ: Lawrence Erlbaum.

———. 1991. Tradition in animals: Field observations and laboratory analyses. In: *Interpretation and Explanation in the Study of Behavior*, vol. 1: *Interpretation, Intentionality and Communication* (Ed. by M. Bekoff and D. Jamieson), pp. 74–95. Boulder, CO: Westview Press.

———. 1992. The question of animal culture. *Human Nature*, 3, 157–178.

———. 1995. Why behaviour patterns that animals learn socially are locally adaptive. *Animal Behaviour*, 49, 1325–1334.

——. 1996. Traditions in animals: Field observations and laboratory analyses. In: *Readings in Animal Cognition* (Ed. by M. Bekoff and D. Jamieson), pp. 91–106. Cambridge, MA: MIT Press.

——. 2003a. "Traditional" foraging behaviors of brown and black rats (*Rattus norvegicus* and *Rattus rattus*). In: *The Biology of Traditions: Models and Evidence* (Ed. by D. M. Fragaszy and S. Perry), pp. 159–186. Cambridge: Cambridge University Press.

——. 2003b. Social learning: Promoter or inhibitor of innovation? In: *Animal Innovation* (Ed. by S. M. Reader and K. N. Laland), pp. 137–152. Oxford: Oxford University Press.

——. 2004. Approaches to the study of traditional behaviors of free-living animals. *Learning and Behavior*, 32, 53–61.

Galef, B. G., Jr., and Allen, C. 1995. A model system for studying animal traditions. *Animal Behaviour*, 50, 705–717.

Galef, B. G., Jr., and Giraldeau, L.-A. 2001. Social influences on foraging in vertebrates: Causal mechanisms and adaptive functions. *Animal Behaviour*, 61, 3–15.

Galef, B. G., Jr., and Heyes, C. M. 2004a. Introduction. *Learning and Behaviour*, 32, 1–3.

——. 2004b. Social learning and imitation. Special Issue of *Learning and Behavior*, 32.

Galef, B. G., Jr., and Laland, K. N. 2005. Social learning in animals: Empirical studies and theoretical models. *Bioscience*, 55, 489–499.

Galef, B. G., Jr., Manzig, L. A., and Field, R. M. 1986. Imitation learning in budgerigars: Dawson and Foss (1965) revisited. *Behavioural Processes*, 13, 191–202.

Galef, B. G., Jr., and Whiskin, E. E. 2001. Interaction of social and individual learning in food preferences of Norway rats. *Animal Behaviour* 62, 41–46.

Galef, B. G., Jr., Whiskin, E. E., and Dewar, G. 2005. A new way to study teaching in animals: Despite demonstrable benefits, rat dams do not teach their young what to eat. *Animal Behaviour*, 70, 91–96.

Galef, B. G., Jr., and White, D. J. 2000. Evidence of social effects on mate choice in vertebrates. *Behavioural Processes*, 51, 167–175.

Gannon, D. P., and Waples, D. M. 2004. Diets of coastal bottlenose dolphins from the US mid-Atlantic coast differ by habitat. *Marine Mammal Science*, 20, 527–545.

Gazda, S. K., Connor, R. C., Edgar, R. K., and Cox, F. 2005. A division of labour with role specialization in group-hunting bottlenose dolphins *(Tursiops truncatus)* off Cedar Key, Florida. *Proceedings of the Royal Society of London, Series B*, 272, 135–140.

Geisel, T. 1984. *The Butter Battle Book*. New York: Random House.

Gergely, G., and Csibra, G. 2006. Sylvia's recipe: The role of imitation and pedagogy in the transmission of human culture. In: *Roots of Human Sociality: Culture, Cognition, and Human Interaction* (Ed. by N. J. Enfield and S. C. Levinson), pp. 229–255. Oxford: Berg Publishers.

Gibson, Q. A., and Mann, J. 2008a. Early social development in wild bottlenose dolphins: Sex differences, individual variation and maternal influence. *Animal Behaviour*, 76, 375–387.

———. 2008b. The size and composition of wild bottlenose dolphin (*Tursiops* sp.) mother-calf groups in Shark Bay, Australia. *Animal Behaviour*, 76, 389–405.

Gibson, R. M., and Bachman, G. C. 1992. The costs of female choice in a lekking bird. *Behavioral Ecology* 3(4): 300–309.

Gibson, R. M., Bradbury, J. W., and Vehrencamp, S. L. 1991. Mate choice in lekking sage grouse revisited: The roles of vocal display, female site fidelity, and copying. *Behavioral Ecology*, 2, 165–180.

Gil-White, F. J. 2001. Are ethnic groups biological "species" to the human brain? Essentialism in our cognition of some social categories. *Current Anthropology*, 42, 515–554.

Ginsberg, J. R., and Young, T. P. 1992. Measuring association between individuals or groups in behavioural studies. *Animal Behaviour*, 44, 377–379.

Gintis, H. 2000. Strong reciprocity and human sociality. *Journal of Theoretical Biology*, 206, 169–179.

Giraldeau, L.-A., Valone, T. J., and Templeton, J. J. 2002. Potential disadvantages of using socially-acquired information. *Philosophical Transactions of the Royal Society, Series B*, 357, 1559–1566.

Glazier, A. M., Nadeau, J. H., and Aitman, T. J. 2002. Finding genes that underlie complex traits. *Science*, 298, 2345–2349.

Goldberg, T. L., and Ruvolo, M. 1997. Molecular phylogenetics and historical biogeography of East African chimpanzees. *Biological Journal of the Linnean Society*, 61, 301–324.

Goodale, E., and Kotagama, S. W. 2006. Context-dependent vocal mimicry in a passerine bird. *Proceedings of the Royal Society, Series B*, 273, 875–880.

Goodall, J. 1986. *The Chimpanzees of Gombe*. Cambridge, MA: Harvard University Press.

Goossens, B., Chikhi, L., Jalil, M. F., Ancrenaz, M., Lackman-Ancrenaz, I., Mohamed, M., Andau, P., and Bruford, M. W. 2005. Patterns of genetic diversity and migration in increasingly fragmented and declining orang-utan (*Pongo pygmaeus*) populations from Sabah, Malaysia. *Molecular Ecology*, 14, 441–456.

Goren-Inbar, N., Sharon, G., Melamed, Y., and Kislev, M. 2002. Nuts, nut cracking, and pitted stones at Gesher Benot Ya'agov, Israel. *Proceedings of the National Academy of Sciences of the United States of America*, 99, 2455–2460.

Gould, S. J., and Lewontin, R. C. 1979. The spandrels of San Marco and the Panglossian paradigm: A critique of the adaptationist programme. *Proceedings of the Royal Society of London, Series B*, 205, 581–598.

Grant, P. R. 1986. *Ecology and Evolution of Darwin's Finches*. Princeton, NJ: Princeton University Press.

Grant, P. R., Grant, R., and Petren, K. 2000. The allopatric phase of speciation: The sharp-beaked ground finches (*Geospiza difficilis*) on the Galápagos Islands. *Biological Journal of the Linnean Society*, 69, 287–317.

Griffin, A. S. 2004. Social learning about predators: A review and prospectus. *Learning and Behavior,* 32, 131–140.

Gros-Louis, J., Perry, S., and Manson, J. H. 2003. Violent coalitionary attacks and intraspecific killing in wild capuchin monkeys *(Cebus capucinus). Primates,* 44, 341–346.

Groves, C. P. 2001. *Primate Taxonomy.* Washington, DC: Smithsonian Institution Press.

Guglielmino, C. R., Viganotti, C., Hewlett, B., and Cavalli-Sforza, L. L. 1995. Cultural variation in Africa: Role of mechanisms of transmission and adaptation. *Proceedings of the National Academy of Sciences of the United States of America,* 92, 585–589.

Guinet, C., Barrett-Lennard, L. G., and Loyer, B. 2000. Co-ordinated attack behavior and prey sharing by killer whales at Crozet Archipelago: Strategies for feeding on negatively-buoyant prey. *Marine Mammal Science,* 16, 829–834.

Guinet, C., and Bouvier, J. 1995. Development of intentional stranding hunting techniques in killer whale *(Orcinus orca)* calves at Crozet Archipelago. *Canadian Journal of Zoology,* 73, 27–33.

Gurven, M. 2004. To give and to give not: The behavioral ecology of human food transfers. *Behavioral and Brain Sciences,* 27, 543–583.

Hall, K. R. L., and Schaller, G. B. 1964. Tool-using behavior of the California sea otter. *Journal of Mammalogy,* 45, 287–298.

Hannah, A., and McGrew, W. 1987. Chimpanzees using stones to crack open oil palm nuts in Liberia. *Primates,* 28, 31–46.

Hare, R. D. 1993. *Without Conscience: The Disturbing World of the Psychopaths among Us.* New York: Simon and Schuster.

———. 2003. *The Psychopathy Checklist—Revised.* 2nd ed. Toronto: Multi-Health Systems.

Harley, H. E., Putman, E. A., and Roitblat, H. L. 2003. Bottlenose dolphins perceive object features through echolocation. *Nature,* 424, 667–669.

Harpending, H. C., and Sobus, J. 1987. Sociopathy as an adaptation. *Ethology and Sociobiology,* 8, 63S–72S.

Hart, B. L., Hart, L. A., McCoy, M., and Sarath, C. R. 2001. Cognitive behaviour in Asian elephants: Use and modification of branches for fly switching. *Animal Behaviour,* 62, 839–847.

Hauser, M. D. 1992. Costs of deception: Cheaters are punished in rhesus monkeys *(Macaca mulatta). Proceedings of the National Academy of Sciences of the United States of America,* 89, 12137–12139.

Hauser, M. D., Chen, M. K., Chen, F., and Chuang, E. 2003. Give unto others: Genetically unrelated cotton-top tamarin monkeys preferentially give food to those who altruistically give food back. *Proceedings of the Royal Society of London, Series B,* 270, 2363–2370.

Hauser, M. D., Chomsky, N., and Fitch, W. T. 2002. The faculty of language: What it is, who has it, and how did it evolve? *Science,* 298, 1569–1579.

Haviland, W. A. 1996. *Cultural Anthropology.* Fort Worth: Harcourt Brace College Publishers.

Hayes, K., and Hayes, C. 1952. Imitation in a home-raised chimpanzee. *Journal of Comparative and Physiological Psychology*, 45, 450–459.

Heithaus, M. R., and Dill, L. M. 2002. Food availability and tiger shark predation risk influence bottlenose dolphin habitat use. *Ecology*, 83, 480–491.

———. 2006. Does tiger shark predation risk influence foraging habitat use by bottlenose dolphins at multiple spatial scales? *Oikos*, 114, 257–264.

Helfman, G. S., and Schultz, E. T. 1984. Social tradition of behavioural traditions in a coral reef fish. *Animal Behaviour*, 32, 379–384.

Hemelrijk, C. K. 1990. Models of, and tests for, reciprocity, unidirectionality and other social interaction patterns at a group level. *Animal Behaviour*, 39, 1013–1029.

Henrich, J. 2004a. Inequity aversion in capuchins? *Nature*, 428, 139.

———. 2004b. Cultural group selection, coevolutionary processes and large-scale coöperator. *Journal of Economic Behavior and Organization*, 53, 3–35.

Henrich, J., and Boyd, R. 1998. The evolution of conformist transmission and the emergence of between-group differences. *Evolution and Human Behavior*, 19, 215–241.

Henrich, N., and Henrich J. 2006. Why humans cooperate. Oxford: Oxford University Press.

Henshilwood, C. S., and Marean, C. W. 2002.The origin of modern human behavior: A review and critique of models and test implications. *Current Anthropology*, 44, 627–651.

Herman, L. M. 2002a. Exploring the cognitive world of the bottlenose dolphin. In: *The Cognitive Animal* (Ed. by M. Bekoff, C. Allen, and G. M. Burghardt), pp. 275–283. Cambridge, MA: MIT Press.

———. 2002b. Vocal, social, and self-imitation by bottlenose dolphins. In: *Imitation in Animals and Artifacts* (Ed. by K. Dautenhahn, and C. L. Nehaniv), pp. 63–108. Cambridge, MA: MIT Press.

Hewlett, B. S., and Cavalli-Sforza, L. L. 1986. Cultural transmission among Aka pygmies. *American Anthropologist*, 88, 922–934.

Heyes, C. M. 1993. Imitation, culture and cognition. *Animal Behaviour*, 46, 999–1010.

Heyes, C. M., and Galef, B. G., Jr., eds. 1996. *Social Learning in Animals: The Roots of Culture*. San Diego: Academic Press.

Heyes, C. M., Jaldow, E., and Dawson, G. R. 1993. Observational extinction: Observation of nonreinforced responding reduces resistance to extinction in rats. *Animal Learning and Behavior*, 21, 221–225.

Hill, K. 2002. Altruistic cooperation during foraging by the Ache, and the evolved human predisposition to cooperate. *Human Nature*, 13, 105–128.

Hill, K., Barton, M., and Hurtado, A. M. In press. The emergence of human uniqueness. *Evolutionary Anthropology*.

Hirata, S., Watanabe, K., and Kawai, M. 2002. "Sweet-potato washing" revisited. In: *Primate Origins of Human Cognition and Behavior* (Ed. by T. Matsuzawa), pp. 487–508. Tokyo: Springer.

Hockett, C. F. 1960. The origin of speech. *Scientific American*, 203(9), 89–96.

Hohmann, G., and Fruth, B. 2003. Culture in bonobos? Between-species and within-species variation in behavior. *Current Anthropology,* 44, 563–571.

Holbrook, S. J., and Schmitt, R. J. 1992. Causes and consequences of dietary specialization in surfperches: Patch choice and intraspecific competition. *Ecology,* 73, 402–412.

Hopper, L. M., Spiteri, A., Lambeth, S. P., Schapiro, S. J., Horner, V., and Whiten, A. 2007. Experimental studies of traditions and underlying transmission processes in chimpanzees. *Animal Behaviour,* 73, 1021–1032

Hoppitt, W. J. E., Brown, G. R., Kendal, R. L., Rendell, L., Thornton, A., Webster, M., and Laland, K. N. 2008. Lessons from Animal Teaching. *Trends in Ecology and Evolution,* 23, 486–493.

Horner, V., and Whiten, A. 2005. Causal knowledge and imitation/emulation switching in chimpanzees *(Pan troglodytes)* and children *(Homo sapiens).* *Animal Cognition,* 8, 164–181.

Horner, V., Whiten, A., and de Waal, F. B. M. 2006. Faithful replication of foraging techniques along cultural transmission chains by chimpanzees and children. *Proceedings of the National Academy of Sciences of the United States of America,* 103, 13878–13883.

Huffman, M. A. 1996. Acquisition of innovative cultural behaviors in nonhuman primates: A case study of stone handling, a socially transmitted behavior in Japanese macaques. In: *Social Learning in Animals: The Roots of Culture* (Ed. by C. M. Heyes and B. G. Galef Jr.), pp. 267–289. San Diego: Academic Press.

Huffman, M. A., and Hirata, S. 2004. An experimental study of leaf swallowing in captive chimpanzees. *Primates,* 45, 113–118.

Humle, T. 2006. Ant-dipping in chimpanzees: An example of how microecological variables, tool use, and culture reflect the cognitive abilities of chimpanzees. In *Cognitive Development in Chimpanzees* (Ed. by T. Matsuzawa, M. Tomonaga, and M. Tanaka), pp. 452–475. Tokyo: Springer-Verlag.

Humle, T., and Matsuzawa, T. 2002. Ant-dipping among the chimpanzees of Bossou, Guinea, and some comparisons with other sites. *American Journal of Primatology,* 58, 133–148.

Hunt, G. R. 1996. Manufacture and use of hook-tools by New Caledonian crows. *Nature,* 379, 249–251.

———. 2000. Human-like, population-level specialization in the manufacture of pandanus tools by New Caledonian crows, *Corvus moneduloides.* *Proceedings of the Royal Society of London, Series B,* 267, 403–413.

Hunt, G. R., and Gray, R. D. 2003. Diversification and cumulative evolution in tool manufacture by New Caledonian crows. *Proceedings of the Royal Society of London, Series B,* 270, 867–874.

Imanishi, K. 1952. The evolution of human nature. In: *Ningen* (Ed. by K. Imanishi), pp. 36–94. Tokyo: Mainichi-Shinbunsha. (Japanese)

Ingold, T. 2001. The use and abuse of ethnography. *Behavioral and Brain Sciences,* 24, 337.

Inoue-Nakamura, N., and Matsuzawa, T. 1997. Development of stone tool use by wild chimpanzees *(Pan troglodytes).* *Journal of Comparative Psychology,* 111, 159–173.

Itani, J., and Nishimura, A. 1973. The study of infrahuman culture in Japan: A review. In: *Symposia of the Fourth International Congress of Primatology*, vol. 1: *Precultural Primate Behavior* (Ed. by E. W. Menzel), pp. 26–50. Basel: Karger.

Jaeggi, A. 2006. The role of social learning in the acquisition of foraging skills in wild Bornean orangutans *(Pongo pygmaeus)*. Diploma thesis, Anthropological Institute, University of Zürich.

Janik, V. M. 2000a. Food-related bray calls in wild bottlenose dolphins *(Tursiops truncatus)*. *Proceedings of the Royal Society of London, Series B*, 267, 923–927.

———. 2000b. Whistle matching in wild bottlenose dolphins *(Tursiops truncatus)*. *Science*, 289, 1355–1357.

———. 2001. Is cetacean social learning unique? *Behavioral and Brain Sciences*, 24, 337–338.

Janik, V. M., Sayigh, L., and Wells, R. 2006. Signature whistle shape conveys identity information to bottlenose dolphins. *Proceedings of the National Academy of Sciences of the United States of America*, 103, 8293–8297.

Janik, V. M., and Slater, P. J. B. 1997. Vocal learning in mammals. *Advances in the Study of Behavior*, 26, 59–99.

———. 2003. Traditions in mammalian and avian vocal communication. In: *The Biology of Traditions: Models and Evidence* (Ed. by D. M. Fragaszy and S. Perry), pp. 213–235. Cambridge: Cambridge University Press.

Jensen, K., Call, J., and Tomasello, M. 2007. Chimpanzees are rational maximizers in an ultimatum game. *Science*, 317, 107–109.

Johnson, D. H. 1999. The insignificance of statistical significance testing. *Journal of Wildlife Management*, 63, 763–772.

Joyce, R. 2005. *Evolution of Morality*. Cambridge, MA: MIT Press.

Kanthaswamy, S., Kurushima, J. D., and Smith, D. G. 2006. Inferring *Pongo* conservation units: A perspective based on microsatellite and mitochondrial DNA analyses. *Primates*, 47, 310–321.

Kaplan, H., and Hill, K. 1985. Food sharing among Ache foragers: Tests of explanatory hypotheses. *Current Anthropology*, 26, 223–245.

Kawai, M. 1965. Newly acquired pre-cultural behavior of the natural troop of Japanese monkeys on Koshima Islet. *Primates*, 6, 1–30.

Kawamura, S. 1959. The process of sub-culture propagation among Japanese macaques. *Primates*, 2, 43–54.

Kelemen, D. 1999. The scope of teleological thinking in preschool children. *Cognition*, 70, 241–272.

Kendal, R. L., Kendal, J., and Laland, K. N. 2007. Quantifying and modeling social learning processes in monkey populations. *International Journal of Psychology and Psychological Therapy*, 7(2), 123–138.

Kenward, B., Weir, A. A. S., Rutz, C., and Kacelnik, A. 2005. Tool manufacture by naïve juvenile crows. *Nature*, 433, 121.

Kirkpatrick, M., and Dugatkin, L. A. 1994. Sexual selection and the evolutionary effects of copying mate choice. *Behavioral Ecology and Sociobiology*, 34, 443–449.

Kleinbaum, D. G., Kupper, L. L., and Muller, K. E. 1988. *Applied Regression Analysis and Other Multivariable Methods*. 2nd ed. Boston: PWS-Kent.

Koehler, W. 1925. *The Mentality of Apes*. New York: Harcourt, Brace and World.

Koster, F., and Koster, H. 1983. Twelve days among the "vampire-finches" of Wolf Island. *Noticias de Galápagos*, 38, 4–10.

Krebs, C. J. 1989. *Ecological Methodology*. New York: Harper and Row.

Kroeber, A. L. 1928. Sub-human cultural beginnings. *Quarterly Review of Biology*, 3, 325–342.

Kroeber, A. L., and Kluckhohn, C. 1952. *Culture: A Critical Review of Concepts and Definitions*. Cambridge, MA: Harvard University Press.

Krützen, M., Barre, L. M., Connor, R. C., Mann, J., and Sherwin, W. B. 2004. "O father: where art thou?"—Paternity assessment in an open fission-fusion society of wild bottlenose dolphins (*Tursiops* sp.) in Shark Bay, Western Australia. *Molecular Ecology*, 13, 1975–1990.

Krützen, M., Mann, J., Heithaus, M. R., Connor, R. C., Bejder, L., and Sherwin, W. B. 2005. Cultural transmission of tool use in bottlenose dolphins. *Proceedings of the National Academy of Sciences of the United States of America*, 102, 8939–8943.

Krützen, M., Sherwin, W. B., Berggren, P., and Gales, N. 2004. Population structure in an inshore cetacean revealed by microsatellite and mtDNA analysis: Bottlenose dolphins (*Tursiops* sp.) in Shark Bay, Western Australia. *Marine Mammal Science*, 20, 28–47.

Krützen, M., van Schaik, C., and Whiten, A. 2007. Response to Laland and Janik: The animal cultures debate. *Trends in Ecology and Evolution*, 22, 6.

Kummer, H. 1971. *Primate Societies: Group Techniques of Ecological Adaptation*. Chicago: Aldine-Atherton.

———. 1995. *In Quest of the Sacred Baboon: A Scientist's Journey*. Princeton, NJ: Princeton University Press.

Kummer, H., and Goodall, J. 1985. Conditions of innovative behavior in primates. *Philosophical Transactions of the Royal Society of London. Series B, Biological Sciences*, 308, 203–214.

Kuper, A. 1999. *Culture: The Anthropologists' Account*. Cambridge, MA: Harvard University Press.

Kurzban, R., and Houser, D. 2005. Experiments investigating cooperative types in humans: A complement to evolutionary theory and simulations. *Proceedings of the National Academy of Sciences of the United States of America*, 102, 1803–1807.

Lachlan, R. F., and Slater, P. J. B. 1999. The maintenance of vocal learning by gene-culture interaction: The cultural trap hypothesis. *Proceedings of the Royal Society of London, Series B*, 266, 701–706.

Lack, D. 1969. Subspecies and sympatry in Darwin's finches. *Evolution*, 23, 252–263.

Laland, K. N. 1994. Sexual selection with a culturally transmitted mating preference. *Theoretical Population Biology*, 45, 1–15.

321

————. 2001. Imitation, social learning, and preparedness as mechanisms of bounded rationality. In: *Bounded Rationality: The Adaptive Toolbox* (Ed. by G. Gigerenzer and R. Selton), pp. 233–248. Cambridge, MA: MIT Press.

————. 2004. Social learning strategies. *Learning and Behavior*, 32, 4–14.

Laland, K. N., and Hoppitt, W. 2003. Do animals have culture? *Evolutionary Anthropology*, 12, 150–159.

Laland, K. N., and Janik, V. M. 2006. The animal cultures debate. *Trends in Ecology and Evolution*, 21, 542–547.

Laland, K. N., and Kendal, J. R. 2003. What the models say about animal social learning. In *Traditions in Nonhuman Primates: Models and Evidence* (Ed. by D. Fragaszy and S. Perry), pp. 33–55. Chicago: University of Chicago Press.

Laland, K. N., Odling-Smee, F. J., and Feldman, M. W. 2001. Cultural niche construction and human evolution. *Journal of Evolutionary Biology*, 14, 22–33.

Laland, K. N., and Plotkin, H. C. 1990. Social learning and social transmission of foraging information in Norway rats *(Rattus noregicus)*. *Animal Learning and Behavior*, 18, 246–251.

————. 1993. Social transmission of food preferences amongst Norway rats by marking of food sites, and by gustatory contact. *Animal Learning and Behavior*, 21, 35–41.

Laland, K. N., Richerson, P. J., and Boyd, R. 1996. Developing a theory of animal social learning. In: *Social Learning in Animals: The Roots of Culture* (Ed. by C. M. Heyes and B. G. Galef Jr.), pp. 129–154. San Diego: Academic Press.

Laland, K. N., and Williams, K. 1997. Shoaling generates social learning of foraging information in guppies. *Animal Behaviour*, 53, 1161–1169.

————. 1998. Social transmission of maladaptive information in the guppy. *Behavioral Ecology*, 9, 493–499.

Landova, E., Horacek, I., and Frynta, D. 2006. Independent origins of the culturally transmitted technique of pine cone opening: Black rats on Cyprus Island behave like their conspecifics in Israel. *Israel Journal of Ecology*, 52, 151–158.

Lanjouw, A. 2002. Tool use in Tongo chimpanzees. In: *Behavioral Diversity in Chimpanzees and Bonobos* (Ed. by C. Boesch, G. Hohmann, and L. F. Marchant), pp. 52–60. Cambridge: Cambridge University Press.

Lawick-Goodall, J. van. 1973. Cultural elements in a chimpanzee community. In *Precultural Primate Behavior, Symposia of the Fourth International Congress of Primatology* (Ed. by E. W. Menzel), pp. 144–184. Basel: Karger.

LeDuc, R. G., Perrin, W. F., and Dizon, A. E. 1999. Phylogenetic relationships among the delphinid cetaceans based on full cytochrome B sequences. *Marine Mammal Science*, 15, 619–648.

Lefebvre, L. 1986. Cultural diffusion of a novel food-finding behaviour in urban pigeons: An experimental field test. *Ethology*, 71, 295–304.

————. 1995a. Culturally transmitted feeding behaviour in primates: Evidence for accelerating learning rates. *Primates*, 36, 227–239.

————. 1995b. The opening of milk-bottles by birds: Evidence for accelerating learning rates, but against the wave-of-advance model of cultural transmission. *Behavioural Processes,* 34, 43–53.

Lefebvre, L., and Bouchard, J. 2003. Social learning about food in birds. In: *The Biology of Traditions: Methods and Theory* (Ed. by D. M. Fragaszy and S. Perry), pp. 94–126. Cambridge: Cambridge University Press.

Lefebvre, L., Marino, L., Sol, D., Lemieux-Lefebvre, S., and Arshad, S. 2006. Large brains and lengthened life history periods in odontocetes. *Brain, Behavior and Evolution,* 68, 218–228.

Lefebvre, L., Nicolakakis, N., and Boire, D. 2002. Tools and brains in birds. *Behaviour,* 139, 939–973.

Lefebvre, L., and Palameta, B. 1988. Mechanisms, ecology, and population diffusion of socially learned, food-finding behaviour in feral pigeons. In: *Social Learning: Psychological and Biological Perspectives* (Ed. by T. R. Zentall and B. G. Galef Jr.), pp. 141–164. Hillsdale, NJ: Lawrence Erlbaum.

Lefebvre, L., Templeton, J., Brown, K., and Koelle, M. 1997. Carib grackles imitate conspecific and Zenaida dove tutors. *Behaviour,* 134, 1003–1017.

Legendre, P., and Legendre, L. 1998. *Numerical Ecology.* 2nd ed. Amsterdam: Elsevier.

Lenski, G., Nolan, P., and Lenski, J. 1995. *Human Societies.* 7th ed. New York: McGraw-Hill.

LeVine, R. A. 1984. Properties of culture: An ethnographic view. In: *Culture Theory: Essays on Mind, Self and Emotion* (Ed. by R. A. Schweder and R. A. LeVine), pp. 67–87. Cambridge: Cambridge University Press.

Levinson, S. C. 2006. Introduction: The evolution of culture in a microcosm. In *Evolution and Culture* (Ed. by S. C. Levinson and P. Jaisson), pp. 1–41. Cambridge, MA: MIT Press.

Lonsdorf, E. V. 2005. Sex differences in the development of termite-fishing skills in the wild chimpanzees, *Pan troglodytes schweinfurthii,* of Gombe National Park, Tanzania. *Animal Behaviour,* 70, 673–683.

Lonsdorf, E. V., Eberly, L. E., and Pusey, A. E. 2004. Sex differences in learning in chimpanzees. *Nature,* 428, 715–716.

Lumsden, C., and Wilson, E. O. 1981. *Genes, Mind, and Culture.* Cambridge, MA: Harvard University Press.

Lusseau, D. 2003. The emergence of cetaceans: Phylogenetic analysis of male social behaviour supports the Cetartiodactyla clade. *Journal of Evolutionary Biology,* 16, 531–535.

Lusseau, D., and Newman, M. E. J. 2004. Identifying the role that animals play in their social networks. *Proceedings of the Royal Society of London, Series B: Biological Sciences,* 271, S477–S481.

Lycett, S.J., Collard, M., & McGrew, W.C. 2007. Phylogenetic analyses of behavior support existence of culture among wild chimpanzees. Proceedings of the National Academy of Sciences of the United States of America, 104(45), 17588–17592.

Maestripieri, D., and Whitham, J. 2001. Teaching in marine mammals? Anecdote versus science. *Behavioral and Brain Sciences,* 24, 342–343.

Manly, B. F. J. 1997. *Randomization, Bootstrap and Monte Carlo Methods in Biology.* London: Chapman and Hall.

Mann, J. 2001. Cetacean culture: Definitions and evidence. *Behavioral and Brain Sciences,* 24, 343.

————. 2006. Establishing trust: Sociosexual behaviour and the development of male-male bonds among Indian Ocean bottlenose dolphin calves. In: *Homosexual Behaviour in Animals: An Evolutionary Perspective* (Ed. by P. Vasey and V. Sommer), pp. 107–121. Cambridge: Cambridge University Press.

Mann, J., and Barnett, H. 1999. Lethal tiger shark *(Galeocerdo cuvier)* attack on bottlenose dolphin *(Tursiops* sp.) calf: Defense and reactions by the mother. *Marine Mammal Science,* 15, 568–575.

Mann, J., Connor, R. C., Barre, L. M., and Heithaus, M. R. 2000. Female reproductive success in bottlenose dolphins *(Tursiops* sp.): Life history, habitat, provisioning, and group-size effects. *Behavioral Ecology,* 11, 210–219.

Mann, J., and Sargeant, B. L. 2003. Like mother, like calf: The ontogeny of foraging traditions in wild Indian Ocean bottlenose dolphins *(Tursiops* sp.). In: *The Biology of Traditions: Models and Evidence* (Ed. by D. M. Fragaszy and S. Perry), pp. 236–266. Cambridge: Cambridge University Press.

Mann, J., Sargeant, B. L., and Minor, M. 2007. Calf inspections of fish catches in bottlenose dolphins *(Tursiops* sp.): Evidence for oblique social learning? *Marine Mammal Science,* 23, 197–202

Mann, J., and Smuts, B. 1999. Behavioral development in wild bottlenose dolphin newborns *(Tursiops* sp.). *Behaviour,* 136, 529–566.

Mann, J., and Watson-Capps, J. J. 2005. Surviving at sea: Ecological and behavioural predictors of calf mortality in Indian Ocean bottlenose dolphins, *Tursiops* sp. *Animal Behaviour,* 69, 899–909.

Mann, N. I., Dingess, K. A., and Slater, P. J. B. 2006. Antiphonal four-part synchronized chorusing in a neotropical wren. *Biology Letters,* 2, 1–4.

Marcoux, M., Rendell, L., and Whitehead, H. 2007. Indications of fitness differences among vocal clans of sperm whales. *Behavioural Ecology and Sociobiology,* 61, 1093–1098.

Marean, C. W., Bar-Mathews, M., Bernatchez, J., Fisher, E., Goldberg, P., Herries, A. I. R., Jacobs, Z., Jerardino, A., Karkanas, P., Minichillo, T., Nilssen, P. J., Thompson, E., Watts, I., and Williams, H. M. 2007. Early human use of marine resources and pigment in South Africa during the Middle Pleistocene. *Nature,* 449, 905–909.

Marino, L. 1998. A comparison of encephalization between odontocete cetaceans and anthropoid primates. *Brain, Behavior, and Evolution,* 51, 230–238.

Marino, L., McShea, D. W., and Uhen, M. D. 2004. Origin and evolution of large brains in toothed whales. *Anatomical Record,* 281A, 1247–1255.

Marks, J. 2002. *What It Means to Be 98% Chimpanzee: Apes, People, and Their Genes.* Berkeley: University of California Press.

Marler, P. 1952. Variation in the song of the chaffinch *Fringilla coelebs. Ibis,* 94, 458–472.

Marler, P., and Tamura, M. 1964. Culturally transmitted patterns of vocal behavior in sparrows. *Science,* 146, 1483–1486.

Marshall, A., Ancrenaz, M., Brearly, F. Q., Fredriksson, G. M., Ghaffar, N., Heydon, M., Husson, S., Leighton, M., McConkey, K. R., Morrogh-Bernard, H., Proctor, J., van Schaik, C. P., Yeager, C., and Wich, S. A. In press. The effects of forest phenology and floristics on populations of Bornean and Sumatran orangutans: Are Sumatran forests better òrangutan habitat than Bornean forests? In: *Orangutans: Geographic Variation Behavioral Ecology and Conservation* (Ed. by S. Wich, S. Suci Utami Atmoko, T. Mitra Setia, and C. P. van Schaik). Oxford: Oxford University Press.

Marshall, A. J., Wrangham, R. W., and Arcadi, A. C. 1999. Does learning affect the structure of vocalizations in chimpanzees? *Animal Behaviour*, 58, 825–830.

Marshall-Pescini, S., and Whiten, A. 2008. Chimpanzees (*Pan troglodytes*) and the question of cumulative culture: An experimental approach. *Animal Cognition*, 11, 449–456.

Mason, W. A. 1985. Experiential influences on the development of expressive behaviors in rhesus monkeys. In: *The Development of Expressive Behavior: Biology-Environment Interactions* (Ed. by G. Zivin), pp. 117–152. Orlando, FL: Academic Press.

Matsusaka, T., Nishie, H., Shimada, M., Kutsukake, N., Zamma, K., Nakamura, M., and Nishida, T. 2006. Tool-use for drinking water by immature chimpanzees of Mahale: Prevalence of an unessential behavior. *Primates*, 47, 113–122.

Matsuzawa, T. 1994. Field experiments on the use of stone tools in the wild. In: *Chimpanzee Cultures* (Ed. by R. W. Wrangham, W. C. McGrew, F. B. M. de Waal, and P. G. Heltne), pp. 351–370. Cambridge, MA: Harvard University Press.

———. 2001. Primate foundations of human intelligence: A view of tool use in nonhuman primates and fossil hominids. In: *Primate Origins of Human Cognition and Behavior* (Ed. by T. Matsuzawa), pp. 3–25. Tokyo: Springer.

Matsuzawa, T., Biro, D., Humle, T., Inoue-Nakamura, N., Tonooka, R., and Yamakoshi, G. 2001. Emergence of culture in wild chimpanzees: Education by master apprenticeship. In: *Primate Origins of Human Cognition and Behavior* (Ed. by T. Matsuzawa), pp. 557–574. Tokyo: Springer.

Matsuzawa, T., and Yamakoshi, G. 1996. Comparison of chimpanzee material culture between Bossou and Nimba, West Africa. In: *Reaching into Thought: The Minds of the Great Apes* (Ed. by A. Russon, K. Bard, and S. Parker), pp. 211–232. Cambridge: Cambridge University Press.

McBeath, N. M., and McGrew, W. C. 1982. Tools used by wild chimpanzees to obtain termites at Mt. Assirik, Senegal: The influence of habitat. *Journal of Human Evolution*, 11, 65–72.

McBrearty, S., and Stringer, C. 2007. Palaeoanthropology: The coast in colour. *Nature*, 449, 793–794.

McElreath, R., Boyd, R., and Richerson, P. J. 2003. Shared norms and the evolution of ethnic markers. *Current Anthropology*, 44, 122–129.

McGregor, A., Saggerson, A., Pearce, J., and Heyes, C. 2006. Blind imitation in pigeons, *Columba livia*. *Animal Behaviour*, 72, 287–296.

McGrew, W. C. 1974. Tool use by wild chimpanzees in feeding upon driver ants. *Journal of Human Evolution*, 3, 501–508.

————. 1992. *Chimpanzee Material Culture: Implications for Human Evolution.* Cambridge: Cambridge University Press.

————. 1998. Culture in nonhuman primates. *Annual Review of Anthropology,* 27, 301–308.

————. 2001. The nature of culture: Prospects and pitfalls of cultural primatology. In: *Tree of Origin: What Primate Behavior Can Tell Us about Human Social Evolution* (Ed. by F. B. M. de Waal), pp. 229–254. Cambridge, MA: Harvard University Press.

————. 2003. Ten dispatches from the chimpanzee culture wars. In: *Animal Social Complexity: Intelligence, Culture, and Individualized Societies* (Ed. by F. B. M. de Waal and P. Tyack), pp. 419–443. Cambridge, MA: Harvard University Press.

————. 2004. *The Cultured Chimpanzee: Reflections on Cultural Primatology.* Cambridge: Cambridge University Press.

McGrew, W. C., Ham, R. M., White, L. T. J., Tutin, C. E. G., and Fernandez, M. 1997. Why don't chimpanzees in Gabon crack nuts? *International Journal of Primatology,* 18, 353–374.

McGrew, W. C., Marchant, L. F., and Nishida, T., eds., 1996. *Great Ape Societies.* Cambridge: Cambridge University Press.

McGrew, W. C., Marchant, L. F., Scott, S. E., and Tutin, C. E. G. 2001. Intergroup differences in social custom of wild chimpanzees: The grooming-hand-clasp of the Mahale Mountains. *Current Anthropology,* 42, 148–153.

McGrew, W. C., and Tutin, C. E. G. 1978. Evidence for a social custom in wild chimpanzees. *Man,* 13, 234–251.

McLaughlin, R. L., Ferguson, M. M., and Noakes, D. L. G. 1999. Adaptive peaks and alternative foraging tactics in brook charr: Evidence of short-term divergent selection for sitting-and-waiting and actively searching. *Behavioral Ecology and Sociobiology,* 45, 386–395.

Mead, J. G., and Potter, C. W. 1990. Natural history of bottlenose dolphins along the Central Atlantic coast of the United States. In: *The Bottlenose Dolphin* (Ed. by S. Leatherwood and R. R. Reeves), pp. 165–195. San Diego: Academic Press.

Mealey, L. 1995. The sociobiology of sociopathy: An integrated evolutionary model. *Behavioral and Brain Sciences,* 18, 523–599.

Mellars, P., and Stringer, C. 1989. *The Human Revolution.* Edinburgh: Edinburgh University Press

Meltzoff, A. 1988. The human infant as "homo imitans." In: *Social Learning: Psychological and Biological Perspectives* (Ed. by T. Zentall and B. G. Galef Jr.). Hillsdale, NJ: Lawrence Erlbaum.

Menzel, E., ed. 1973a. *Precultural Primate Behavior.* Vol. 1. Basel: Karger.

————. 1973b. Further observations on the use of ladders in a group of young chimpanzees. *Folia Primatologica,* 19, 450–457.

Menzel, E. W., Davenport, R. K., and Rogers, C. M. 1972. Protocultural aspects of chimpanzees' responsiveness to novel objects. *Folia Primatologica,* 17, 161–170.

Mercader, J., Panger, M., and Boesch, C. 2002. Excavation of a chimpanzee stone tool site in the African rainforest. *Science,* 296, 1452–1455.

Mercado, I. E., Murray, S. O., Uyeyama, R. K., Pack, A. A., and Herman, L. M. 1998. Memory of recent actions in the bottlenosed dolphin *(Tursiops truncatus)*: Replication of arbitrary behaviors using an abstract rule. *Animal Learning and Behavior*, 26, 210–218.

Merrill, M. Y. 2004. Orangutan cultures? Tool use, social transmission and population differences. PhD dissertation, Department of Biological Anthropology and Anatomy, Duke University.

Mesoudi, A., Whiten, A., and Dunbar, R. I. M. 2006. A bias for social information in human cultural transmission. *British Journal of Psychology*, 97, 405–423.

Mesoudi, A., Whiten, A., and Laland, K. N. 2006. Towards a unified science of cultural evolution. *Behavioral and Brain Sciences*, 29, 329–383.

Miles, H. L., Mitchell, R. W., and Harper, S. E. 1996. Simon says: The development of imitation in an enculturated orangutan. In: *Reaching into Thought: The Minds of the Great Apes* (Ed. by A. E. Russon, K. A. Bard, and S. T. Parker), pp. 278–299. Cambridge: Cambridge University Press.

Milinski, M. 1997. How to avoid the seven deadly sins in the study of behavior. *Advances in the Study of Behavior*, 26, 159–180.

Miller, P. J. O., Shapiro, A. D., Tyack, P. L., and Solow, A. R. 2004. Call-type matching in vocal exchanges of free-ranging resident killer whales, *Orcinus orca. Animal Behaviour*, 67, 1099–1107.

Mineka, S., and Cook, M. 1988. Social learning and the acquisition of snake fear in monkeys. In: *Social Learning: Psychological and Biological Perspectives* (Ed. by T. R. Zentall and B. G. Galef Jr.), pp. 51–74. Hillsdale, NJ: Lawrence Erlbaum.

Mobius, Y., Boesch, C., Koops, K., Matsuzawa, T., and Humle, T. 2008. Cultural differences in army ant predation by West African chimpanzees? A comparative study of microecological variables. *Animal Behaviour*, 76, 37–45.

Möller, L. M., Beheregaray, L. B., Harcourt, R., and Krützen, M. 2001. Alliance membership and kinship in wild male bottlenose dolphins *(Tursiops aduncus)* of southeastern Australia. *Proceedings of the Royal Society of London, Series B*, 268, 1941–1947.

Morgan, B., and Abwe, A. 2006. Chimpanzees use stone hammers in Cameroon. *Current Biology*, 16, R632–R633.

Morgan, C. L. 1896a. *Habit and Instinct*. London: Arnold.

———. 1896b. Of modification and variation. *Science*, 4, 733–739.

Morgan, L. H. 1868. *The American Beaver and His Works*. Philadelphia: Lippincott.

Morin, P. A., Moore, J. J., Chakraborty, R., Jin, L., Goodall, J., and Woodruff, D. S. 1994. Kin selection, social structure, gene flow and the evolution of chimpanzees. *Science*, 265, 1193–1201.

Moritz, C. 1994. Defining 'Evolutionarily Significant Units' for conservation. *Trends in Ecology and Evolution*, 9, 373–375.

Morrell, L. J., Croft, D. P., Dyer, J. R. G., Chapman, B. B., Kelley, J. L., Laland, K. N., and Krause J. In Press. Association patterns and foraging behaviour in natural and artificial guppy shoals. *Animal Behaviour*.

Muir, C. C., Galdikas, B. M. F., and Beckenbach, A. T. 2000. MtDNA sequence diversity of orangutans from the islands of Borneo and Sumatra. *Journal of Molecular Evolution*, 51, 471–480.

Mundinger, P. C. 1980. Animal cultures and a general theory of cultural evolution. *Ethology and Sociobiology,* 1, 182–223.

Myowa-Yamakoshi, G., and Matsuzawa, T. 1999. Factors influencing imitation of manipulatory actions in chimpanzees. *Journal of Comparative Psychology,* 113, 128–136.

Nagell, K., Olguin, K., and Tomasello, M. 1993. Processes of social learning in the tool use of chimpanzees *(Pan troglodytes)* and human children *(Homo sapiens). Journal of Comparative Psychology,* 107, 174–186.

Nakamichi, M., Kato, E., Kejima, Y., and Itoigawa, N. 1998. Carrying and washing of grass roots by free-ranging Japanese macaques at Katsuyama. *Folia primatologica,* 69, 35–40.

Nakamura, M. 2002. Grooming-hand-clasp in Mahale M group chimpanzees: Implications for culture in social behaviours. In: *Behavioural Diversity in Chimpanzees and Bonobos* (Ed. by C. Boesch, G. Hohmann, and L. F. Marchant), pp. 71–89. Cambridge: Cambridge University Press.

Nakamura, M., McGrew, W. C., Marchant, L. F., and Nishida, T. 2000. Social scratch: Another custom in wild chimpanzees? *Primates,* 41, 237–248.

Nakamura, M., and Uehara, S. 2004. Proximate factors of different types of grooming hand-clasp in Mahale chimpanzees: Implications for chimpanzee social customs. *Current Anthropology,* 45, 108–114.

Nakayama, K. 2004. Observing conspecifics scratching induces a contagion of scratching in Japanese monkeys *(Macaca fuscata). Journal of Comparative Psychology,* 118, 20–24.

Newman, M. E. J. 2004. Analysis of weighted networks. *Physical Review E,* 70, 056131.

———. 2006. Modularity and community structure in networks. *Proceedings of the National Academy of Sciences of the United States of America,* 103, 8577–8582.

Nicol, C. J., and Pope, S. J. 1996. The maternal feeding display of domestic hens is sensitive to perceived chick error. *Animal Behaviour,* 52, 767–774.

Nielsen, M. 2006. Copying actions and copying outcomes: Social learning through the second year. *Developmental Psychology,* 42, 555–565.

———. In press. The imitative behaviour of children and chimpanzees: A window on the transmission of cultural traditions. *Primatologie.*

Nihei, Y., and Higuchi, H. 2001. When and where did crows learn to use automobiles as nutcrackers? *Tohoku Psychologica Folia,* 60, 93–97.

Nishida, T. 1987. Local traditions and cultural transmission. In: *Primate Societies* (Ed. by B. B. Smuts, D. L. Cheney, R. M. Seyfarth, R. W. Wrangham, and T. T. Struhsaker), pp. 462–474. Chicago: University of Chicago Press.

Nishida, T., Mitani, J. C., and Watts, D. P. 2004. Variable grooming behaviours in wild chimpanzees. *Folia Primatologica,* 75, 31–36.

Noad, M. J., Cato, D. H., Bryden, M. M., Jenner, M. N., and Jenner, K. C. S. 2000. Cultural revolution in whale song. *Nature,* 408, 537.

Noren, S. R., Lacave, G., Wells, R. S., and Williams, T. M. 2002. The development of blood oxygen stores in bottlenose dolphins *(Tursiops truncatus):* Implications for diving capacity. *Journal of Zoology,* 258, 105–113.

Noren, S. R., Williams, T. M., Pabst, D. A., McLellan, W. A., and Dearolf, J. L. 2001. The development of diving in marine endotherms: Preparing the skeletal muscles of dolphins, penguins, and seals for activity during submergence. *Journal of Comparative Physiology B: Biochemical, Systemic and Environmental Physiology,* 171, 127–134.

Norris, K. S., and Schilt, C. R. 1988. Cooperative societies in three-dimensional space: On the origins of aggregations, flocks and schools, with special reference to dolphins and fish. *Ethology and Sociobiology,* 9, 149–179.

Norton-Griffiths, M. 1967. Some ecological aspects of the feeding behaviour of the oystercatcher *Haematopus ostralegus* on the edible mussel *Mytilus edulis. Ibis,* 109, 412–424.

Nowacek, D. P. 2002. Sequential foraging behaviour of bottlenose dolphins, *Tursiops truncatus,* in Sarasota Bay, FL. *Behaviour,* 139, 1125–1145.

Odling-Smee, F. J., Laland, K. N., and Feldman, M. W. 2003. *Niche Construction: The Neglected Process in Evolution.* Princeton, NJ: Princeton University Press.

O'Malley, R. C., and Fedigan, L. M. 2004. Evaluating social influences on food-processing behavior in white-faced capuchins *(Cebus capucinus). American Journal of Physical Anthropology,* 127, 481–491.

Opie, I., and Opie, P. 1987. *The Language and Lore of School Children.* New York: Oxford University Press.

Osborne, H. F. 1896. A mode of evolution requiring neither natural selection nor the inheritance of acquired characteristics. *Transactions of the New York Academy of Sciences,* 15, 141–148.

Ottoni, E. B., de Resende, B. D., and Izar, P. 2005. Watching the best nutcrackers: What capuchin monkeys *(Cebus apella)* know about others' tool-using skills. *Animal Cognition,* 8, 215–219.

Owen, E. C. G., Wells, R. S., and Hofmann, S. 2002. Ranging and association patterns of paired and unpaired adult male Atlantic bottlenose dolphins, *Tursiops truncatus,* in Sarasota, Florida, provide no evidence for alternative male strategies. *Canadian Journal of Zoology,* 80, 2072–2089.

Palameta, B., and Lefebvre, L. 1985. The social transmission of a food-finding technique in pigeons: What is learned? *Animal Behaviour,* 33, 892–896.

Panger, M. A., Perry, S., Rose, L., Gros-Louis, J., Vogel, E., Mackinnon, K. C., and Baker, M. 2002. Cross-site differences in foraging behavior of white-faced capuchins *(Cebus capuchinus). American Journal of Physical Anthropology,* 119, 52–56.

Parsons, K. M., Durban, J. W., and Claridge, D. E. 2003. Kinship as a basis for alliance formation between male bottlenose dolphins, *Tursiops truncatus,* in the Bahamas. *Animal Behaviour,* 66, 185–194.

Partridge, L., and Green, P. 1985. Intraspecific feeding specializations and population dynamics. In: *Behavioural Ecology: Ecological Consequences of Adaptive Behaviour* (Ed. by R. M. Sibly and R. H. Smith), pp. 207–226. Oxford: Blackwell Scientific Publications.

Payne, K. 1999. The progressively changing songs of humpback whales: A window on the creative process in a wild animal. In: *The Origins of Music* (Ed.

by N. L. Wallin, B. Merker, and S. Brown), pp. 135–150. Cambridge, MA: MIT Press.

Payne, K., and Payne, R. 1985. Large scale changes over 19 years in songs of humpback whales in Bermuda. *Zeitschrift für Tierpsychologie*, 68, 89–114.

Payne, R., and McVay, S. 1971. Songs of humpback whales. *Science*, 173, 587–597.

Perry, S. 1996. Intergroup encounters in wild white-faced capuchins, *Cebus capucinus*. *International Journal of Primatology*, 17, 309–330.

———. 2006. What cultural primatology can tell anthropologists about the evolution of culture. *Annual Review of Anthropology*, 35, 171–190.

Perry, S., Baker, M., Fedigan, L., Gros-Louis, J., Jack, K., MacKinnon, K. C., Manson, J. H., Panger, M., Pyle, K., and Rose, L. 2003. Social conventions in wild white-faced capuchin monkeys: Evidence for traditions in a neotropical primate. *Current Anthropology*, 44, 241–268.

Perry, S., Barrett, H. C., and Manson, J. M. 2004. White-faced capuchin monkeys exhibit triadic awareness in their choice of allies. *Animal Behaviour*, 67, 165–170.

Perry, S., and Manson, J. 2003. Traditions in monkeys. *Evolutionary Anthropology*, 12, 71–81.

Perry, S., and Ordoñez Jiménez, J. C. 2006. The effects of food size, rarity, and processing complexity on white-faced capuchins' visual attention to foraging conspecifics. In: *Feeding Ecology in Apes and Other Primates* (Ed. by G. Hohmann, M. Robbins, and C. Boesch), pp. 203–234. Cambridge: Cambridge University Press.

Perry, S., Panger, M., Rose, L., Baker, M., Gros-Louis, J., Jack, K., MacKinnon, K., Manson, J., Fedigan, L., and Pyle, K. 2003. Traditions in wild white-faced capuchin monkeys. In: *The Biology of Traditions: Models and Evidence* (Ed. by D. Fragaszy and S. Perry), pp. 391–425. Cambridge: Cambridge University Press.

Pitchford, I. 2001. The origins of violence: Is psychopathy an adaptation? *Human Nature Review*, 1, 28–36.

Polly pachyderm. 2006. *Science*, 314, 29.

Poole, J. H., Tyack, P. L., Stoeger-Horwath, A. S., and Watwood, S. B. 2005. Elephants are capable of social learning. *Nature*, 434, 455–456.

Premack, D., and Premack, A. J. 1994. Why animals have neither culture nor history. In: *Companion Encyclopedia of Anthropology* (Ed. by T. Ingold), pp. 350–365. London: Routledge.

Preston, S. D., and de Waal, F. B. M. 2002. Empathy: Its ultimate and proximate bases. *Behavioral and Brain Sciences*, 25, 1–72.

Preuschoft, S., and van Hooff, J. A. R. A. M. 1995. Homologizing primate facial displays: A critical review of methods. *Folia Primatologica*, 65, 121–137.

Price, T. 1987. Diet variation in a population of Darwin's finches. *Ecology*, 68, 1015–1028.

Pruetz, J. D., and Bertolani, P. 2007. Savanna Chimpanzees *Pan troglodytes verus*, Hunt with Tools. *Current Biology*, 17, 412–417.

Queller, D. C., Strassman, J. E., and Hughes, C. R. 1993. Microsatellites and kinship. *Trends in Ecology and Evolution*, 8, 285–288.

Radcliffe-Brown, A. R. 1922. *The Andamanese Islanders*. Cambridge: Cambridge University Press.

Rakoczy, H., Warneken, F., and Tomasello, M. 2008. The sources of normativity: Young children's awareness of the normative structure of games. *Developmental Psychology*, 44, 875–881.

Ramsey, G., Bastian, M. L., and van Schaik, C. P. 2007. How to study innovation in the wild. *Behavioral and Brain Sciences*, 30, 393–437.

Reader, S. M. 2004. Distinguishing social and asocial learning using diffusion dynamics. *Learning and Behavior*, 32, 90–104.

Reader, S. M., Kendal, J. R., and Laland, K. N. 2003. Social learning of foraging sites and escape routes in wild Trinidadian guppies. *Animal Behaviour*, 66, 729–739.

Reader, S. M., and Laland, K. N. 2002. Social intelligence, innovation, and enhanced brain size in primates. *Proceedings of the National Academy of Sciences of the United States of America*, 99, 4436–4441.

———. 2003. *Animal Innovation*. Oxford: Oxford University Press.

Reiss, D., and Marino, L. 2001. Mirror self-recognition in the bottlenose dolphin: A case of cognitive convergence. *Proceedings of the National Academy of Sciences of the United States of America*, 98, 5937–5942.

Rendell, L., and Whitehead, H. 2001. Culture in whales and dolphins. *Behavioral and Brain Sciences*, 24, 309–324.

———. 2003. Vocal clans in sperm whales *(Physeter macrocephalus)*. *Proceedings of the Royal Society of London, Series B*, 270, 225–231.

Richerson, P., and Boyd, R. 2005. *Not by Genes Alone: How Culture Transformed Human Evolution*. Chicago: University of Chicago Press.

Ridley, M. 1986. *Evolution and Classification: The Reformulation of Cladism*. London: Longman.

Riesch, R., Ford, J. K. B., and Thomsen, F. 2006. Stability and group specificity of stereotyped whistles in resident killer whales, *Orcinus orca*, off British Columbia. *Animal Behaviour*, 71, 79–91.

Ripoll, T., and Vauclair, J. 2001. Can culture be inferred from the absence of genetic and environmental factors? *Behavioral and Brain Sciences*, 24, 355–356.

Robinson, B. W., Wilson. D. S., Margosian, A. S., and Lotto, P. T. 1993. The ecological and morphological differentiation of pumpkinseed sunfish in lakes without bluegill sunfish. *Evolutionary Ecology*, 7, 451–464.

Rodseth, L., Wrangham, R. W., Harrigan, A. M., and Smuts, B. B. 1991. The human community as a primate society. *Current Anthropology*, 32, 221–254.

Roitblat, H. L. 1988. A cognitive action theory of learning. In: *Systems with Learning and Memory Abilities* (Ed. by J. Delacour and J. C. Levy), pp. 13–26. Amsterdam: North- Holland.

———. 1991. Cognitive action theory as a control architecture. In: *From Animals to Animats: Proceedings of the First International Conference on Simulation of Adaptive Behavior* (Ed. by J. A. Meyer and S. W. Wilson), pp. 444–450. Cambridge, MA: MIT Press.

Romanes, G. J. 1882. *Animal Intelligence*. London: Kegan, Paul, Trench and Co.

———. 1884. *Mental Evolution in Animals*. New York: Appleton and Co.

Roper, T. J. 1986. Cultural evolution of feeding behaviour in animals. *Science Progress*, 70, 571–583.

Roughgarden, J. 1974. Niche width: Biogeographic patterns among *Anolis* lizard populations. *American Naturalist*, 108, 429–442.

Rousch, R. S., and Snowdon, C. T. 2000. Quality, quantity, distribution and audience effects on food calling in cotton-top tamarins. *Ethology*, 106, 673–690.

———. 2001. Food transfer and development of feeding behavior and food-associated vocalizations in cotton-top tamarins. *Ethology*, 107, 415–429.

Russon, A. E. 2003. Developmental perspectives on great ape traditions. In: *The Biology of Traditions: Models and Evidence* (Ed. by D. M. Fragaszy and S. Perry), pp. 329–364. Cambridge: Cambridge University Press.

Russon, A. E., and Galdikas, B. 1993. Imitation in free-ranging rehabilitant orang-utans. *Journal of Comparative Psychology*, 107, 147–160.

Sanfey, A. G., Rilling, J. K., Aronson, J. A., Nystrom, L. E., and Cohen, J. D. 2003. The neural basis of economic decision-making in the ultimatum game. *Science*, 300, 1755–1758.

Sanz, C., Morgan, D., and Gulick, S. 2004. New insights into chimpanzees, tools, and termites from the Congo Basin. *American Naturalist*, 164, 567–581.

Sapolsky, R. M. 1994. The physiological relevance of glucocorticoid endangerment of the hippocampus. *Annals of the New York Academy of Sciences*, 746, 294–304.

———. 2006. Social cultures among non-human primates. *Current Anthropology*, 47, 641–656.

Sapolsky, R. M., and Share, L. J. 2004. A pacific culture among wild baboons: Its emergence and transmission. *Public Library of Science Biology*, 2, 534–541.

Sargeant, B. L. 2005. Foraging development and individual specialization in wild bottlenose dolphins (*Tursiops* sp.). PhD thesis, Georgetown University.

Sargeant, B. L., Mann, J., Berggren, P., and Krützen, M. 2005. Specialization and development of beach hunting, a rare foraging behavior, by wild Indian Ocean bottlenose dolphins (*Tursiops* sp.). *Canadian Journal of Zoology*, 83, 1400–1410.

Sargeant, B. L., Wirsing, A. J., Heithaus, M. R., and Mann, J. 2007. Can environmental heterogeneity explain individual foraging variation in wild bottlenose dolphins (*Tursiops* sp.)? *Behavioral Ecology and Sociobiology*, 5, 679–688.

Savage-Rumbaugh, E. S. 1990. Language as a cause-effect communication system. *Philosophical Psychology*, 3, 55–76.

———. 1998. Scientific schizophrenia with regard to the language act. In: *Piaget, Evolution, and Development* (Ed. by J. Langer and M. Killen), pp. 145–169. Mahwah, NJ: Lawrence Erlbaum Associates.

Sayigh, L. S., Tyack, P. L., Wells, R. S., and Scott, M. D. 1990. Signature whistles of free-ranging bottlenose dolphins, *Tursiops truncatus*: Stability and mother-offspring comparisons. *Behavioral Ecology and Sociobiology*, 26, 247–260.

Schiller, P. H. 1952. Innate constituents of complex responses in primates. *Psychological Review*, 59, 177–191.

————. 1957. Innate motor actions as a basis of learning. In: *Instinctive Behavior: The Development of a Modern Concept* (Ed. by C. H. Schiller), pp. 264–287. New York: International Universities Press.

Schluter, D., and Grant, P. R. 1984. Ecological correlates of morphological evolution in a Darwin's finch, *Geospiza difficilis. Evolution,* 38, 856–869.

Schnell, G. D., Watt, D. J., and Douglas, M. E. 1985. Statistical comparison of proximity matrices: Applications in animal behaviour. *Animal Behaviour,* 33, 239–253.

Schöning, C., Ellis, D., Fowler, A., and Sommer, V. 2007. Army ant availability and consumption by chimpanzees *(Pan troglodytes vellerosus)* at Gashaka (Nigeria). *Journal of Zoology,* 271, 125–133.

Schöning, C., Humle, T., Moebius, Y., and McGrew, W. C. 2008. The nature of culture: Technical variation in chimpanzee predation on army ants revisited. *Journal of Human Evolution,* 55, 48–59.

Scott, E. M., Mann, J., Watson-Capps, J. J., Sargeant, B. L., and Connor, R. C. 2005. Aggression in bottlenose dolphins: Evidence for sexual coercion, male-male competition, and female tolerance through analysis of tooth-rake marks and behaviour. *Behaviour,* 142, 21–44.

Scupin, R. 1992. *Cultural Anthropology: A Global Perspective.* Englewood Cliffs, NJ: Prentice Hall.

Seppanen, J. T., Forsman, J. T., Monkkonen, M., and Thompson, R. L. 2007. Information use is a process across time, space & ecology, reaching heterospecifics. *Ecology,* 88, 1622–1633.

Seyfarth, R. M., and Cheney, D. L. 2003. The structure of social knowledge in monkeys. In: *Animal Social Complexity* (Ed. by F. B. M. de Waal and P. L. Tyack), pp. 207–229. Cambridge, MA: Harvard University Press.

Shane, S. H. 1990. Behavior and ecology of the bottlenose dolphin at Sanibel Island, Florida. In: *The Bottlenose Dolphin* (Ed. by S. Leatherwood and R. R. Reeves), pp. 245–265. San Diego: Academic Press.

Shapiro, A. D., Slater, P. J. B., and Janik, V. M. 2004. Call usage learning in gray seals *(Halichoerus grypus). Journal of Comparative Psychology,* 118, 447–454.

Sherry, D. F., and Galef, B. G., Jr. 1984. Cultural transmission without imitation: Milk bottle opening by birds. *Animal Behaviour,* 32, 937–938.

Shettleworth, S. J. 1998. *Cognition, Evolution, and Behavior.* Oxford: Oxford University Press.

Silber, G. K., and Fertl, D. 1995. Intentional beaching by bottlenose dolphins *(Tursiops truncatus)* in the Colorado River Delta, Mexico. *Aquatic Mammals,* 21, 183–186.

Silk, J. B., Brosnan, S. F., Vonk, J., Henrich, J., Povinelli, D. J., Richardson, A. S., Lambeth, S. P., Mascaro, J., and Shapiro, S. J. 2005. Chimpanzees are indifferent to the welfare of unrelated group members. *Nature,* 437, 1357–1359.

Singleton, I. S., and van Schaik, C. P. 2002. The social organisation of a population of Sumatran orang-utans. *Folia Primatologica,* 73, 1–20.

Slater, P. J. B. 1985. *An Introduction to Ethology.* Cambridge: Cambridge University Press.

———. 2001. There's CULTURE and "Culture." *Behavioral and Brain Sciences,* 24, 356–357.

Slocombe, K. E., and Zuberbühler, K. 2005a. Agonistic screams in wild chimpanzees *(Pan troglodytes schweinfurthii)* vary as a function of social role. *Journal of Comparative Psychology,* 119, 67–77.

———. 2005b. Functionally referent communication in a chimpanzee. *Current Biology,* 15, 1779–1784.

Smith, J. 1977. *The Behavior of Communicating.* Cambridge, MA: Harvard University Press.

Smith, J. D., Shields, W. E., and Washburn, D. A. 2003. The comparative psychology of uncertainty monitoring and metacognition. *Behavioral and Brain Sciences,* 26, 317–373.

Smolker, R., and Pepper, J. W. 1999. Whistle convergence among allied male bottlenose dolphins (Delphinidae, *Tursiops* sp.). *Ethology,* 105, 595–617.

Smolker, R. A., Richards, A. F., Connor, R. C., Mann, J., and Berggren, P. 1997. Sponge-carrying by Indian Ocean bottlenose dolphins: Possible tool-use by a delphinid. *Ethology,* 103, 454–465.

Smolker, R. A., Richards, A. F., Connor, R. C., and Pepper, J. W. 1992. Sex differences in patterns of association among Indian Ocean bottlenose dolphins. *Behaviour,* 123, 38–69.

Smouse, P. E., Long, J. C., and Sokal, R. R. 1986. Multiple regression and correlation extensions of the Mantel test of matrix correspondence. *Systematic Zoology,* 35, 627–632.

Sokal, R. R., and Rohlf, F. J. 1994. *Biometry.* 3rd ed. New York: W. H. Freeman.

Soltis, J., Boyd, R., Richerson, P. J. 1995. Can group-functional behaviors evolve by cultural group selection? *Current Anthropology,* 36, 473–483.

Spalding, D. 1873. Instinct, with original observations on young animals. *Macmillan's Magazine,* 27, 282–293.

Stanford, C. B. 1998. *Chimpanzee and Red Colobus: The Ecology of Predator and Prey.* Cambridge, MA: Harvard University Press.

———. 1999. The social behavior of chimpanzees and bonobos: Empirical evidence and shifting assumptions. *Current Anthropology,* 39, 399–420.

Sterelny, K. 2006. The evolution and evolvability of culture. *Mind and Language,* 21, 137–165.

Stout, M. 2005. *The Sociopath Next Door.* New York: Broadway.

Strassmann, J. E., Zhu, Y., and Queller, D. C. 2000. Altruism and social cheating in the social amoeba *Dictyostelium discoideum. Nature,* 408, 965–967.

Strum, S. C. 1975. Primate predation: Interim report on the development of a tradition in a troop of olive baboons. *Science,* 187, 755–757.

Sugiyama, Y. 1995. Tool-use for catching ants by chimpanzees at Bossou and Monts-Nimba. *Primates,* 36, 193–205.

Sumita, K., Kitahara-Frisch, J., and Norikoshi, K. 1985. The acquisition of stone tool use in captive chimpanzees. *Primates,* 26, 168–181.

Sumpter, D. J. T. 2006. The principles of collective animal behaviour. *Philosophical Transactions of the Royal Society of London, Series B: Biological Sciences,* 361, 5–22.

Svanbäck, R., and Persson, L. 2004. Individual diet specialization, niche width and population dynamics: Implications for trophic polymorphisms. *Journal of Animal Ecology,* 73, 973–982.

Szathmáry, E. 2006. Darwin for all seasons. *Science,* 313, 306–307.

Taglialatela, J. P., Savage-Rumbaugh, E. S., and Baker, L. A. 2003. Vocal production by a language-competent *Pan paniscus. International Journal of Primatology,* 24, 1–17.

Tanaka, I. 1998. Social diffusion of modified louse egg-handling techniques during grooming in free-ranging Japanese macaques. *Animal Behaviour,* 56, 1229–1236.

Tattersall, I. 2002. *The Monkey in the Mirror: Essays on the Science of What Makes Us Human.* New York: Harcourt.

Tayler, C. K., and Saayman, G. S. 1973. Imitative behaviour by Indian Ocean bottlenose dolphins *(Tursiops aduncus)* in captivity. *Behaviour,* 44, 286–298.

Tebbich, S., Taborsky, M., Fessl, B., and Blomqvist, D. 2001. Do woodpecker finches acquire tool use by social learning? *Proceedings of the Royal Society of London, Series B,* 268, 2189–2193.

Tennie, C., Call, J., and Tomasello, M. 2006. Push or pull: Emulation versus imitation in great apes and human children. *Ethology,* 112, 1159–1169.

Terkel, J. 1996. Cultural transmission of feeding behavior in the black rat *(Rattus rattus).* In: *Social Learning in Animals: The Roots of Culture* (Ed. by C. M. Heyes and B. G. Galef Jr.), pp. 17–48. San Diego: Academic Press.

Thieme, H. 1997. Lower Paleolithic hunting spears from Germany. *Nature,* 385, 807–810.

Thorndike, E. L. 1898. Animal intelligence: An experimental study of the associative processes in animals. *Psychological Reviews Monograph,* 2, 551–553.

———. 1911. *Animal Intelligence.* New York: Hafner.

Thornton, A., and McAuliffe, K. 2006. Teaching in wild meerkats. *Science,* 313, 227–229.

Thorpe, W. H. 1963. *Learning and Instinct in Animals.* Cambridge, MA: Harvard University Press.

Thouless, C. R., Fanshawe, J. H., and Bertram, B. C. R. 1989. Egyptian vultures, *Neophon percnopterus,* and ostrich, *Struthio camelus,* eggs: The origins of stone throwing behavior. *Ibis,* 131, 9–15.

Tinbergen, N. 1963. On aims and methods of ethology. *Zeitschrift für Tierpsychologie,* 20, 410–433.

Tinker, M. T. 2004. Sources of variation in the foraging behavior and demography of the sea otter, *Enhydra lutris.* PhD thesis, University of California at Santa Cruz.

Tomasello, M. 1990. Cultural transmission in the tool use and communicatory signalling of chimpanzees? In: *"Language" and Intelligence in Monkeys and Apes: Comparative Developmental Perspectives* (Ed. by S. Parker and K. Gibson), pp. 274–311. Cambridge: Cambridge University Press.

———. 1992. The social bases of language acquisition. *Social Development,* 1, 67–87.

———. 1994. The question of chimpanzee culture. In: *Chimpanzee Cultures* (Ed.

by R. W. Wrangham, W. C. McGrew, F. B. M. de Waal, and P. G. Heltne), pp. 301–317. Cambridge, MA: Harvard University Press.

———. 1996. Do apes ape? In *Social Learning in Animals: The Roots of Culture* (Ed. by C. M. Heyes and B. G. Galef Jr.), pp. 319–346. San Diego: Academic Press.

———. 1998. Emulation learning and cultural learning. *Behavior and Brain Sciences*, 21, 703–704.

———. 1999a. *The Cultural Origins of Human Cognition*. Cambridge, MA: Harvard University Press.

———. 1999b. The human adaptation for culture. *Annual Review of Anthropology*, 28, 509–529.

———. 2000. Two hypotheses about primate cognition. In: *The Evolution of Cognition* (Ed. by C. M. Heyes and L. Huber), pp. 165–183. Cambridge, MA: MIT Press.

———. 2003. *Constructing a Language: A Usage-Based Theory of Language Acquisition*. Cambridge, MA: Harvard University Press.

Tomasello, M., and Call, J. 1997. *Primate Cognition*. Oxford: Oxford University Press.

Tomasello, M., Call, J., Nagell, K., Olguin, R., and Carpenter, M. 1994. The learning and use of gestural signals by young chimpanzees: A transgenerational study. *Primates*, 37, 137–154.

Tomasello, M., Call, J., Warren, J., Frost, G. T., Carpenter, M., and Nagell, K. 1997. The ontogeny of chimpanzee gestural signals: A comparison across groups and generations. *Evolution of Communication*, 1, 223–259.

Tomasello, M., and Carpenter, M. 2005. The emergence of social cognition in three young chimpanzees. *Monographs of the Society for Research in Child Development*, 70, 1 (Serial no. 279).

Tomasello, M., Davis-Dasilva, M., Camak, L., and Bard, K. 1987. Observational learning of tool use by young chimpanzees and enculturated chimpanzees. *Human Evolution*, 2, 175–183.

Tomasello, M., George, B., Kruger, A., Farrar, J., and Evans, E. 1985. The development of gestural communication in young chimpanzees. *Journal of Human Evolution*, 14, 175–186.

Tomasello, M., Gust, D., and Frost, T. 1989. A longitudinal investigation of gestural communication in young chimpanzees. *Primates*, 30, 35–50.

Tomasello, M., Kruger, A. C., and Ratner, H. H. 1993. Cultural learning. *Behavioral and Brain Sciences*, 16, 495–552.

Tomasello, M., Savage-Rumbaugh, E. S., and Kruger, A. C. 1993. Imitative learning of actions on objects by children, chimpanzees, and enculturated chimpanzees. *Child Development*, 64, 1688–1705.

Tuttle, R. 2001. Culture and traditional chimpanzees. *Current Anthropology*, 42, 407–409.

Tyack, P. L. 1997. Development and social functions of signature whistles in bottlenose dolphins *Tursiops truncatus*. *Bioacoustics*, 8, 21–46.

Tyack, P. L., and Sayigh, L. S. 1997. Vocal learning in cetaceans. In: *Social*

*Influences on Vocal Development* (Ed. by C. T. Snowdon and M. Hausberger), pp. 208–233. Cambridge: Cambridge University Press.

Tylor, E. B. 1871. *Primitive Culture.* London: Murray

Urbani, B. 1998. An early report on tool use by neotropical primates. *Neotropical Primates*, 6, 123–124.

Uzgiris, I. C. 1981. Two functions of imitation during infancy. *International Journal of Behavioral Development*, 4, 1–12.

van Schaik, C. P. 2003. Local traditions in orangutans and chimpanzees: Social learning and social tolerance. In: *The Biology of Traditions: Models and Evidence* (Ed. by D. M. Fragaszy and S. Perry), pp. 297–328. Cambridge: Cambridge University Press.

———. 2004. *Among Orangutans: Red Apes and the Rise of Human Culture.* Cambridge, MA: Belknap Press of Harvard University Press.

———. 2006. Why are some animals so smart? *Scientific American*, 294(4), 64–71.

van Schaik, C. P., Ancrenaz, M., Borgen, G., Galdikas, B., Knott, C., Singleton, I., Suzuki, A., Utami, S. S., and Merrill, M. 2003. Orangutan cultures and the evolution of material culture. *Science*, 299, 102–105.

van Schaik, C. P., Ancrenaz, M., Djojoasmoro, R., Knott, C. D., Morrogh-Bernard, H., Nuzuar, Odom, K., Atmoko, S. S. U., & van Noordwijk, M. A. In press. Orangutan cultures revisited. In: *Orangutans: Geographic Variation Behavioral Ecology and Conservation.* (Ed. by S. Wich, S. Suci Utami Atmoko, T. Mitra Setia, & C. P. van Schaik). Oxford University Press.

van Schaik, C. P., Deaner, R. O., and Merrill, M. Y. 1999. The conditions for tool use in primates: Implications for the evolution of material culture. *Journal of Human Evolution*, 36, 719–741.

van Schaik, C. P., Fox, E. A., and Fechtman, L. T. 2003. Individual variation in the rate of use of tree-hole tools among wild orang-utans: Implications for hominin evolution. *Journal of Human Evolution*, 44, 11–23.

van Schaik, C. P., and Knott, C. D. 2001. Geographic variation in tool use on *Neesia* fruits in orangutans. *American Journal of Physical Anthropology*, 114, 331–342.

van Schaik, C. P., and Pradhan, G. R. 2003. A model for tool-use traditions in primates: Implications for the evolution of culture and cognition. *Journal of Human Evolution*, 44, 645–664.

van Schaik, C. P., van Noordwijk, M. A., and Wich, S. A. 2006. Innovation in wild Bornean orangutans *(Pongo pygmaeus wurmbii). Behaviour*, 143, 839–876.

Vekua, A., Lordkipanidze, D., Rightmire, G. P., Agusti, J., Ferring, R., Maisuradze, G., Mouskhelishvili, A., Nioradze, M., Ponce de Leon, M., Tappen, M., Tvalchrelidze, M., and Zollikofer, C. 2002. A new skull of early Homo from Dmanasi, Georgia. *Science*, 297(5578), 85–89.

Visalberghi, E., and Fragaszy, D. 1990. Do monkeys ape? In: *Language and Intelligence in Monkeys and Apes: Comparative Developmental Perspectives* (Ed. by S. Parker and K. Gibson), pp. 247–273. Cambridge: Cambridge University Press.

Voight, B. F., Kudaravalli, S., Wen, X., and Pritchard, J. K. 2006. A map of recent positive selection in the human genome. *Public Library of Science Biology,* 4, 446–458.

Vonk, J., Brosnan, S., Silk, J. B., Henrich, J., Richarson, A. S., Lambeth, S. P., Schapiro, S. J., and Povinelli, D. J. 2008. Chimpanzees do not take advantage of very low cost opportunities to deliver food to unrelated group members. *Animal Behaviour,* 75, 1757–1770.

Wallace, A. R. 1870. *Contributions to the Theory of Natural Selection.* London: Macmillan.

Wang, E. T., Kodama, G., Baldi, P., and Moyzis, R. K. 2006. Global landscape of recent inferred Darwinian selection for *Homo sapiens. Proceedings of the National Academy of Sciences of the United States of America,* 103, 135–140.

Warneken, F., Chen, F., Tomasello, M. 2006. Cooperative activities in young children and chimpanzees. *Child Development,* 77, 640–663.

Warner, R. R. 1988. Traditionality of mating-site preferences in a coral reef fish. *Nature,* 335, 719–721.

———. 1990. Male versus female influences on mating-site determination in a coral-reef fish. *Animal Behaviour,* 39, 540–548.

Washburn, S. L., and Benedict, B. 1979. Non-human primate culture. *Man,* 14, 163–164.

Watanabe, K. 1989. Fish: A new addition to the diet of Japanese macaques on Koshima Island. *Folia Primatologica,* 52, 124–131.

———. 1994. Precultural behavior of Japanese macaques: Longitudinal studies of the Koshima troops. In: *The Ethological Roots of Culture* (Ed. by R. A. Gardner, A. B. Chiarelli, B. T. Gardner, and F. X. Plooij), pp. 81–94. Dordrecht, Netherlands: Kluwer.

Watwood, S. L., Tyack, P. L., and Wells, R. S. 2004. Whistle sharing in paired male bottlenose dolphins, *Tursiops truncatus. Behavioral Ecology and Sociobiology,* 55, 531–543.

Weaver, A., and de Waal, F. B. M. 2003. The mother-offspring relationship as a template in social development: Reconciliation in captive brown capuchins *(Cebus apella). Journal of Comparative Psychology,* 117, 101–110.

Wells, R. S. 1991. The role of long-term study in understanding the social structure of a bottlenose dolphin community. In: *Dolphin Societies: Discoveries and Puzzles* (Ed. by K. Pryor and K. S. Norris), pp. 198–225. Berkeley: University of California Press.

Werner, T. K., and Sherry, T. W. 1987. Behavioral feeding specialization in *Pinarolaxias inornata,* the "Darwin's Finch" of Cocos Island, Costa Rica. *Proceedings of the National Academy of Sciences of the United States of America,* 84, 5506–5510.

West, M. J., and King, A. P. 1996. Social learning: Synergy and songbirds. In: *Social Learning in Animals: The Roots of Culture* (Ed. by C. M. Heyes and B. G. Galef Jr.), pp. 155–178. San Diego: Academic Press.

West-Eberhard, M. J. 2003. *Developmental Plasticity and Evolution.* New York: Oxford University Press.

What are the roots of human culture? 2005. *Science*, 309, 99.

Wheatley, B. P. 1999. *The Sacred Monkeys of Bali*. Prospect Heights, IL: Waveland Press.

White, J. 2006. Early and profound human impact. *Science*, 311, 472.

Whitehead, H. 1998. Cultural selection and genetic diversity in matrilineal whales. *Science*, 282, 1708–1711.

———. 2003a. *Sperm Whales: Social Evolution in the Ocean*. Chicago: University of Chicago Press.

———. 2003b. Society and culture in the deep and open ocean: The sperm whale. In: *Animal Social Complexity: Intelligence, Culture, and Individualized Societies* (Ed. by F. B. M. de Waal and P. L. Tyack), pp. 444–464. Cambridge, MA: Harvard University Press.

———. 2005. Genetic diversity in the matrilineal whales: Models of cultural hitch-hiking and group-specific non-heritable demographic variation. *Marine Mammal Science*, 21, 58–79.

———. 2008. Analyzing animal societies: Quantitative methods for vertebrate social analysis. University of Chicago Press, Chicago.

Whitehead, H., Dillon, M., Dufault, S., Weilgart, L., and Wright, J. 1998. Non-geographically based population structure of South Pacific sperm whales: Dialects, fluke-markings and genetics. *Journal of Animal Ecology*, 67, 253–262.

Whitehead, H., and Dufault, S. 1999. Techniques for analyzing vertebrate social structure using identified individuals: Review and recommendations. *Advances in the Study of Behavior*, 28, 33–74.

Whitehead, H., and Mann, J. 2000. Female reproductive strategies of cetaceans. In: *Cetacean Societies* (Ed. by J. Mann, R. Connor, P. L. Tyack, and H. Whitehead), pp. 219–246. Chicago: University of Chicago Press.

Whitehead, H., and Rendell, L. 2004. Movements, habitat use and feeding success of cultural clans of South Pacific sperm whales. *Journal of Animal Ecology*, 73, 190–196.

Whitehead, H., Richerson, P. J., and Boyd, R. 2002. Cultural selection and genetic diversity in humans. *Selection*, 3, 115–125.

Whiten, A. 1998. Imitation of the sequential structure of actions by chimpanzees *(Pan troglodytes)*. *Journal of Comparative Psychology*, 112, 270–281.

———. 1999. Parental encouragement in Gorilla in comparative perspective: Implications for social cognition. In: *The Mentality of Gorillas and Orangutans* (Ed. by S. T. Parker, H. L. Miles, and R. W. Mitchell), pp. 342–366. Cambridge: Cambridge University Press.

———. 2000. Primate culture and social learning. *Cognitive Science*, 24, 477–508.

———. 2005. The second inheritance system of chimpanzees and humans. *Nature*, 437, 52–55.

Whiten, A., and Custance, D. M. 1996. Studies of imitation in chimpanzees and children. In: *Social Learning in Animals: The Roots of Culture* (Ed. by C. M. Heyes and B. G. Galef Jr.), pp. 291–318. San Diego: Academic Press.

Whiten A., Custance, D. M., Gomez, J., Teixidor, P., and Bard, K. 1996. Imitative learning of artificial fruit processing in children *(Homo sapiens)* and chimpanzees *(Pan troglodytes)*. *Journal of Comparative Psychology*, 110, 3–14.

Whiten, A., Goodall, J., McGrew, W. C., Nishida, T., Reynolds, V., Sugiyama, Y., Tutin, C. E. G., Wrangham, R. W., and Boesch, C. 1999. Cultures in chimpanzees. *Nature, 399,* 682–685.

———. 2001. Charting cultural variation in chimpanzees. *Behaviour,* 138, 1481–1516.

Whiten, A., and Ham, R. 1992. On the nature and evolution of imitation in the animal kingdom: Reappraisal of a century of research. *Advances in the Study of Behavior, 21,* 239–283.

Whiten, A., Horner, V., and de Waal, F. B. M. 2005. Conformity to cultural norms of tool use in chimpanzees. *Nature, 437,* 737–740.

Whiten, A., Horner, V., Litchfield, C., and Marshall-Pescini, S. 2004. How do apes ape? *Learning and Behaviour, 32,* 36–52.

Whiten, A., Horner, V., and Marshall-Pescini, S. 2003. Cultural Panthropology. *Evolutionary Anthropology, 12,* 92–105.

Whiten, A., and van Schaik, C. P. 2007. The evolution of animal 'cultures' and social intelligence. *Philosophical Transactions of the Royal Society of London, Series B, 362,* 603–620.

Whiten, A., Spiteri, A., Horner, V., Bonnie, K. E., Lambeth, S. P., Schapiro, S. J., and de Waal, F. B. M. 2007. Transmission of multiple traditions within and between chimpanzee groups. *Current Biology, 17,* 1038–1043.

Wich, S. A., Marshall, A. J., Frederiksson, G., Ghaffar, N., Leighton, M., Yeager, C., Brearley, F. Q., Proctor, J., Heydon, M., and van Schaik, C. P. In press. A comparison of forest fruit production between Sumatra and Borneo. *Journal of Tropical Ecology.*

Wich, S. A., Utami-Atmoko, S. S., Setia, T. M., Djoyosudharmo, S., and Geurts, M. L. 2006. Dietary and energetic responses of *Pongo abelii* to fruit availability fluctuations. *International Journal of Primatology, 27,* 1535–1550.

Wilson, A. C. 1985. The molecular basis of evolution. *Scientific American,* 253(4), 164–173.

Wilson, D. R. B. 1995. The ecology of bottlenose dolphins in the Moray Firth, Scotland: A population at the northern extreme of the species' range. Ph.D. diss., University of Aberdeen.

Wilson, M. L., Hauser, M. D., and Wrangham, R. W. 2001. Does participation in intergroup conflict depend on numerical assessment, range location, or rank for wild chimpanzees? *Animal Behaviour, 61,* 1203–1216.

Witte, K., and Noltemeier, B. 2002. The role of information in mate-choice copying in female sailfin mollies *(Poecilia latipinna)*. *Behavioral Ecology and Sociobiology, 52,* 194–202.

Wittemyer, G., Douglas-Hamilton, I., and Getz, W. M. 2005. The socio-ecology of elephants: Analysis of the processes creating multi-tiered social structures. *Animal Behaviour, 69,* 1357–1371.

Wood, D., Bruner, J., and Ross, G. 1976. The role of tutoring in problem solving. *Journal of Child Psychology and Psychiatry, 17,* 89–100.

Woolfenden, G. E., and Fitzpatrick, J. W. 1984. *The Florida Scrub Jay.* Princeton, NJ: Princeton University Press.

Worden, B. D., and Papaj, D. R. 2005. Flower choice copying in bumble bees. *Biology Letters,* 1, 504–507.

Wrangham, R. W. 1999. Evolution of coalitionary killing. *Yearbook of Physical Anthropology,* 42, 1–30.

———. 2006. Chimpanzees: The culture-zone concept becomes untidy. *Current Biology,* 16, R634–R635.

Wrangham, R. W., de Waal F. B. M., and McGrew, W. C. 1994. The challenge of behavioral diversity. In: *Chimpanzee Cultures* (Ed. by R. Wrangham, W. C. McGrew, F. B. M. de Waal, and P. G. Heltne), pp. 1–18. Cambridge, MA: Harvard University Press.

Wrangham, R. W., McGrew, W., de Waal, F. B. M., and Heltne, P., eds. 1994. *Chimpanzee Cultures.* Cambridge, MA: Harvard University Press.

Wyles, J. S., Kunkel, J. G., and Wilson, A. C. 1983. Birds, behavior, and anatomical evolution. *Proceedings of the National Academy of Sciences of the United States of America,* 80, 4394–4397.

Wylie, P. 1942. *The Generation of Vipers.* New York: Rinehart.

Wynn, T., and Coolidge, F. L. 2004. The expert Neandertal mind. *Journal of Human Evolution,* 46, 467–487.

Yamakoshi, G., and Sugiyama, Y. 1995. Pestle-pounding behavior of wild chimpanzees at Bossou, Guinea: A newly observed tool-using behavior. *Primates,* 36, 489–500.

Yurk, H., Barrett-Lennard, L., Ford, J. K. B., and Matkin, C. O. 2002. Cultural transmission within maternal lineages: Vocal clans in resident killer whales in southern Alaska. *Animal Behaviour,* 63, 1103–1119.

Zentall, T. R. 2004. Action imitation in birds. *Learning and Behavior,* 32, 15–23.

Zentall, T. R., and Galef, B. G., Jr., eds. 1988. *Social Learning: Psychological and Biological Perspectives.* Hillsdale, NJ: Lawrence Erlbaum.

Zhu, R. X., Potts, R., Xie, F., Hoffman, K. A., Deng, C. L., Shi, C. D., Pan, Y. X., Wang, H. Q., Shi, R. P., Wang, Y. C., Shi, G. H., and Wu, N. Q. 2004. New evidence on the earliest human presence at high northern platitudes in northeast Asia. *Nature,* 431, 559–561.

Zohar, O., and Terkel, J. 1992. Acquisition of pine cone stripping behaviour in black rats *(Rattus rattus). International Journal of Comparative Psychology,* 5, 1–6.

Zuberbühler, K. 2005. The phylogenetic roots of language—Evidence from primate communication and cognition. *Current Directions in Psychological Science,* 14, 126–130.

KRISTIN E. BONNIE, Emory University

FRANS B. M. DE WAAL, Emory University

BENNETT G. GALEF, McMaster University

KIM HILL, Arizona State University

JEREMY R. KENDAL, Durham University

RACHEL L. KENDAL, Durham University

KEVIN N. LALAND, University of St. Andrews

JANET MANN, Georgetown University

W. C. MCGREW, University of Cambridge

SUSAN PERRY, University of California, Los Angeles

BROOKE L. SARGEANT, Florida International University

KIM STERELNY, Victoria University of Wellington and the Australian
National University

MICHAEL TOMASELLO, Max Planck Institute for Evolutionary Anthropology,
Leipzig

CAREL P. VAN SCHAIK, University of Zürich

HAL WHITEHEAD, Dalhousie University

ANDREW WHITEN, University of St. Andrews